MODAL LOGIC

AND ITS APPLICATIONS

MODAL LOGIC
AND
ITS APPLICATIONS

D. PAUL SNYDER
Temple University

Van Nostrand Reinhold Company

New York • Cincinnati • Toronto • London • Melbourne

Van Nostrand Reinhold Company Regional Offices:
New York Cincinnati Chicago Millbrae Dallas

Van Nostrand Reinhold Company International Offices:
London Toronto Melbourne

Library of Congress Catalog Card Number 72-126043

Published by Van Nostrand Reinhold Company
450 West 33rd Street, New York, N.Y. 10001

Published simultaneously in Canada by
Van Nostrand Reinhold Ltd.

8 7 6 5 4 3 2 1

TO

Carl A. Eisdorfer
John H. Wilms
David Rubinstein

PREFACE

This book is addressed to philosophers and students of philosophy who want to use formal logic as a tool of their trade. It presupposes only the familiarity with truth-functions and quantifiers that is gained in a one-semester introduction to symbolic logic. It is designed to present in as clear and straightforward a way as possible a set of techniques for manipulating formal systems of modal logic and the related interpretative devices, and for developing further systems of modal logic to meet specific contextual needs.

Considerable work is being done currently by philosophers who employ systems of modal logic as an analytic tool for the clarification of a number of different contexts. To most students and many professionals, however, modal logic remains something of a mystery because its techniques are complex and its interpretations difficult to grasp. Although many people have been interested in the field, relatively few have been able to master its techniques or its theory, since these usually presuppose considerable advanced work in formal logic.

Two recent developments in formal logic make it possible to treat modal logic on an elementary level without limiting oneself to the juggling of mysterious object language primitives. The first development has to do with the remantic interpretation of systems of modal logic, and the second with techniques of proof. The semantic techniques of Saul Kripke and Jaakko Hintikka provide a clear and straightforward way of interpreting formal systems of modal logic. Hintikka's model system technique in particular seems to be the simplest and most direct way yet devised for interpreting formal systems. It is a fertile technique in that the model is easy to manipulate and modify, and a progression of formal systems beginning with elementary logic and running through simple systems of modal logic to the most complex systems of quantified modal logic with entailment can be interpreted on the same basic model structure. At the same time, Hintikka's technique allows the treatment of variant formal systems that are adaptable to deontic, epistemic, temporal and other interpretations.

The proof technique employed throughout the book was devised by

R. W. Binkley and R. L. Clark. The technique, called "cancellation", is a mechanically abbreviated reduction of the Gentzen type, which has been extended to multiple quantification and modal logic by its originators. It is simple, precise, flexible, easy to teach and easy to learn. Its presentation here is self-contained, in the sense that no prior familiarity with reductive techniques is presupposed, and the technique is presented in such a way as to constitute a review of the principles of truth-functional and quantified elementary logic before it is applied to modal logic. The cancellation technique mechanizes the business of determining whether or not a formula is a theorem of a given formal system. The time and effort that are generally devoted to running proofs can be turned to the comparison of formal systems and the development of variant formal systems to meet specific situations, as well as the spelling out in a clear way of the philosophical considerations relevant to the use of modal logic.

The combination of the Binkley-Clark proof techniques and the Hintikka semantic technique is very powerful. Neither technique is wholly new, of course. Each has its roots in the work of other people, and each has appeared in the professional journals. They are brought together here because each of the two techniques seems to provide the most simple and straightforward way of doing what it does. The two techniques mesh in such a way that for each modification in either the object language or the model, the corresponding modification in the other is obvious.

Further, the techniques converge at two places. First, the Hintikka model system serves to develop the interpretation for the Binkley-Clark versions of the classical formal systems of modal logic. Second, and most important, the Binkley-Clark formulation of elementary logic is used to mechanize the Hintikka semantic technique so that, in effect, formulas containing modal operators can be interpreted on formulas that do not contain modal operators, and *both* can be subjected to mechanical proof techniques. One important outgrowth of this second convergence is the virtual elimination of metaproofs of the usual sort from the body of the text. The relation between formal system and interpretation is immediately clear in all cases, but the actual demonstration of completeness and soundness properties can be set aside and treated in an Appendix to the main body of the work.

No attempt has been made to provide a full historical survey of modal logic, although the bulk of the classical systems of modal logic are treated. The aim throughout is to have the student work with systems of modal logic, cope with their interpretation and application and be aware of the philosophical problems attendant to their use. Likewise, no attempt is made to resolve all the philosophical issues that arise. The techniques used make the issues clear in virtually every case, but a full philosophical treatment of the issues belongs elsewhere. The philosophical prejudices of the author may be obvious, but every attempt has been made to avoid turning philosophical prejudice into received doctrine.

Acknowledgments. On the occasion of one's first book, there are many people to thank. The debts to family, friends and teachers are too difficult to describe and too numerous to list. My formal debt to Romane Clark has already been acknowledged, although not sufficiently. My personal debt to him is something else again. As my advisor and friend during my graduate study at Duke University he prodded, encouraged, criticized and taught with incredible incisiveness and good grace. Robert Binkley, who was also on the faculty at Duke during my graduate study, was a constant source of ideas. Many things suggested over coffee by Binkley six years ago have ended up in this book in one form or another, and it has not always been possible to give credit for specific ideas. Professional friends who have helped with criticism and suggestions include Charles E. Dyke, John Fisher, John Heintz, David L. Hull, C. Frederick Koehler, S.J., Hugues Leblanc, Jay Rosenberg, Robert Sheehan and William A. Wisdom.

Finally, to my students at Temple University who studied from early versions of this material for three successive years, my sincere thanks.

Donald Paul Snyder

CONTENTS

Introduction

THE SCOPE OF MODAL LOGIC

Modal logic is not the Philosopher's Stone. It will not answer the riddles of the universe, nor will it provide a quick and easy answer to basic philosophical questions. What modal logic does provide is a powerful, flexible and adaptable instrument for the analysis of a variety of different contexts of human discourse and reasoning.

For some years, it was felt that there was precisely one correct system of formal logic which could be used to impose clarity and rigor on all significant verbal contexts. This is, quite simply, false. There is not precisely one correct system of logic. There are many. Which system is the appropriate one to use for a given task is, to a large extent, determined by the nature of that task. In the following chapters we will discuss literally hundreds of distinct systems of formal logic. Each of these systems has desirable formal properties. Many of them will be suggested as appropriate for the analysis of specific sorts of verbal contexts. But more important than the presentation of a variety of formal systems is the development of techniques for *devising* formal systems to meet specific needs. This, as much as the laying down of systems already devised, will be the task of this book.

1 MODAL LOGIC AND ELEMENTARY LOGIC

The core of all the formal systems to be discussed in the following chapters is a version of quantified first-order logic which we will call *elementary logic*. In one form or another, it is elementary logic that is covered in most introductory courses in formal logic. A technique for detecting theorems of elementary logic will be presented in the following two chapters.

The basic content-bearing unit of both elementary logic and modal logic is the *statement*. The simplest well-formed formulas of a formal system represent complete statements, either in the form of statement variables or constants (usually 'p', 'q', 'r', etc.) or in the form of predicates and individual variables or constants put together in such a way as to represent complete statements (usually 'Fx', 'Gx', 'Rxy', etc.). A statement is assumed to be precisely one of *true* or *false*. The logical operators or logical constants of the formal system operate on complete statements in such a way as to form new, more complex statements.

The logical constants of elementary logic fall into two groups, *truth-functional operators* and *quantifiers*. Truth-functional operators operate on one or more (usually two) complete statements in such a way as to form new statements. They are called "truth-functional" because the truth or falsity of the new statement formed by the operator is a function of—i.e., is uniquely determined by—the truth or falsity of the statements operated on. Usually, a system of elementary logic will include truth-functional operators representing alternation (either—or—), conjunction (both—and—), material implication (if—then—) and material equivalence (—if and only if—), as well as negation (it is not the case that—). All of these operators form complex statements whose truth values are calculable from the actual truth-values of their components. The alternation "*Either* Susie will go to the party *or* her brother will go," for example, is true just in case at least one of the contained statements is true, and false if both contained statements are false. The material implication, "*If* Susie goes to the party, *then* her brother will stay home," claims simply that it is not the case that the first statement is true and the second false; i.e., the material implication is false when the "if" clause is true and the "then" clause is false, but is true under all other combinations of truth-value of its components. Even in the case of material implication and material equivalence, then, truth-functional operators combine statements in such a way as to make claims about what *is* the case, not about what must be, or could be the case.

The quantifiers of elementary logic allow us to manipulate the notions "all" and "some" in combination with the truth-functional operators. Quantifiers are always coupled with individual variables, and there are usually two quantifiers, the existential (there is at least one individual, x, such that —) and the universal (for every individual, x, —). Quantified statements in elementary logic make claims about what is the case with respect to every existing individual or some existing individual.

Modal logic, as distinct from elementary logic, has to do with the structure of statements which are in grammatical moods other than the indicative. While it is easy enough in other languages to identify modal contexts thus, the spelling out of exactly which contexts these are is rather difficult in contemporary English because the subjunctive has all but disappeared from the language. Where we might once have said "It is necessary that there be precisely one even prime number," contemporary English calls for "It is necessary that there is precisely one even prime number." The subjunctive form has been replaced by the indicative, although its force remains. The force of the subjunctive is to take a statement away from assertions about what simply is the case, and allow discussion of what must be, or ought to be, or might be, or is believed to be, hoped to be, and so on.

Modal contexts can be detected by the presence of verbal cues that there is a modal statement operator present. One formal feature of modal operators is that they range over complete statements, forming new statements in such a way that the truth or falsity of the new statement is *not* a function of the truth or falsity of the contained statement. For example, "*It is not the case that* today is Tuesday" is a *non*-modal statement. The truth or falsity of this statement is entirely determined by the truth or falsity of the contained statement "Today is Tuesday." "*It is possible that* today is Tuesday," on the other hand, is a modal statement. Its truth-value is "independent" of the truth-value of "Today is Tuesday," in the sense that the truth or falsity of one does not entirely determine the truth or falsity of the other. "It is possible that p" is true when 'p' is true, but it can also be true when 'p' is false. When 'p' is false, "It is possible that p" may be either true or false. In general, we will say that a context is modal if there is present a statement operator which forms a new statement having a truth-value independent from that of the statement operated on.

This account of modality may seem excessively broad. It *is* quite broad, and deliberately so. This reflects a recent trend in the development of formal systems. Traditionally, modal logic has been restricted to the consideration of the contexts "It is possible that—" and "It is necessary that—" and their negations (impossible, unnecessary). In the twentieth century, most work in modal logic has focused on attempts to develop "if-then" operators which are more satisfactory than the truth-functional material implication. Such attempts have used possibility and necessity in combination with truth-functional operators for the most part, although there have been some recent departures that we will take note of shortly. Very recently, however, philosophers have attempted to formalize other statement operators, like "It is known that—," "It is believed that—," "It is obligatory that—," "It is permissible that—," "It is sometimes the case that—," and even "desirable," "hoped for," and so on. All of these qualify as *modal* under the description of modality given here, and all of them are subject to some formal clarification using the same *sort* of logical machinery as is used for possibility and necessity. Modality, in its semantic interpretation as well as its verbal usage, involves considerations of states of affairs other than the one which actually exists.

Several of the contexts mentioned are sometimes regarded as neither truth-functional nor truly modal. "It is known that—," for example, has been analyzed as a *heterogeneous relation*—a relation involving items of different category—between a person on the one hand and a statement on the other.[1] In its basic form, the relation appears as "Smith knows that p," and the apparent modal statement "It is known that p" involves a quantification: "There is an x such that x knows that p." Other intentional notions such as wishing and hoping, and even such notions as obligation and permission, are subject to this more sophisticated treatment. But even in the form of heterogeneous relations, such notions bear strong similarities both formally and semantically to the traditional modalities—strong enough that they bear consideration in this text.

1.1 Monadic and Dyadic Modality. Dissatisfaction with the formalization of implication provided by truth-functional logic has led to a number of attempts to develop systems of modal logic which provide a strong "if-then" operator. Some of these, as noted in the preceding section, have involved a combination of the material implication operator with the modal operator "It is necessary that—." Other attempts to formalize "if—then—" have taken a different line of attack: the introduction of a modal operator which ranges over *pairs* of statements. Such two-place operators are called *dyadic* modal operators, as distinct from the *monadic* modal operators which range over (simple or complex) statements one at a time. The dyadic modal operator for the "if-then" relation is usually called *entailment*, to distinguish it from the weaker implication operators. Ideally, a formal system of entailment should guarantee that whenever a statement, 'p', entails another, 'q', it can be shown that 'q' is a relevant consequence of 'p', and that the entailment relation does not hold between two statements simply on the strength of the truth-values of the statements.

If we think of entailment as the "dyadic necessity" operator, then the "dyadic possibility" operator is *compatibility*. Compatibility is also relational and carries the sense that, whatever the truth-values of two statements, they are compatible if there is nothing about one of them that prevents the other's being true.

[1] Cf. [1], also Chapter VI.

Note: Bibliographical references are made with respect to chapter bibliographies. '[II.2]', for example, refers to item 2 in the bibliography of Chapter II. Section references within this book are made by citing the chapter number and section number or, within a given chapter, by citing the section number alone. '§II.2', occurring in a chapter other than II, refers to Section 2, Chapter II. '§2' occurring in Chapter II, refers to the same section.

2 SPECIES AND VARIETIES OF MODALITY

There are certain surface similarities among the various sorts of modal operators we have mentioned with respect to the way in which they behave in verbal contexts. Generally, within each sort of modality there is a strong operator (necessity, obligation, knowledge) and a weaker one (possibility, permission, belief). In most cases, the strong and weak operators are related definitionally. To say that a statement is necessary, for example, is equivalent (by definition) to saying that the negation of that statement is impossible.

Some of the features of each sort of modality become clear immediately, and it is worthwhile to consider modality informally by way of introduction to the formal work to follow. For the present, we will adopt the convention of abbreviating the modal operators in the following way: "It is necessary that p" will be written "[necessary] p", and, in general, "It is—that p" will be written "[—] p."

2.1 Alethic Modality. Modal logic which is concerned with possibility, necessity, entailment and compatibility is called *alethic*, from the Greek word for truth, because it has to do with the way in which statements are, or are not, true. On some views, alethic modality has to do with the *strength* of assertions that statements are, or are not, true.

Typically, one of the monadic alethic operators is taken as primitive and the other operator introduced by definition. We may, for example, take "[possible] p" as primitive and define "[necessary] p" in the following way:

$$\text{[necessary] p} =_{\text{df}} \text{not [possible] not-p.}$$

In the same sense that logical intuitions demand certain things of the operators of truth-functional logic, they demand certain things of modal operators. In truth-functional logic, for example, we would not be inclined to accept any formal version of "Both—and—" for which it did not hold that whenever "Both p and q" is true then 'p' must be true and so must 'q'. In alethic modality, the following are demanded of any adequate formal system that includes the monadic operators:

(1) If [necessary] p, then [possible] p.
(2) If [necessary] p, then p.
(3) If p, then [possible] p.

These three formulas simply give the relative strength of necessity, truth and possibility. If it is necessary that all winged creatures be vertebrates, then it must also be true that they are. If it is true that all winged creatures are vertebrates, then it must be at least possible that they are.

The contrapositives of the three statements above must also be true for any adequate system of alethic modality:

(4) If not [possible] p, then not [necessary] p.
(5) If not-p, then not [necessary] p.
(6) If not [possible] p, then not-p.

Given the definitional relation between possibility and necessity, further interrelations can be spelled out.

(7) [Possible] not-p if and only if not [necessary] p.
(8) [Necessary] not-p if and only if not [possible] p.

Further, we can finish out the interrelations of the operators in an obvious way by considering how they relate to a statement and its negation. Precisely one of 'p' or 'not-p' must be *true*, but both of them might be *possible*. "There is a brick house on the corner of Morris Street" is true, but not necessarily true. "There is *no* brick house on the corner of Morris Street" must then be false, but not necessarily so. But if neither a statement nor its negation is necessary, both of them are possible, whichever happens to be contingently true. But a statement and its negation cannot both be necessary, of course.

(9) Either [possible] p or [possible] not-p (or both).
(10) Not both [necessary] p and [necessary] not-p.

The *dyadic* alethic modalities will be discussed only after a formal foundation has been laid. But certain negative criteria can be laid down for them. Entailment is taken to be relational and carries the sense that the entailed statement is a *relevant consequence* of the entailing statement, whatever the truth-value or modal status of the statements taken individually. This means that we should never be able to prove that 'p' entails 'q' from the statement that 'p' is false or impossible, or from the statement that 'q' is true or even necessary. Compatibility is also relational and should not be affected by anything that can be said about the statements individually. If both 'p' and 'q' are true, then they must be compatible—there could be nothing about one of them that prevented the other from being true—but two statements which are impossible might well be compatible with each other. "There are seven hundred whirling dervishes in the hall closet" is, in an obvious sense, impossible, but it is compatible with "There are six hundred whirling dervishes in the hall closet."

Apart from meeting informal criteria and some more sophisticated criteria to be discussed later, any adequate account of modality must provide a way of articulating the various senses of possibility, necessity, entailment and compatibility. What, after all, does it mean to say that a statement is necessarily true? Under what conditions are we justified in saying so? Any or all of the following statements might be called necessary in one sense or another:

Either it will rain tomorrow or it won't.
Koala bears are marsupials.
Bachelors are unmarried.
Copper is a conductor of electricity.
Bicyclists are two-legged.
Sugar dissolves in warm water.
For every action there is an equal and opposite reaction.
If Smith passed Philosophy 1, then he registered for Philosophy 1.
There is precisely one even prime number.
If Agnes is a female astronaut, then she is a female.

Clearly not all of these statements are necessary in the same sense or for the same reasons. Some, like the first statement, are *logically* necessary—necessarily true because of sheer structure. Others are necessarily true because of interrelations between non-logical features: bachelor—unmarried; koala bear—marsupial; etc. Still others might be taken to be "theoretically" or "physically" or "causally" necessary. While the numbered statements above may hold *generically* for all the senses of "necessary" and "possible," there will be some features of logical necessity that will not be shared by physical necessity.

In a similar way, there is more than one species under the genus *entailment*, and more than one species of *compatibility*. "Alice punched Bill in the mouth" is *logically* compatible with "Alice and Bill have never been within fifty miles of each other." There is nothing in the logical structure of one to prevent the other's being true. But clearly the two statements are not compatible in another sense. Part of what is involved in developing alethic modal logics that will be useful for analytic purposes is the consideration of the various species of alethic modality, how they differ, and how they are related to each other.

2.2 Temporal Modality. Numerous attempts have been made to formalize temporal relations, and not all attempts have been overtly modal. Indeed, merely making explicit the time reference of statements and attempting to systematize logically the ways in which statements about "tensed" events relate to each other seems not to be a modal matter at all. Considerations of what happens at ordered points in time can be carried out in modal terms, however, and most of the attempts to formalize temporal relations bear strong formal similarities to modal logic.

As will be the case with most non-alethic modalities, we will consider temporal modalities in simple and complex forms. The simplest temporal logic—so simple as to be almost gross—merely introduces operators representing "It is always the case that—" and "It is sometimes the case that—" into a system of elementary logic. Formulas that do not fall under modal operators may be regarded as present tense ("now"). In such a temporal logic, the strong operator [always] will behave precisely like the alethic operator [necessary], and the weaker operator [sometimes] will behave like [possible]. This is easy enough to see with respect to the very basic truths of modal logic given in the preceding section, and it will also be the case when alethic modal logic is developed further in Chapter III. The two operators are interdefinable in the usual way:

$$\text{[always] p} =_{\text{df}} \text{not [sometimes] not-p}$$

(11) If [always] p, then [sometimes] p.
(12) If [always] p, then p.
(13) If p, then [sometimes] p.
(14) If not [sometimes] p, then not [always] p.
(15) If not p, then not [always] p.
(16) If not [sometimes] p, then not-p.
(17) [Sometimes] not-p if and only if not [always] p.
(18) [Always] not-p if and only if not [sometimes] p.
(19) Either [sometimes] p or [sometimes] not-p (or both).
(20) Not both [always] p and [always] not-p.

More complex temporal modality takes several forms. First, there are variants that are somewhat ordered, like "It is or has been the case that," coupled with "It will be the case that," and other variants where past, present and future can be accommodated. Then there are more complicated systems of temporal logic where time variables are introduced along with individual variables. In such a system, 'Fx,t' might represent "x is a freshman at (time) t." Quantification over the time variables is, of course, impossible to resist in such a system. Finally, there are fully ordered systems of temporal logic where the time variables and constants represent arbitrarily divided instants or units of time. Only the more simple treatments of temporal modality will be considered in this text, but the formal wherewithal for developing the more sophisticated systems is built into the machinery we will develop for modal logics, and it is not too difficult to make the extension.

2.3 Deontic Modality. The logic of obligation and permission is called *deontic* logic. Typically, in serious attempts to develop deontic logics, there is an attempt to limit the sort of statement than can fall under the operator "It is obligatory that—." The limitation might take the form of requiring that the deontic operators range over "act-descriptions" rather than statements, in order to rule out such apparent nonsense as "It is obligatory that today be Tuesday," and "It is permissible that copper be a metal." We will make no attempt at such a limitation here, however, and will consider the deontic operators as if they were statement operators of the same sort as the alethic and temporal operators.

The definitional relation between the strong and weak operators that holds in the alethic and temporal modes holds also in the deontic modes. To say that something is obligatory is equivalent to saying that its opposite is not permissible:

$$\text{[obligatory] p} =_{\text{df}} \text{not [permissible] not-p.}$$

Deontic modality will differ from alethic and temporal modality, however, with respect to the ten truths of modal logic that relate the strong and weak modalities with truth. Whether or not something happens (is true) has no bearing on whether it is obligatory or permissible from a moral point of view. The principles, "If [obligatory] p, then p" and "If p, then [permissible] p," are not acceptable. The following, however, should be included in any simple deontic logic:

(21) If [obligatory] p, then [permissible] p. (cf. (1) and (11))
(22) If not [permissible] p, then not [obligatory] p. (cf. (4) and (14))
(23) [Permissible] not-p if and only if not [obligatory] p. ((7) and (17))
(24) [Obligatory] not-p if and only if not [permissible] p. ((8) and (18))
(25) Either [permissible] p or [permissible] not-p. ((9) and (19))
(26) Not both [obligatory] p and [obligatory] not-p. ((10) and (20))

A reasonable system of quantified deontic logic can be developed around the notions of obligation and permission. In addition, logicians have recently considered systems that attempt to formalize what is involved in practical reasoning—reasoning on moral or other grounds which leads to action.

Deontic logic, like alethic modal logic, has a place for *dyadic* modality. From some ethical points of view, no action is ever absolutely obligatory, but obligation is always relative: one action or state of affairs obligates us to do (morally entails) some other action. Making promises obligates one to keep them; being a member of a certain profession obligates one to perform certain actions or to maintain certain standards of behavior, and so on.

2.4 Epistemic Modality. Epistemic logic is concerned with knowledge, belief and related concepts. In its simpler forms, epistemic logic simply involves introducing the apparent modal operators "It is known that—" and "It is believed that—" along with certain other related operators. More complex forms of epistemic logic involve considerations of what is known by particular individuals at particular moments of time. In the more sophisticated forms, epistemic logic appears less like a "true" modality.

The two simple operators [known] and [believed] differ from the other modal operators considered in that they are not interdefinable. "It is not believed that not-p" is not equivalent to "It is known that p." Nobody may have the foggiest notion whether 'p' is the case or 'not-p' is the case. Then neither is believed, and surely neither is known.

Logical intuitions are not as clear with respect to epistemic operators as they are with respect to the other modal operators. There are many senses of the terms "know" and "believe," and any formalization must select one sense from among the many available. On any reasonable account, however, the following principles seem to be called for:

(27) If [known] p, then [believed] p.
(28) If not [believed] p, then not [known] p.
(29) Not both [known] p and [known] not-p.

With respect to the generalized operator "It is believed that—," it can of course be the case that both of 'p' and 'not-p' are believed. When belief is treated as a heterogeneous relation, however, there seems to be strong argument for requiring that a given person cannot at one and the same time believe both a statement and its negation. But it is psychologically possible to do so, especially if it "never occurs to" the person in question that he believes one while sincerely proclaiming the other. For this reason, attempts to formalize knowledge and belief are often restricted to "responsible knowledge" and "rational belief." What, exactly, might be meant by "responsible" and "rational" in this context we leave to a later chapter.

2.5 "Intentional" Modality. While little work has been done in the logic of such intentional notions as wishing, hoping, desiring and intending, certain attempts to clarify such notions have drawn on the formal machinery of modal logic. There will be no attempt in this text to develop a formal calculus of intention, but it will be possible to suggest, at least, ways in which the formal and semantic machinery might be used in the analysis of intentional notions.

Some recent work in value theory has also drawn upon the sort of formal machinery we will develop, and appropriate notice will be taken of such work when the machinery is before us.

2.6 Further By-Products of Modal Logic. Along with all the various contexts we have called modal, there are other sorts of context that can be examined in the light of the formal resources developed for modal logic. We have noted that modal logic has to do with consideration of states of affairs other than the one which happens to be the case. Alethic modality is concerned with what must be the case in *every* possible state of affairs (necessity) or in *some* possible state of affairs (possibility), if not the actual one. Deontic modality has to do with morally idealized states of affairs: what is obligatory is what happens in a morally governed universe, and what is permissible is what could happen in a morally governed universe—surely such a discussion is not about this world. Since our formal and semantic machinery will allow us to deal with states of affairs other than the actual one, it can be extended to deal with states of affairs that involve different groups of individuals as well.

Fictional entities like Pegasus and Sherlock Holmes have come under discussion in the literature of (non-modal) quantification theory, because there is no reasonable way of treating them in a logic that can deal only with matters of fact. Some of the machinery of modal logic suggests a way of handling statements like "Sherlock Holmes caught Professor Moriarty" without committing ourselves to the (actual) existence of either the detective or the criminal. In a similar way, the machinery we will develop for modal logic will suggest a way of handling statements—or quasi-statements—which are neither blatant nonsense nor the sort of thing we would want to claim to be true or false. "The present King of France is wise" has been called simply false by some philosophers, but others have objected, saying that to call this statement false is to say that "The present King of France is *not* wise" must be true, leaving us with a true statement about a non-existent king. Whichever position one chooses, it is interesting to note that formal systems can be developed which allow us to include statements like "The present King of France is wise" and "The present King of France is *not* wise," and treat them *both* as false without (really) tampering with the Law of Excluded Middle. One such system is a by-product of one of the formal systems of logical entailment to be considered in the final chapters.

3 THE ENTERPRISE

The task we have undertaken is an ambitious one. To develop machinery that will allow us to formalize a number of modal and quasi-modal notions and to provide semantic interpretations of the formalizations, and in addition to develop some interesting logical by-products like those mentioned in the preceding section, is a formidable task indeed. The task would be considerably more difficult were it not for some very recent developments in formal logic and semantics, which make the manipulation of formal systems considerably easier than it has been heretofore. Such developments are never the work of one or two people, but draw ultimately on the work of many people.

The formal techniques to be used are essentially mechanical. Proving theorems within a given system of logic involves following a straightforward mechanical procedure with paper and pencil that could as well be done by a computer (and, as a matter of fact, *has* been done by an IBM 7074). The high adventure of seeking clever strategies for deductive proofs, and the concomitant satisfaction of finding such proofs and being able to claim new theorems, are lost to us in the present set of formal systems. Instead, the adventure of doing logic in the way we will be doing it lies in the development of a variety of systems of logic for a variety of tasks. The basic proof technique is easy enough to master. Once this is done, almost endless finagling is possible in order to construct new formal systems and variants of old ones.

BIBLIOGRAPHY—INTRODUCTION

[1] Hintikka, Jaakko. *Knowledge and Belief.* (Cornell University Press, Ithaca, N.Y., 1962)

[2] Hintikka, Jaakko. "A Program and A Set of Concepts for Philosophical Logic," *The Monist* **51** (1967), pp. 69–92.

[3] Rescher, Nicholas. "Recent Developments in Philosophical Logic," in his *Topics in Philosophical Logic* (Dordrecht, 1968), pp. 1–13.

[4] von Wright, G. H. *An Essay in Modal Logic.* (North Holland, Amsterdam, 1951)

Chapter I

NOTATION AND SYMBOLIZATION

There are as many as seven or eight different notations in current use by people who do work in modal logic or in related logical theory. The principal notations are shown in Table 3 at the end of §4 below, and on the endpapers of this book, along with the notation selected for this book.

The notation to be used here is one that seems most clear and typographically convenient. The reductive proof techniques used throughout the book demand a notation that follows the Polish order, where the principal operator of a formula appears to the left of the subformulas on which it operates. It is assumed that the student has had occasion to use at least one notation of the Polish type or one notation like that of *Principia Mathematica*. The Polish notation, once mastered, is simple, straightforward and unambiguous.

1 TRUTH-FUNCTIONAL LOGIC

In Polish notation, truth-functional operators are written before the formulas on which they operate. The notation to be used in this book differs only slightly from standard Polish notation. Both will be described in this section.

Mastering a new notation takes a certain amount of practice. To the student who knows only a notation of the Polish type or only a notation of the *Principia* type, the other notation looks strange and unreasonable at first. A few exercises in symbolization and a reasonable amount of work with proof techniques in the unfamiliar notation will generally overcome the strangeness.

1.1 Vocabulary of Polish Notation. The vocabulary of Polish notation for truth-functional logic consists of statement constants and logical constants. The logical constants are represented by capital letters, statement constants by lower-case letters. Specifically, our vocabulary consists of the following:

Statement constants, a countable stock: p, q, r, s, ···, p′, q′, ···, p″, ···
Logical constants

A Alternation		Apq	Either p or q.
K Conjunction		Kpq	Both p and q.
C Conditional		Cpq	If p, then q.
E Biconditional		Epq	p if and only if q.
N Negation		Np	Not-p.

1.2 Formation Rules for Polish Notation. The simplest well-formed formula in a formal system is called an *atomic formula*. An atomic formula of a logical system represents the simplest statement expressible in that system. In truth-functional logic, an atomic formula is a statement constant ('p', 'q', etc.).

A *basic formula* is an atomic formula or the negation of an atomic formula.

Any string of the logical constants or statement constants listed above is a *formula*. Some are not put together in such a way that we can interpret them in a logical system, and are called *ill-formed*. The formulas that are interpreted in a logical system are called *well-formed formulas*. A well-formed formula in Polish notation is one of the following:

(i) an atomic formula ('p', 'q', 'r', etc.)
(ii) the result of writing 'K', 'A', 'C' or 'E' to the left of a pair of well-formed formulas. ('Kpq', 'Cpq', etc., and, since these are well-formed, 'ArKpq', 'CKpqCpq', 'EKpqApq', etc.)
(iii) The result of writing 'N' to the left of a well-formed formula. ('Np', 'NKpq', 'NArKpq', etc.)

All of the following are well-formed formulas. The "translations" given follow a style that makes the reading of Polish notation fairly easy: where a prefix word is available, use it. 'Kpq' is best read "Both p and q," not just "p and q."

Np	Not-p.
KpAqr	Both p and either q or r.
KpNq	Both p and not-q.
CKpqr	If both p and q, then r.
KpNAqNr	Both p and not either q or not-r; Both p and neither q nor not-r.

Unlike *Principia* notation, Polish notation requires no parentheses or brackets for punctuation.

1.3 The Notation of This Book. Now that the basic rules of Polish notation have been laid out, we will immediately change them. For the proof techniques to be used in this book, it is most convenient to use negation in a somewhat different form than the one used in standard Polish notation. The logical constant for negation is as follows:

$\bar{}$ Negation $\bar{p}$ Not-p.

The formation rule which replaces rule (iii) above is the following:

(iii) The result of writing "$\bar{}$" over the left-most sign of a well-formed formula (is a well-formed formula). ('$\bar{p}$', '$\bar{K}pq$', '$\bar{A}rKpq$', etc.)

The negation bar is easy to handle with the two-place operators 'A', 'K' and 'C' (the operator 'E' will not be used). The last formula in the list of the preceding section, '$KpNAqNr$'—"Both p and neither q nor not-r," is now written '$Kp\bar{A}q\bar{r}$'. If two negation operators occur together, they are written one over the other; the negation of '$\bar{A}pq$' is '$\bar{\bar{A}}pq$'.

1.4 Polish Notation Compared with Principia Notation. In Table 1, formulas in the variant of Polish notation to be used in this book are written next to corresponding formulas in one of the many variants of *Principia* notation.

TABLE 1

Polish	Principia
$KApqCrs$	$(p \vee q) . (r \supset s)$
$CCCpqrs$	$(((p \supset q) \supset r) \supset s)$
$CpKqr$	$p \supset (q . r)$
$C\bar{A}pq\bar{A}\bar{r}s$	$\sim(p \vee q) \supset \sim(\sim r \vee s)$
$\bar{C}\bar{A}pq\bar{A}\bar{r}s$	$\sim(\sim(p \vee q) \supset \sim(\sim r \vee s))$

2 ELEMENTARY LOGIC (QUANTIFIED FIRST-ORDER LOGIC)

We will use the term *elementary logic* to designate a system of formal logic which contains both truth-functional operators and quantifiers.

2.1 Vocabulary. In addition to the vocabulary of truth-functional logic, the vocabulary of elementary logic contains constants which represent predicates—properties of individuals—and variables which range over individuals, as well as quantifiers which allow the construction of formulas which can be interpreted as reflecting the notions "all" and "some."

Predicates that apply to individuals one at a time are called predicates of *degree one*. "—is a frog," "—is red," and other such predicates that can be applied to one individual are predicates of degree one. Predicates like "—is larger than—" and "—loves—," which require the designation of *two* individuals in order to form a complete statement, are called predicates of *degree two*. A predicate of *degree three* would be "—is between—and—," where three individuals must be designated in order for a complete statement to be formed. The vocabulary of elementary logic includes all of the items mentioned in §1.1 above, in addition to the following:

Predicate constants, a countable stock, of varying degrees: F, G, J, F′, G′, J′, F″, ···
Individual variables, a countable stock: x, y, z, x′, y′, z′, x″, ···
Logical constants, the two quantifiers:

∃	Existential Quantifier	∃x Fx	Something is an F.
∀	Universal Quantifier	∀x Fx	Everything is an F.

2.2 Formation Rules for the New Vocabulary. An *atomic formula* is still the simplest well-formed formula that can be constructed in a formal system to represent a complete statement. Now that our vocabulary has been enriched to deal with individuals and predicates that are attached to individuals, the notion of atomic formula is broadened. In elementary logic, an atomic formula is either a statement constant (as before) *or* a predicate of degree *n* followed by *n* individual variables. If 'G' is a predicate of degree two, for example, 'Gxy' is an atomic formula.

With the notion of *atomic formula* broadened, there is no need to change the formation rules (i), (ii) and (iii) above. We simply add to them a fourth formation rule to cover quantifiers. A well-formed formula of elementary logic in Polish notation is one of (i), (ii) or (iii) in §1.2 above, or

(iv) the result of writing '∃' or '∀', followed by an individual variable, to the left of a well-formed formula. ('∃x Fx', '∃x K Fx Jx', '∀x C Fx Jx', '∀x C Fx ∃y Gxy', etc.)

The following well-formed formulas involving quantifiers are shown with verbal translations that reflect the various locutions that may be used to interpret quantified formulas.

∃x K Fx Gx	Some F's are G's; or There is something that is both an F and a G.
∀x C Fx Gx	All F's are G's; or For all x, if x is an F, then x is a G.
∀x C Fx K Gx $\bar{\text{J}}$x	All F's are both G's and non-J's.
$\bar{\exists}$x K Fx Gx	No F's are G's; or There is nothing that is both an F and a G.
$\bar{\forall}$x C Fx Gx	Not all F's are G's.
∀x C Fx ∃y Gxy	Every F is a G to some y; or For all x, if x is an F, then there is a y such that x G's y.

2.3 Polish and Principia. Once again, in Table 2, we compare formulas in Polish notation with the corresponding formulas in *Principia* notation.

TABLE 2

Polish	Principia
∀x C Fx K Gx Jx	(x)(Fx ⊃ (Gx . Jx))
∃x K Fx K Gx Jx	(∃x)(Fx . (Gx . Jx))
∀x C ∃y Fxy ∃z Gzx	(x)((∃y)(Fxy) ⊃ (∃z)(Gzx))
∀x $\bar{\text{K}}$ Gx K $\bar{\text{F}}$x J x	(x) ∼ (Gx . (∼Fx . Jx))
C ∃x Fx ∀x C Gx Fx	((∃x)(Fx)) ⊃ (x)(Gx ⊃ Fx)

3 SYNTACTICAL VARIABLES

In discussing the formal systems to be presented, it will be helpful to have on hand a stock of variables to represent elements of the formal systems. Such variables are called *syntactical variables*, since they are used to represent elements of the syntax of the formal systems in the stating of rules. There will be two principal sorts of syntactical variable used in the text.

Syntactical variables representing well-formed formulas: α, β, γ. These variables are used to represent any formula which meets the conditions (i) through (iv) above. 'K$\alpha\beta$' represents any conjunction, say 'K p q', or 'K A p q C r K p s', or 'K ∃x K Fx Gx ∀x C Fx Jx'.

Syntactical variables representing well-formed *contexts* for individual variables: Φ, Ψ. These variables are always used with one or more individual variables to represent well-formed formulas. 'Φx' represents a well-formed formula which contains one or more occurrences of 'x' which are not picked up by a quantifier, perhaps ‘K Fx Gx’, or ‘C Fx ∃y Gxy’. ‘∃x Φx’ might then be used to represent ‘∃x K Fx Gx’, or ‘∃x K Fx ∀y C Fy Gxy’.

4 ADDITIONAL VOCABULARY

Other items will be added to the vocabulary of elementary logic as they are needed in the presentation of systems of modal logic. Such items as modal operators, individual constants, and special signs used in the proof techniques of the text will not be needed for the exercises which follow this chapter. The various modal operators are shown in Table 3, however, so that the table may be used for reference.

The following abbreviations will be used throughout the book.

“$=_{df}$” for “—is equivalent by definition to—”
“wff” for “well-formed formula”
“wffs” for “well-formed formulas”
“bsf” for “basic subformula”
“bsfs” for “basic subformulas”

5 REVIEW: RELATIONS AMONG THE LOGICAL OPERATORS

Each of the truth-functional operators can be explicated in terms of the conditions under which it forms a true statement. An alternation, for example, is true if and only if at least one of its members is true. A conjunction is true if and only if both of its members are true. A negation is true if and only if the formula negated is false. The relationships among the truth-functional operators are definitional; that is, they are stipulated and built into the rules of the formal systems. We say that any two formulas that are true or false under precisely the same conditions are *equivalent*, and when such formulas contain different truth-functional operators, we say that they are *equivalent by definition*. Any formula containing one of the operators ‘A’, ‘K’ or ‘C’ is equivalent by definition to a formula containing one of the remaining operators and negation. Following are three sets of formulas that are equivalent by definition. We use the sign “$=_{df}$” to indicate the equivalences.

$$\mathrm{A}\alpha\beta =_{df} \bar{\mathrm{K}}\bar{\alpha}\bar{\beta} =_{df} \mathrm{C}\bar{\alpha}\beta$$
$$\mathrm{K}\alpha\beta =_{df} \bar{\mathrm{C}}\alpha\bar{\beta} =_{df} \bar{\mathrm{A}}\bar{\alpha}\bar{\beta}$$
$$\mathrm{C}\alpha\beta =_{df} \mathrm{A}\bar{\alpha}\beta =_{df} \bar{\mathrm{K}}\alpha\bar{\beta}$$

TABLE 3. *Principcl Logical Notations*

	POLISH ORDER			PRINCIPIA ORDER			
	This Text	Lukasiewicz	Clark & Welsh	Peano-Russell	Quine	von Wright	Other Variants
Negation	$\bar{\alpha}$	$N\alpha$	$-\alpha, \bar{\alpha}$	$\sim\alpha$	$-\alpha$	$\sim\alpha$	$\neg\alpha, \bar{\alpha}$
Conjunction	$K\alpha\beta$	$K\alpha\beta$	$\&\alpha\beta$	$\alpha\,.\,\beta$	$\alpha\beta$	$\alpha\,\&\,\beta$	$\alpha\wedge\beta$
Alternation	$A\alpha\beta$	$A\alpha\beta$	$\vee\,\alpha\beta$	$\alpha\vee\beta$	$\alpha\vee\beta$	$\alpha\vee\beta$	$\alpha\beta$
Conditional	$C\alpha\beta$	$C\alpha\beta$	$\supset\alpha\beta$	$\alpha\supset\beta$	$\alpha\supset\beta$	$\alpha\rightarrow\beta$	
Biconditional	—	$E\alpha\beta$	$\equiv\alpha\beta$	$\alpha\equiv\beta$	$\alpha\equiv\beta$	$\alpha\leftrightarrow\beta$	$\alpha\sim\beta$
Universal Quantifier	$\forall x\,\Phi x$	$\Pi x\,\Phi x$	$\forall x\,\Phi x$	$(x)(\Phi(x))$	$(x)(\Phi x)$	$(x)(\Phi x)$	$(\forall x)(\Phi x)$
Existential Quantifier	$\exists x\,\Phi x$	$\Sigma x\,\Phi x$	$\exists x\,\Phi x$	$(\exists x)(\Phi(x))$	$(\exists x)(\Phi x)$	$(\exists x)(\Phi x)$	$\exists x(\Phi x)$

MODAL OPERATORS	This Text	Lukasiewicz	Lewis	von Wright	Anderson-Belnap
Monadic Possibility	$M\alpha$	$M\alpha$	$\Diamond\,\alpha$	$M\alpha$	$P\alpha$
Monadic Necessity	$L\alpha$	$L\alpha$	$\Box\,\alpha$	$N\alpha$	$N\alpha$
Strict Implication	$LC\alpha\beta$	$LC\alpha\beta$	$a \dashv\beta$	$N(\alpha\supset\beta)$	$\alpha\dashv\beta$
Dyadic Possibility (Compatibility)	$Z\alpha\beta$	—	—	$M(\beta/\alpha)$	—
Dyadic Necessity (Entailment)	$Y\alpha\beta$	—	—	$N(\beta/\alpha)$	$\alpha\rightarrow\beta$

The material conditional, 'C', is somewhat more complicated than alternation and negation, in that the circumstances under which a conditional is true or false must be described with reference to the order in which the component formulas fall. The first component of a conditional, the "if" clause, we call the *antecedent*, and the second component, the "then" clause, we call the *consequent*. A material conditional may then be said to be true if and only if either its antecedent is false or its consequent is true.

Negated truth-functional operators are sometimes difficult to interpret verbally. '$\bar{A}$' easily enough becomes the English construction "Neither—nor—." '$\bar{K}$' poses only slight difficulty if it is read simply "Not both—and—." The negated conditional, however, has no easy and natural English counterpart based on "If—, then—." While it is easy enough to put the words together and read '$\bar{C}pq$' as "Not if p then q," it is difficult to make sense of the construction or to have a sense for when such a statement is true. The definitional equivalences among truth-functional operators are helpful in such a case. Extending the equivalences cited above, we can show that every formula containing truth-functional operators, negated or not, is equivalent to a formula containing 'A' or 'K', with negations falling only on component formulas:

$$\bar{A}\alpha\beta =_{df} K\bar{\alpha}\bar{\beta}$$
$$\bar{K}\alpha\beta =_{df} A\bar{\alpha}\bar{\beta}$$
$$C\alpha\beta =_{df} A\bar{\alpha}\beta$$
$$\bar{C}\alpha\beta =_{df} K\alpha\bar{\beta}$$

'$\bar{C}pq$' is equivalent to "Both p and not-q." The formulas to the right of the sign '$=_{df}$' we call constructions with *confined negation*. Occurrences of the negation sign, including the occurrences that are implicit in 'C', are confined to component formulas. While we will never use definitions as a means of operating on formulas in the text, it is often helpful to break down a difficult construction into an equivalent construction with confined negation, for the sake of clarity.

The two quantifiers can also be explicated in terms of the claims for truth that they make, and there are some definitional equivalences among constructions involving quantifiers that are worth keeping in mind when reading and writing formulas.

The existential quantifier, coupled with an individual variable and written before a formula, as in '∃x K Fx Gx', makes the claim that the formula it introduces is true for at least one individual. '∃x K Fx Gx' is true if and only if there is at least one individual, say i, such that 'K Fi Gi' is true.

The universal quantifier, on the other hand, makes the claim that the formula it introduces is true of every individual, but it does not claim that there are any individuals at all. We usually interpret universally quantified formulas in such a way that it is understood that the universe isn't empty, but this understanding is not built into the quantifier formally or semantically. '$\forall x\, C\, Fx\, Gx$' is true if and only if for every individual, i, '$C\, Fi\, Gi$' is true, assuming that there is at least one individual.

The definitional equivalences between the two quantifiers are as follows:

$$\exists x\, \Phi x =_{df} \overline{\forall x}\, \bar{\Phi} x$$
$$\forall x\, \Phi x =_{df} \overline{\exists x}\, \bar{\Phi} x$$

"Something is Φ" makes the same claim as "Not everything is non-Φ," and "Everything is Φ" makes the same claim as "Nothing is non-Φ."

The following confinement equivalences, based upon those given for truth-functional operators and the two just given for quantifiers, are helpful in understanding complex formulas.

$$\overline{\exists x}\, K\, Fx\, Gx =_{df} \forall x\, C\, Fx\, \bar{G}x$$
$$\overline{\forall x}\, C\, Fx\, Gx =_{df} \exists x\, K\, Fx\, \bar{G}x$$

"No F's are G's" is equivalent to "All F's are non-G's," and "Not all F's are G's" is equivalent to "Some F's are non-G's."

6 READING AND WRITING FORMULAS OF ELEMENTARY LOGIC

We conclude with some things to remember in reading and writing formulas of truth-functional logic, and a device for generating endless exercises in reading and writing such formulas.

It is important to keep in mind that the truth-functional operators and quantifiers only make claims for truth and falsity; they do not make any claim for temporal sequence or causal sequence or even for relevance. "Susie ate an apple on Monday and died on Tuesday" suggests, in ordinary English, that there is some connection between the apple and the death. The suggestion does not come over into truth-functional "Both—and—." 'K p q', as a symbolization of the statement, is true just in case each of 'p' and 'q' is true, whether or not 'p' and 'q' are in any sense relevant to each other. Even the "If—, then—" construction has no relevance claim, and no claim for sequence. "If the doorbell rang, then somebody pushed the button," is just 'C p q'; there is no claim in the conditional that the antecedent precedes the consequent in time or in causal sequence.

The universal quantifier, in that it almost always ranges over a material conditional, is sometimes a source of confusion. "All freshmen are girls," symbolized '∀x C Fx Gx', does not make the claim that there are any freshmen at all. It is true if, for every individual, i, 'C Fi Gi' is true; that is, if every individual is either *not* a freshman or else is a girl. If there are no freshmen at all, then 'C Fi Gi' is true for every individual, and '∀x C Fx Gx' is true.

A device that will be used from time to time to generate examples will serve for a review of reading and writing formulas of elementary logic. It involves statements in a limited vocabulary about a limited number of objects. Objects are placed in "worlds" in a "miniature universe."

There are three kinds of object in the miniature universe: dots, stars and wavy lines. They are placed in vertical lines within their worlds, in the following fashion:

1 2 3 4 5

There are five "worlds" above, each of which is populated with dots, stars and/or wavy lines. We will use the following vocabulary to construct statements about the worlds:

> Let 'Dx' represent "x is a dot";
> Let 'Sx' represent "x is a star";
> Let 'Wx' represent "x is a wavy line";
> Let 'Oxy' represent "x is over y." (somewhere over, not immediately over)

The statement '∃x Sx'—"There is a star," or "There is at least one object, x, such that x is a star," is true of worlds 1, 2 and 4, but false of worlds 3 and 5.

'∃x K Sx ∃y K Wy Oxy' is a little more complex: "There is a star, x, and there is a wavy line, y, such that x is over y." This is true only of worlds 1 and 2.

Constructions involving the universal quantifier are a little more tricky, because this quantifier almost always ranges over a material conditional. "All stars are over wavy lines," symbolized "∀x C Sx ∃y K Wy Oxy', is clearly true of worlds 1 and 2 and false of 4. It is not so obvious how this statement fares with respect to worlds 3 and 5, where there are no stars at all. The equivalences given above are a help here. The formula in question is equivalent to '∀x A $\bar{\text{S}}$x ∃y K Wy Oxy'—"For every object, x, *either* x is a non-star, *or else* there is a y such that y is a line and x is above y." Every object in worlds 3 and 5 is a non-star. The statement is true of worlds 3 and 5. In such cases, the material conditional is said to be *trivially true*.

In Table 4 are three statements, with their truth-values indicated for

TABLE 4

Symbolized Statements	1	2	3	4	5
∃x K Wx ∃y K Dy Oxy	T	F	T	F	F
∀x C Dx ∃y K Wy Oyx	T	T	T	T	F
∀x∀y C K Sx Wy Oxy	T	T	T	T	T

each world. Note that the third statement is true for worlds 3, 4 and 5 *because* there is no combination of objects that makes the antecedent true—i.e., no 'x' and 'y' such that the first is a star and the second a wavy line.

Determine whether the following statements are true or false with respect to each of the five worlds.

(a) ∀x C Wx ∃y K Dy Oxy
(b) ∃x ∃y ∃z K K K Dx Sy Wz K Oxy Oyz
(c) ∃x K Sx∀y C Wy Oxy
(d) ∀x C Dx∀y C Wy Oyx
(e) ∀x C Sx∀y C Wy A Oxy Oyx

6.1 "Counting" objects. There is no way with the present notation to indicate directly whether or not two quantifiers range over distinct objects. '∃x ∃y K Dx Dy' does not claim that there are two dots; 'x' and 'y' could be the same object, since each of the existential quantifiers simply claims that there is at least one object that meets the description following the quantifier. To make it clear that there is more than one object, we need to add an explicit disclaimer: "x and y are not the same object." We add the following predicate:

Let 'Ixy' represent "x and y are the same object."

Now, a number of claims can be symbolized that could not be symbolized before:

∃x ∃y K K Dx Dy $\bar{\text{I}}$xy	There are at least two dots. (There is an x, and there is a y, such that x is a dot and y is a dot, and x and y are not the same object.)

∃x K Dx∀y C Dy Ixy — There is precisely one dot. (There is an x such that x is a dot, and anything that is a dot is the same object as x.)

∀x∀y∀z C K K Dx Dy Dz A A Ixy Iyz Ixz — There are at most two dots. (For any x, y and z such that x, y and z are all dots, some two of them are the same object.)

In Table 5 are three statements, with their truth-values indicated for

TABLE 5

Symbolized Statements	1	2	3	4	5
∀x∀y C K Dx Dy Ixy	T	T	F	T	F
∃x ∃y K K K Sx Sy $\bar{\text{I}}$xy∀z C Sz A Ixz Iyz	F	T	F	F	F
∃x ∃y K K K Sx Sy $\bar{\text{I}}$xy∀z C Wz K Oxz Oyz	F	T	F	T	F

each of the five worlds above. The three statements claim, respectively, "There is at most one dot," "There are precisely two stars," and "There are (at least) two stars that are over every wavy line." Note that the third statement is true of world 4, where there are no wavy lines at all.

Determine whether the following statements are true or false with respect to each of the five worlds.

(a) ∃x K Dx ∀y C Dy Ixy
(b) ∃x ∃y K K K Wx Wy $\bar{\text{I}}$xy∀z C Sz K Ozx Ozy
(c) ∃x ∃y ∃z K K K Sx Sy Sz K K $\bar{\text{I}}$xy $\bar{\text{I}}$yz $\bar{\text{I}}$xz

6.2 A Predicate of Degree Three. Finally, to complicate the examples a little more, we make the following addition to the vocabulary.

Let 'Bxyz' represent "x is between y and z."

The statement '∃x ∃y ∃z K K K Dx Sy Wz Bxyz' claims that there is a dot between a wavy line and a star. It is false of all five worlds. '∀x∀y∀z C K K Sx Sy Bzxy Sz' claims that any object (z) that is between two stars (x and y) is another star. It is true of all five worlds—*trivially* true in all worlds except 4, because 4 is the only world where there are three objects that meet the condition in the antecedent of the conditional.

6.3 Practice in Writing Formulas. The miniature universe can be used to generate endless examples. For a start, construct a formula that is true of world 1, but not true of any other world. Then work your way through, constructing for each of the five worlds a formula that is true of it alone. Then take pairs of worlds, and construct, for a first example, a statement that is true of 1 and 2, but not of 3, 4 or 5; then 1 and 3, and so on. If you feel that you need more practice, create five more worlds and start over.

Chapter II

THE BASIC PROOF TECHNIQUES

1 PROOF REDUCTION AND CANCELLATION

There are a great many formal techniques for dealing with elementary logic. *Deductive* proof techniques, which are usually learned in the first course in logic, begin with axioms or assumed premisses and allow for the derivation of provable formulas by means of rules of inference. *Reductive* proof techniques, on the other hand, work in precisely the opposite way: they begin with a formula which may or may not be provable and allow the application of certain rules to this formula to determine whether or not it is provable. Where deductive techniques rely in large measure on the cleverness of the person manipulating them, reductive techniques are generally simple and mechanical. The basic reductive method for detecting provable formulas has been abbreviated mechanically by R. W. Binkley and R. L. Clark, and extended to quantified formulas and modal formulas. The abbreviated procedure is called *cancellation*.[1]

We have been using the expressions "formal system" and "system of logic" in rather obvious ways so far. We now define a *formal system* as a set of well-formed formulas and a system of rules for operating on these formulas in order to set apart a group of formulas as *theorems* of the formal system. "Theorem" will then be defined relative to the rules of each formal system to be studied.

All the formal systems to be considered in this book are based on reduction and cancellation procedures. The rules by means of which it is decided whether or not a given well-formed formula is a theorem of a given formal system will be reduction and cancellation rules.

[1] Reductive proof techniques which are precise reversals of deductive proof techniques are usually associated with G. Gentzen. The version of Gentzen's reduction technique to be presented here is that of R. L. Clark and Paul Welsh [II.2, pp. 100 ff.]. For cancellation techniques, see R. W. Binkley and R. L. Clark [II.1, pp. 79–97]. The sequel on monadic modal logic is forthcoming.

Formal systems which have different rules are different formal systems. It may happen, however, that two formal systems have different rules but that the rules of each select precisely the same set of well-formed formulas as theorems. We say that two such formal systems are *equivalent.*

In this chapter, reduction and cancellation procedures will be described for elementary logic through quantification. In the following chapters, the cancellation procedures will be used to study systems of modal logic. An advantage of the reductive techniques is that they facilitate not only our attempts to prove things *in* a formal system but also our attempts to prove things *about* the formal systems under study. In this way, the reductive techniques provide insight into the concepts embodied in formal systems and aid in the development of variant formal systems.

2 NOTATION

For reduction and cancellation procedures we use the variant of Polish notation given in Chapter I. In place of the usual sign 'N' for negation, we use a bar '‾' placed over the left-most sign of the formula being negated. The negation of 'p' is '$\bar{p}$'; the negation of 'Kpq' is '$\bar{K}pq$'; the negation of '$\bar{A}pq$' is '$\bar{\bar{A}}pq$'.

2.1 Vocabulary.

A countable stock of *statement constants*: p, q, r, s, ⋯, p′, ⋯, p″, ⋯

The following *logical constants*:

K	Conjunction	(Both—and—)
A	Alternation	(Either—or—)
C	Conditionality	(If—, then—)
‾	Negation	(Not—)

Other items will be added to the vocabulary as needed.

The signs listed above comprise the entire vocabulary of the propositional calculus or statement calculus, with the addition only of the sign 'E' which is usually used to represent the biconditional. For the reduction and cancellation procedures to be described here, the formula 'Epq' will be replaced by the more complicated formula 'KCpqCqp'.

2.2 Formulas. A *formula* is any sequence of the signs above. Some are without sense, like 'pKqrsAc', and some are sensible vehicles for making English-like assertions, like 'Apq'. We say that these latter are *well-formed.* We define a *well-formed formula* (wff) of the statement calculus as one of the following:

(i) A statement constant. (p, q, r, s)
(ii) The result of prefixing 'K', 'A' or 'C' to a pair of wffs. (Kpq, ApKqr, CKpqAqr)
(iii) The result of writing the negation bar '$^{-}$' over the left-most sign in a wff. ($\bar{\text{p}}$, $\bar{\text{K}}$pq, $\bar{\text{A}}$p$\bar{\text{K}}$q$\bar{\text{r}}$, CK$\bar{\text{p}}\bar{\text{q}}\bar{\text{A}}$qr)

A wff that contains no logical constants is an *atomic formula*. A *basic formula* is either an atomic formula or the negation of an atomic formula. A formula which contains one or more occurrences of 'A', 'K' or 'C' is a *complex formula*. Any formula and its negation are said to comprise a *pair of opposites*.

2.3 Sequents. A *sequent* is a finite set (perhaps empty) of wffs. The wffs in a sequent, if there are more than one, are set off by commas. 'p, Kqr, $\bar{\text{s}}$' is a sequent. The wffs are called the *members* of the sequent.

A *true sequent* is a sequent that contains at least one true member.

A sequent that contains a pair of opposites is a *closed sequent*. A closed sequent is always a true sequent by the definitions given above. 'p, Kqr, s, $\bar{\text{p}}$, A$\bar{\text{q}}$s' is a closed sequent because it contains the pair of opposites 'p' and '$\bar{\text{p}}$'.

A sequent that contains only basic formulas is a *terminal sequent*. 'p, $\bar{\text{q}}$, $\bar{\text{r}}$, s' is a terminal sequent.

2.4 Metalinguistic Notation. In describing the reduction technique, it will be convenient to use certain metalinguistic symbols: symbols that are used to talk *about* the formal system as distinct from the symbols we use to express things *within* the system. We use 'α', 'β' and 'γ' to represent wffs of the formal system. 'X' and 'Z' are sequent variables, representing sequents of wffs, in some cases empty sequents or sequents with no members.

3 REDUCTION

A sequent may contain any number of members. To begin the *reduction* of a complex formula, we treat it as a sequent with precisely one member. We then apply reduction rules to that sequent and derive from it a second sequent or pair of sequents. The reduction rules are given in the following section. The members of the derived sequent(s) are always less complex than the members of the original sequent, but the derived sequent(s) may contain more members than the original sequent. The process is repeated by applying appropriate reduction rules to the resulting sequent(s), with further sequents being derived until a terminal sequent is reached. No further reduction rules can be applied to a terminal sequent.

The *active member* of a sequent is the left-most wff in that sequent which is not a basic formula. A reduction rule is always applied to the active member of the sequent being reduced, and the first sign of the active member indicates which reduction rule is to be used. In the sequent 'p, Kqr, $Ap\bar{s}$, s' the wff 'Kqr' is the active member.

3.1 Reduction Rules. In the following rules, the sequent variables 'X' and 'Z' are used to represent wffs that may occur to the left and to the right of the active member of the sequent being reduced. The rules are to be understood as follows: From a sequent of the form shown above the line, derive a sequent or pair of sequents of the form(s) shown below the line.[2]

$$\text{A-rule: } \frac{X, \quad A\alpha\beta, \quad Z}{X, \quad \alpha, \; \beta, \quad Z} \qquad \bar{\text{A}}\text{-rule: } \frac{X, \quad \bar{A}\alpha\beta, \quad Z}{X, \bar{\alpha}, Z \qquad X, \bar{\beta}, Z}$$

$$\text{K-rule: } \frac{X, \quad K\alpha\beta, \quad Z}{X, \alpha, Z \qquad X, \beta, Z} \qquad \bar{\text{K}}\text{-rule: } \frac{X, \quad \bar{K}\alpha\beta, \quad Z}{X, \quad \bar{\alpha}, \; \bar{\beta}, \quad Z}$$

$$\text{C-rule: } \frac{X, \quad C\alpha\beta, \quad Z}{X, \quad \bar{\alpha}, \; \beta, \quad Z} \qquad \bar{\text{C}}\text{-rule: } \frac{X, \quad \bar{C}\alpha\beta, \quad Z}{X, \alpha, Z \qquad X, \bar{\beta}, Z}$$

$$\text{=-rule: } \frac{X, \bar{\bar{\alpha}}, Z}{X, \alpha, Z}$$

Applying the A-, C-, K̄- and =-rules results in a single derived sequent written beneath the original sequent. Applying the Ā-, C̄- and K-rules results in a pair of derived sequents written beneath the original sequent. These latter rules are called *splitting rules*, and the reduction splits into two *branches* when they are applied. The two branches have equal status, and each of them is independently reduced until a terminal sequent is reached.

3.2 Proof. A *reduction proof* of a formula is a reduction of that formula all of whose terminal sequents are closed. A *theorem* of a given reduction system is a formula for which a reduction proof can be constructed within that system.

3.3 Rationale. It is somewhat easier to use the reductive technique and to remember its mechanics if one understands how and why it works. Each of the reduction rules is designed to be applied to a sequent that

[2] A reduction rule for '$E\alpha\beta$' is not included in the list. Such a rule would produce precisely the same results as the reduction of the formula '$KC\alpha\beta C\beta\alpha$'.

contains at least one truth-functional compound, and to generate a sequent or pair of sequents that will be true under precisely the same conditions as the first, but will contain members that are less complicated.

Consider, for example, the A-rule. From a sequent of the form '$X, A\alpha\beta, Z$' we derive a sequent of the form 'X, α, β, Z'. Suppose '$X, A\alpha\beta, Z$' is true. This can only be the case if at least one of its members is true, i.e., if either '$A\alpha\beta$' is true or else some member of the sub-sequents 'X' or 'Z' is true. If it is some member of 'X' or 'Z' that makes the first sequent true, then the second sequent is true also, since it contains all the members of 'X' and 'Z'. If, on the other hand, '$A\alpha\beta$' is the true member that makes the first sequent true, it may not be so obvious that the second sequent is also true. We can make this clear by recalling that an alternation is true if and only if at least one of its members is true: '$A\alpha\beta$' is true just in case at least one of 'α' or 'β' is true. But this is precisely the condition under which the sub-sequent 'α, β' of the derived sequent is true.

Consider now a splitting rule, say the K-rule. From a sequent of the form '$X, K\alpha\beta, Z$' we obtain *two* sequents of the forms 'X, α, Z' and 'X, β, Z'. The claim of the reduction rule is that the original sequent is true if and only if *both* derived sequents are true. Again, if it is some member of 'X' or 'Z' that makes the original sequent true, it is obvious that both derived sequents must be true also. Suppose, however, that the sole true member of the original sequent is '$K\alpha\beta$', and all the members of 'X' and 'Z' are false. A conjunction is true if and only if both conjuncts are true. But then if 'α' and 'β' are both true, both of the derived sequents are true sequents, since each of them contains one of 'α' or 'β'. Working in the opposite direction, it is easy enough to see that if both derived sequents are true, then the original sequent must also be true.[3]

3.4 Tautology. A *tautology* of truth-functional logic is a formula that is true under all possible assignments of truth-value to its component formulas. Clearly, a closed sequent meets this condition. One or the other of any pair of opposites must be true, and any sequent containing a pair of opposites therefore has at least one true member under any assignment of truth-value to its members. A sequent derived by a reduction rule is true under precisely the same conditions as the sequent from which it was derived. If the derived sequent is true under *all* conditions, then so

[3] The A-rule and the K-rule may be viewed as the only basic reduction rules. The other rules, with the exception of the =-rule, may be viewed as derived rules which serve to define the operator 'C' and the negations of all three operators. Refer to the definitions in §5 of Chapter I, where negations were confined. Compare the reductions of formulas that are equivalent by definition; for example, '$\bar{C}pq$' and '$Kp\bar{q}$'.

was the original sequent. Thus, any formula that reduces entirely to closed sequents is a tautology.

A wff is a *contradiction* if and only if its negation is a tautology. A wff is *contingent* if and only if it is neither a tautology nor a contradiction.

3.4.1 Examples. Consider a wff, say, 'CCApqrCqr'. Is this a tautology? If we view it as the sole member of a sequent, that sequent is not closed. But neither is it a terminal sequent: the sole member of this sequent is a complex formula to which reduction rules may be applied. *Which* reduction rule we apply is determined by the form of the sequent. In this case, we have a sequent of the form 'X, Cαβ, Z', where 'X' and 'Z' are both empty:

X, C α β, Z
(empty) CCApqrCqr (empty)

Thus the appropriate rule is the C-rule. We write this reduction step as follows:

$$\text{C}\quad \frac{\text{CCApqrCqr}}{\bar{\text{C}}\text{Apqr},\quad \text{Cqr}}$$

The original sequent above the line is true if and only if the sequent below the line is. To determine whether this must be the case or not, we continue on, attacking the active member of the sequent. In this case, the sequent is of the form 'X, $\bar{\text{C}}$αβ, Z':

X, $\bar{\text{C}}$ α β, Z
(empty) $\bar{\text{C}}$Apqr, Cqr

The appropriate reduction rule is the $\bar{\text{C}}$-rule:

$$\bar{\text{C}}\quad \frac{\bar{\text{C}}\text{Apqr},\quad \text{Cqr}}{\text{Apq},\quad \text{Cqr} \qquad \bar{\text{r}},\quad \text{Cqr}}$$

In this instance, the sequent corresponding to 'X' of the rule is null, since there are no basic formulas to the left of the active member. 'Z' of the rule corresponds in this application to 'Cqr', the *in*active element of the reduction step. Note that the $\bar{\text{C}}$-rule is a splitting rule, and that the reduction has now split into two branches, one containing as members 'Apq' and 'Cqr' and the other containing as members '$\bar{\text{r}}$' and 'Cqr'. Each of these two branches must terminate in closed sequents in order for us to have demonstrated that the original formula was a tautology.

Putting the whole reduction together and continuing on until every branch reaches a terminal sequent, we have the following:

$$\begin{array}{ll} & \mathrm{CCApqrCqr} \\ \mathrm{C} & \overline{} \\ & \bar{\mathrm{C}}\mathrm{Apqr},\ \mathrm{Cqr} \\ \bar{\mathrm{C}} & \overline{} \end{array}$$

$$\begin{array}{llcll} & \mathrm{Apq},\ \mathrm{Cqr} & \qquad & \bar{\mathrm{r}},\ \mathrm{Cqr} & \\ \mathrm{A} & \overline{} & & \overline{} & \mathrm{C} \\ & \mathrm{p},\ \mathrm{q},\ \mathrm{Cqr} & & \bar{\mathrm{r}},\ \bar{\mathrm{q}},\ \mathrm{r} & \\ \mathrm{C} & \overline{} & & & \\ & \mathrm{p},\ \mathrm{q},\ \bar{\mathrm{q}},\ \mathrm{r} & & & \end{array}$$

The rule governing each reduction step is indicated at the side of the reduction. Each of the branches terminates in a closed sequent. The original wff is thus provable, and is a tautology.

Exercises

All the exercises in this group are illustrative of the properties of material implication. Demonstrate by reduction that each of the formulas 1–11 is a tautology.

1.	Cpp	A statement materially implies itself. (Material implication is reflexive.)
2.	CKpqp	A conjunction materially implies each of its members.
3.	CKpqq	
4.	CKCpqCqrCpr	Material implication is transitive.
5.	CpApq	A statement materially implies the alternation of itself and any other statement.
6.	C$\bar{\mathrm{p}}$Cpq	A false statement materially implies every other statement.
7.	CKp$\bar{\mathrm{p}}$q	A contradiction materially implies every statement.
8.	CqCpq	A true statement is materially implied by every statement.
9.	CpAq$\bar{\mathrm{q}}$	A tautology is materially implied by every statement.
10.	ACpqCp$\bar{\mathrm{q}}$	Of any two statements, the first materially implies either the second or the negation of the second.
11.	ACpqC$\bar{\mathrm{p}}$q	Of any two statements, either the first or its negation implies the second.

3.5 Demonstrating Implication, Validity and Equivalence. In introductory logic, the student is concerned primarily with determining whether or not a given formula *implies* another, whether or not arguments are *valid*, and whether or not two formulas are *equivalent*. All of these determinations can of course be made using the reduction system just described. A tautology of truth-functional logic reduces to closed sequents, and tests for implication, validity and equivalence can be carried out by determining whether or not certain formulas are tautologies.

The strongest implication relation that can be expressed in truth-functional logic is material implication, represented by 'C' in our notation. We say that a wff, α, *materially implies* another, β, if '$C\alpha\beta$' is a tautology. To determine, for example, whether or not 'KpAqr' materially implies 'Apr', we construct the conditional 'CKpAqrApr' and use the reduction technique to determine whether or not this conditional is a tautology.

$$
\begin{array}{lll}
\mathrm{C} & \mathrm{CKpAqrApr} & \\
\bar{\mathrm{K}} & \bar{\mathrm{K}}\mathrm{pAqr},\ \mathrm{Apr} & \\
\bar{\mathrm{A}} & \bar{\mathrm{p}},\ \bar{\mathrm{A}}\mathrm{qr},\ \mathrm{Apr} & \\
\mathrm{A} & \bar{\mathrm{p}},\ \bar{\mathrm{q}},\ \mathrm{Apr} \qquad \bar{\mathrm{p}},\ \bar{\mathrm{r}},\ \mathrm{Apr} & \mathrm{A} \\
 & \bar{\mathrm{p}},\ \bar{\mathrm{q}},\ \mathrm{p},\ \mathrm{r} \qquad \bar{\mathrm{p}},\ \bar{\mathrm{r}},\ \mathrm{p},\ \mathrm{r} & \\
 & [\text{closed}] \qquad [\text{closed}] &
\end{array}
$$

The reduction terminates entirely in closed branches; the original formula is a tautology; 'KpAqr' materially implies 'Apr'.

The following reduction would be carried out in order to determine whether or not 'Apq' materially implies 'p'.

$$
\begin{array}{ll}
\mathrm{C} & \mathrm{CApqp} \\
\bar{\mathrm{A}} & \bar{\mathrm{A}}\mathrm{pq},\ \mathrm{p} \\
 & \bar{\mathrm{p}},\ \mathrm{p} \qquad \bar{\mathrm{q}},\ \mathrm{p} \\
 & [\text{closed}] \quad [\text{open}]
\end{array}
$$

The reduction terminates in one closed sequent and one *open* sequent; the original formula is *not* a tautology; 'Apq' does *not* materially imply 'p'.

We say that an argument is *materially valid* if and only if the conjunction of its premises materially implies its conclusion. To test an argument for validity, we construct the material conditional that has as its antecedent the conjunction of all the premises of the argument and has as its consequent the conclusion of the argument, and test this conditional by reduction to determine whether or not it is a tautology. Consider the following argument:

Premise 1. If either the tulips or the daffodils have bloomed, then so have the crocus.
Premise 2. But the crocus have not bloomed.
Conclusion. Therefore the tulips have not bloomed.

Let 'p' represent "The tulips have bloomed;" let 'q' represent "The daffodils have bloomed;" let 'r' represent "The crocus have bloomed."

Symbolized in the usual manner, the argument appears thus:

Premise 1.	CApqr
Premise 2.	$\bar{\text{r}}$
Conclusion.	$\bar{\text{p}}$

We now construct the material conditional that has as its antecedent the conjunction of the premises—'KCApqr$\bar{\text{r}}$'—and as its consequent the conclusion of the argument—'$\bar{\text{p}}$'; and test this material conditional to determine whether or not it is a tautology.

$$
\begin{array}{lcc}
\text{C} & \multicolumn{2}{c}{\text{CKCApqr}\bar{\text{r}}\bar{\text{p}}} \\
\bar{\text{K}} & \multicolumn{2}{c}{\bar{\text{K}}\text{CApqr}\bar{\text{r}},\ \bar{\text{p}}} \\
\bar{\text{C}} & \multicolumn{2}{c}{\bar{\text{C}}\text{Apqr},\ \bar{\bar{\text{r}}},\ \bar{\text{p}}} \\
\text{A} & \text{Apq},\ \bar{\bar{\text{r}}},\ \bar{\text{p}} & \bar{\text{r}},\ \bar{\bar{\text{r}}},\ \bar{\text{p}} \quad = \\
= & \text{p},\ \text{q},\ \bar{\bar{\text{r}}},\ \bar{\text{p}} & \bar{\text{r}},\ \text{r},\ \bar{\text{p}} \\
 & \text{p},\ \text{q},\ \text{r},\ \bar{\text{p}} & [\text{closed}] \\
 & [\text{closed}] &
\end{array}
$$

The reduction terminates entirely in closed branches; the original formula is a tautology; the argument is valid.

Two statements are *materially equivalent* if and only if they materially imply each other. To determine whether or not a symbolized statement α is equivalent to another, β, we construct the conjunction of conditionals 'KC$\alpha\beta$C$\beta\alpha$' and reduce. Note that the reduction must split into two conditionals at the first step, so that the two implications are tested in separate branches of the reduction. To test, for example, whether or not 'Cpq' is equivalent to '$\bar{\text{K}}$p$\bar{\text{q}}$', we construct the conjunction of conditionals and reduce:

$$
\begin{array}{c}
\text{K} \quad \text{KCCpq}\bar{\text{K}}\text{p}\bar{\text{q}}\text{C}\bar{\text{K}}\text{p}\bar{\text{q}}\text{Cpq} \\
\begin{array}{cc}
\begin{array}{c}
\text{C} \quad \text{CCpq}\bar{\text{K}}\text{p}\bar{\text{q}} \\
\bar{\text{C}} \quad \bar{\text{C}}\text{pq},\ \bar{\text{K}}\text{p}\bar{\text{q}} \\
\begin{array}{cc}
\bar{\text{K}} \quad \text{p},\ \bar{\text{K}}\text{p}\bar{\text{q}} & \bar{\text{K}} \quad \bar{\text{q}},\ \bar{\text{K}}\text{p}\bar{\text{q}} \\
\text{p},\ \bar{\text{p}},\ \text{q} & \bar{\text{q}},\ \bar{\text{p}},\ \text{q} \\
[\text{closed}] & [\text{closed}]
\end{array}
\end{array}
&
\begin{array}{c}
\text{C}\bar{\text{K}}\text{p}\bar{\text{q}}\text{Cpq} \quad \text{C} \\
\bar{\bar{\text{K}}}\text{p}\bar{\text{q}},\ \text{Cpq} \quad \text{K}^{4} \\
\begin{array}{cc}
\text{C} \quad \text{p},\ \text{Cpq} & \bar{\text{q}},\ \text{Cpq} \quad \text{C} \\
\text{p},\ \bar{\text{p}},\ \text{q} & \bar{\text{q}},\ \bar{\text{p}},\ \text{q} \\
[\text{closed}] & [\text{closed}]
\end{array}
\end{array}
\end{array}
\end{array}
$$

The reduction terminates entirely in closed branches; the original formula is a tautology; the two formulas are equivalent.

[4]Applications of the $^{=}$rule may be suppressed by simply omitting double negations, as is done later in this same reduction.

Exercises

All the exercises in this group are elementary valid arguments which are often used as rules of inference in deductive proof techniques. Demonstrate by reduction that each argument is valid.

1. Cpq
p
———
q — Modus Ponens

2. Cpq
$\bar{q}$
———
$\bar{p}$ — Modus Tollens

3. Cpq
Cqr
———
Cpr — Hypothetical Syllogism (cf. formula 4, §3.4, p. 32)

4. Apq
$\bar{p}$
———
q — Disjunctive Syllogism

5. Kpq
———
p — Simplification of Conjunction. (An implication may be treated like an argument with one premise.)

6. p
q
———
Kpq — Conjunction

7. p
———
Apq — Addition

8. CpKqr
———
Cpq

9. CApqr
———
Cpr

Each pair of formulas in this group is an equivalent pair. Demonstrate their equivalence by reduction.

10.	$\bar{K}pq$ $A\bar{p}\bar{q}$	De Morgan's Law	11.	$\bar{A}pq$ $K\bar{p}\bar{q}$	De Morgan's Law
12.	Apq Aqp	Commutativity of Alternation	13.	Kpq Kqp	Commutativity of Conjunction
14.	ApAqr AApqr	Associativity of Alternation	15.	KpKqr KKpqr	Associativity of Conjunction
16.	KpAqr AKpqKpr	Distribution	17.	ApKqr KApqApr	Distribution
18.	KCpqCqp $AKpqK\bar{p}\bar{q}$	Material Equivalence	19.	Cpq $C\bar{q}\bar{p}$	Contraposition
20.	Cpq $A\bar{p}q$	Material Implication	21.	Cpq $\bar{K}p\bar{q}$	Material Implication
22.	p App	Tautology	23.	p Kpp	Tautology
24.	CpCqr CKpqr	Exportation			

4 CANCELLATION

Cancellation is a proof technique which abbreviates reductions of the type just described so that one moves more directly from the sequent whose sole member is the original formula to an equivalent set of sequents whose members are all basic formulas. The procedure may be divided into three parts:

- Cancellation of the original formula
- Branch chart construction
- Closure

4.1 Cancellation of the Original Formula. This part of the procedure consists essentially in making certain marks on the original formula which anticipate what would occur if the formula were fully reduced by the procedure of §3. In truth-functional reductions there are three basic things, and several combinations of these, that can happen in application of a reduction rule to a complex formula. First, a truth-functional operator may simply be eliminated, as when the sequent containing 'Apq' is reduced to the sequent containing 'p, q'. This is indicated in cancellation by drawing a diagonal slash '/' through the operator to be eliminated, '$\not{A}pq$'. Second, a truth-functional operator may be eliminated in such a way that two sequents are generated, as when a sequent containing 'Kpq' is reduced to two sequents, one of which contains 'p' and the other of which contains 'q' in place of 'Kpq'. This is indicated by drawing a split marker '└∧┘' so that the vertical line passes through the operator being cancelled, and the two divisions of the horizontal line embrace the subformulas being joined by the operator, 'Kpq'. When drawing a split marker, it is convenient to remember that each half of the horizontal line embraces a single wff. We shall speak later of formulas and other marks as being within the left- or right-hand side of a split marker in an obvious sense. Finally, one or more negations may be introduced when an operator is eliminated. This is indicated by drawing the negation bar over the left-most sign of the formula being negated.

The cancellation rules are as follows. The arrow '→' in these rules indicates that a formula of the form shown at the tail of the arrow is marked so as to produce what is shown at the head of the arrow.

$$A\alpha\beta \to \not{A}\alpha\beta \qquad \bar{A}\alpha\beta \to \bar{\not{A}}\bar{\alpha}\bar{\beta}$$

$$K\alpha\beta \to \not{K}\alpha\beta \qquad \bar{K}\alpha\beta \to \bar{\not{K}}\bar{\alpha}\bar{\beta}$$

$$C\alpha\beta \to \not{C}\bar{\alpha}\beta \qquad \bar{C}\alpha\beta \to \bar{\not{C}}\alpha\bar{\beta}$$

$$\bar{\bar{\alpha}} \to \not{\bar{\bar{\alpha}}}$$

The occurrence of other formulas, subformulas and bits of cancellation debris is suppressed in this presentation of the rules.

It is helpful at this stage of things to do precisely what cancellation makes it unnecessary to do: rewrite the formula after each rule is applied. In the example shown in Figure 1 we work our way through a long formula step by step. In all subsequent examples, only the original formula and the formula as it appears after cancellation will be written.

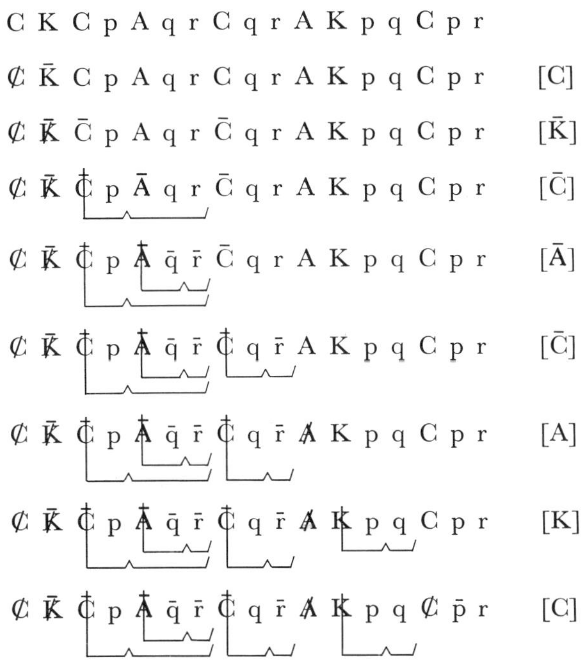

FIGURE 1

For this example only, the rule applied at each step is indicated to the right of the formula. Note that we move always from left to right through a formula, and that split markers within other split markers are written in such a way that the new split marker occurs entirely inside the original one, as with the application of the Ā-rule. Each half of a split marker always embraces a single (basic or complex) wff.

4.2 Branch Chart Construction. The significant marks that remain after cancellation are basic formulas and split markers. The *branch chart* displays the sequents of basic formulas resulting from cancellation.

A *branch space* is the area which contains all the members of a given sequent. A single branch space is assumed to exist below the original formula. This branch space may be divided into two or more branch spaces, which overlap to the left. *Split markers* divide every branch space to which they are relevant into two horizontal branch spaces. Basic formulas are written directly below their original occurrences, in every branch space to which they are relevant. Relevance of marks—basic formulas and split markers—to branch spaces is determined as follows:

(i) Marks which do not occur within a split marker are relevant to every branch space below their original occurrence in the cancelled formula.

(ii) Marks which occur in the left-hand side of a split marker are relevant only to the upper of each pair of branches created by that marker.

(iii) Marks in the right-hand side of a split marker are relevant only to the lower of each pair of branches created by that marker.

If a *split marker* is relevant to a branch according to the three rules of relevance, it splits the branch into two branches. If a *basic formula* is relevant to a branch according to the three rules of relevance, it is written in that branch, directly below its occurrence in the original formula.

For the next several examples, we will number split markers in each cancellation so that we can refer to them for explanation. In Figure 2

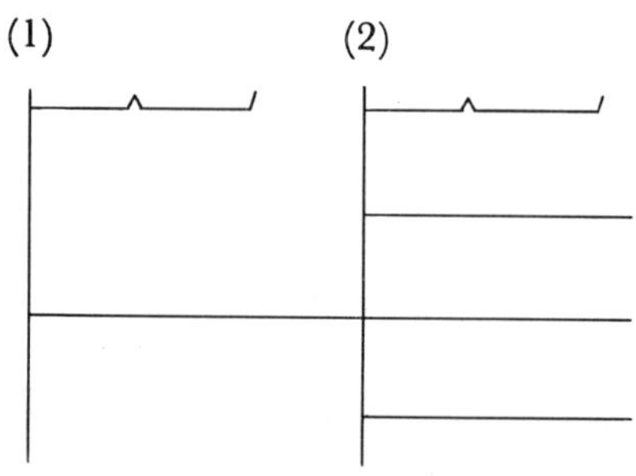

FIGURE 2

the split marker (1) divides the entire branch space into two branches. Split marker (2), by the three rules of relevance above, is relevant to every branch below its occurrence. It therefore divides both branches into two additional branches, making four in all. In the two examples in Figure 3, the second split marker falls within the first split marker. By the rules of relevance, the second split marker is then relevant to only one

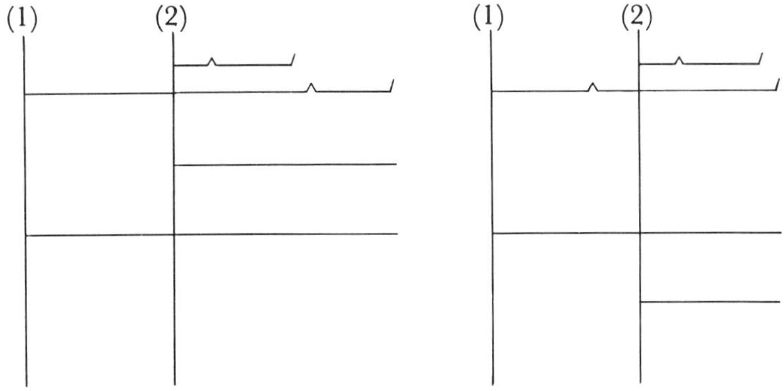

FIGURE 3

of the branches created by the first: the upper branch if it is in the left-hand side of (1), the lower branch if it is in the right-hand side of (1). A split marker divides any branch to which it is relevant. In Figure 4 the split marker (1) divides the entire branch space into two branches. Split marker (2) is within the left-hand side of (1), and therefore divides the upper of the two branches created by (1). Split marker (3) is in the right-hand side of (1) and divides the lower of the two branches created

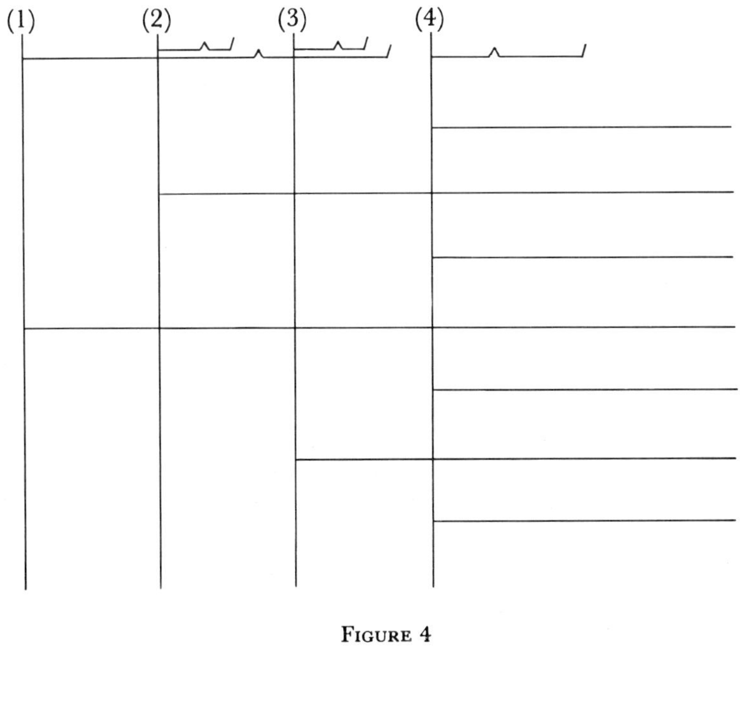

FIGURE 4

by (1). Split marker (4) is not within any other split marker, and is therefore relevant to *every* branch beneath its occurrence. It divides every branch created by previous split markers into two branches.

Finally, following the same rules of relevance, we enter basic formulas in the branch chart. A basic formula is entered *directly under its original occurrence in the cancelled formula* in every branch to which it is relevant. After doing a few problems, one begins to find ways of shortening the procedure further—for example, by writing some basic formulas in the overlap area rather than under their original occurrence when a formula must be entered more than once. *This is not advisable.* The procedure is, in its present form, as compressed as it can be without introducing ambiguity.

Examples C K p C p q q C A p q A p K q r

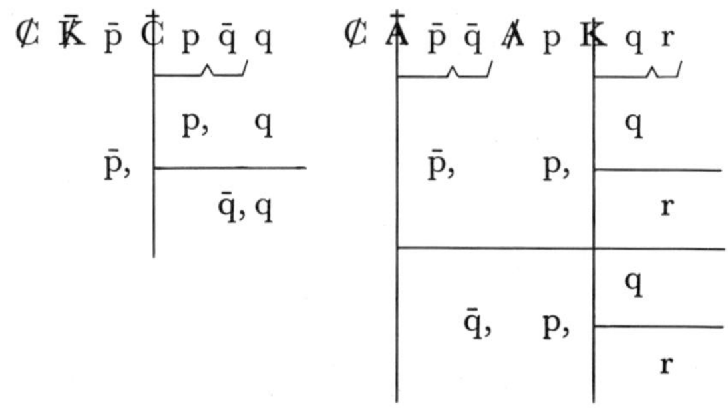

In the first example above, there are two branches, one being the sequent 'p̄, p, q', the other 'p̄, q̄, q'. 'p̄' is a member of both branches because it is in the overlap area at the left of the split marker. In the second example, there are four branches. The sequents are 'p̄, p, q', 'p̄, p, r', 'q̄, p, q', and 'q̄, p, r'.

Finally, in Figure 5 we show the full cancellation and branch chart for the long example of §4.1 (Figure 1). There are twelve branches in the branch chart. Since branches overlap to the left, the topmost branch contains the members 'p', 'q', 'p', 'p̄', 'r'. The ninth branch (from the top) contains the members 'r̄', 'q', 'p', 'p̄', 'r'.

The question to be answered with respect to the branch chart will always be whether or not each branch contains a pair of opposites. The answer is often obvious for major portions of a branch chart without tracing down each branch. Without more than a glance, for example, one can see that the top four branches all contain 'p' in the overlap area and 'p̄' over at the right. The bottom four branches contain 'r̄' in the overlap area and 'r' at the right. A slightly closer inspection reveals that the middle four branches are closed also.

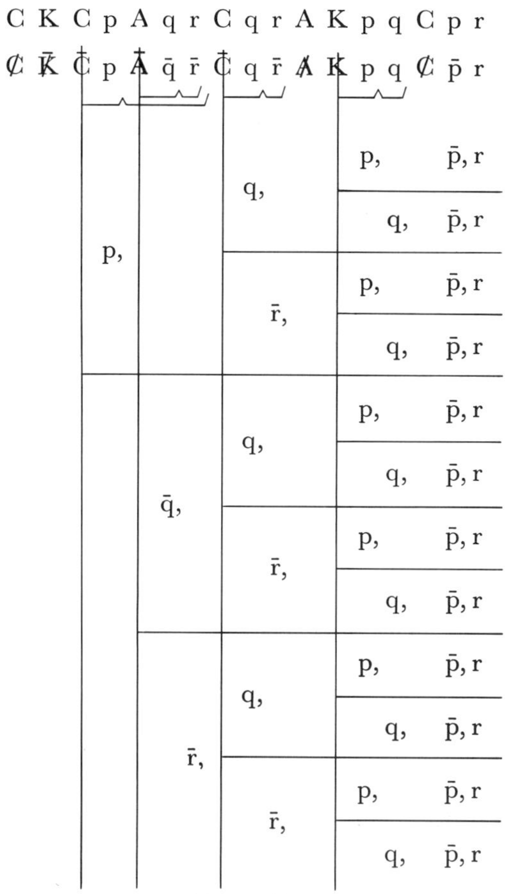

FIGURE 5

4.3 Closure. A *branch* is that sequent whose members occur in a given branch space, including areas of overlap. As with reduction, a branch (sequent) is closed if it contains a pair of opposites.

Exercises

Redo the exercises following §3.4, above, using cancellation instead of reduction.

4.4 Implication, Validity and Equivalence. Since cancellation is merely a mechanically abbreviated version of reduction, it allows us to demonstrate precisely the same things, but with less rewriting of formulas. To demonstrate that 'KpAqr' materially implies 'Apr', we construct the related conditional (cf. §3.5, above) and cancel, obtaining precisely the same terminal sequents (now called "branches") as we did in the reduction:

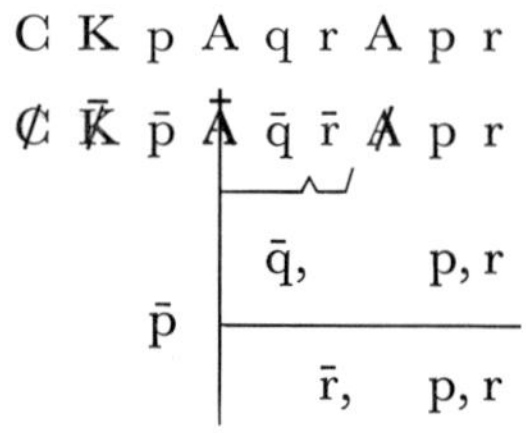

To demonstrate the validity of the argument

$$\begin{array}{l} \mathrm{CApqr} \\ \bar{\mathrm{r}} \\ \hline \bar{\mathrm{p}} \end{array},$$

we construct the conditional that has as its antecedent the conjunction of the premises and as its consequent the conclusion, as before, and cancel, again obtaining the same branches as were obtained in the reduction:

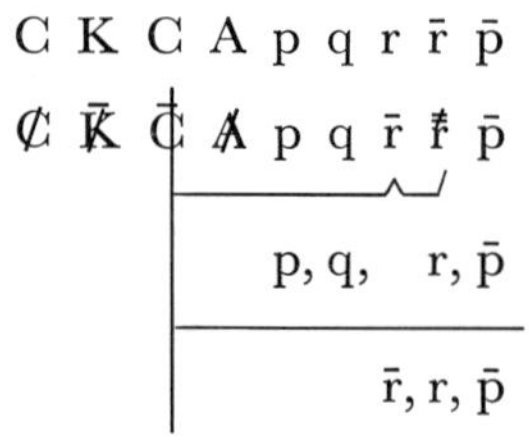

Finally, to demonstrate that two formulas are materially equivalent, we construct the conjunction of conditionals as in §3.5 and cancel:

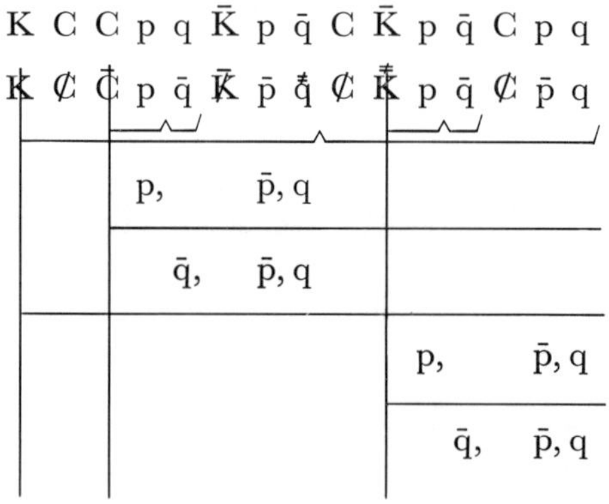

Exercises

Redo the exercises following §3.5 above, using cancellation instead of reduction.

5 CANCELLATION OF QUANTIFIED FORMULAS

We now extend the cancellation procedure in order to operate on formulas containing quantifiers. To the vocabulary of §2, we make the following additions:

∃ Existential Quantifier
∀ Universal Quantifier.
A denumerable stock of *predicates* of varying degrees:
F, G, H, J, F′, G′, ⋯ F″, ⋯
A denumerable stock of *individual variables*:
x, y, z, ⋯ x′, y′, z′, ⋯ x″, ⋯
A denumerable stock of *individual constants*: (the set of positive integers) 1, 2, 3, 4, 5, ⋯
A denumerable stock of *dummy constants*: (lower-case Roman letters, perhaps with primes) a, b, c, d, ⋯ a′, b′, ⋯ a″, ⋯

To the conditions on well-formedness in §2.2 we make the following additions (a wff is one of (i), (ii) or (iii) in §2.2, or one of the following):

(iv) A predicate of degree *n* followed by *n* individual variables or constants.

(v) The result of writing '∃' or '∀', followed by an individual variable, to the left of a wff.

5.1 Quantifier Cancellation Rules. When a quantifier is cancelled, a numerical constant or literal dummy constant is introduced at every occurrence of the individual variable governed by that quantifier, according to the cancellation rules. These dummies and constants are taken in numerical or alphabetical order, so that the first unused dummy or constant is always the one to be used in the application of any given rule. In the following rules, 'd' represents the first unused dummy and 'c' represents the first unused numerical constant. 'Φ' is a syntactical variable representing a well-formed context for an individual variable.

$$\exists x\,\Phi x \rightarrow \overset{d}{\not\exists\!\not x}\,\underset{d}{\Phi x} \qquad \bar{\exists} x\,\Phi x \rightarrow \overset{c}{\not\bar{\exists}\!\not x}\,\underset{c}{\bar{\Phi} x}$$

$$\forall x\,\Phi x \rightarrow \overset{c}{\not\forall\!\not x}\,\underset{c}{\Phi x} \qquad \bar{\forall} x\,\Phi x \rightarrow \overset{d}{\not\bar{\forall}\!\not x}\,\underset{d}{\bar{\Phi} x}$$

Hereafter only the quantifiers will be cancelled.

Examples

∃x K Fx Gx	∀x C Fx Gx	∀x C ∃y Fxy ∃z Gzx
a	1	1 2 a
∄x K̸ Fx Gx	∀̸x C̸ F̄x Gx	∀̸x C̸ ∄y F̄xy ∄z Gzx
a a	1 1	12 a1

5.1.1 The Quantifier Rules. Like the reduction and cancellation rules for truth-functional operators, the quantifier cancellation rules are designed to derive sequents that are true under precisely the same conditions as the sequents from which they were derived. The constants and dummies introduced in quantifier cancellation are ordered, and the cancellation rules require that the first unused constant or dummy be chosen in the application of a given rule. This amounts to an *arbitrary* choice of constants and dummies in cancellation, and the word "arbitrary" is important here.

If '∀x Fx' is true, then 'Fi' must hold for every individual, 'i', no matter how that individual is selected. Whether the cancellation forces us to write 'F1' or 'F7' or 'F143', if '∀x Fx' is true, the result of cancelling it must also be true. Working the other way, if 'Fi' is true for any arbitrarily chosen individual, then it must be true of all. If, for example, 'F7' is true, and we can be sure that it *doesn't matter* that '7' is the constants but that '6' or '148', etc., would do as well, then '∀x Fx' must also be true. The arbitrary choice of ordered constants, and the restrictions involved in closing branch charts, are designed to guarantee that it doesn't matter which particular constant occurs in a basic subformula.

'∃x Fx' requires only that 'Fi' be true for at least one individual, 'i'. It is not necessary that 'Fi' hold for every arbitrarily chosen individual. The use of dummies in the cancellation of '∃' allows us to delay choosing an individual constant at this point in the cancellation procedure. The dummy will be replaced by an appropriately chosen constant later on, and for now it simply represents "some individual." Again, working from the cancelled formula back to the original, if 'Fi' is true for some individual 'i', then '∃x Fx' is also true.[5]

5.2 The Branch Chart. Basic formulas, with their newly acquired dummies and constants, are entered in the branch chart according to the

[5] The ∃-rule and the ∀-rule may be viewed as the only basic quantifier rules, as the A-rule and the K-rule may be viewed with respect to the other truth-functional rules. The $\bar{\exists}$-rule and $\bar{\forall}$-rule may then be viewed as definitional. Referring to the definitional equivalences involving quantifiers in I.5, we may easily enough see that equivalent formulas produce the same results on cancellation. '∀x Fx' and '$\bar{\exists}$x F̄x', for example, produce precisely the same patterns on cancellation.

same rules as for truth-functional cancellation. In addition, all occurrences of dummies and constants written over cancelled quantifiers are entered in the branch chart as if they were bits of formula. See Figure 6 for an example:

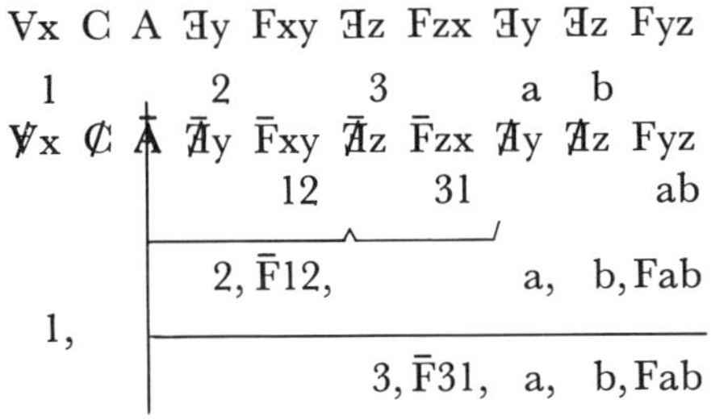

FIGURE 6

5.3 Consolidation and Closure. As was the case with truth-functional cancellation, the branches resulting from the cancellation of quantified formulas are closed if each branch contains a pair of opposites, i.e., some basic formula and its negation. This means that both members of a potential pair of opposites must have precisely the same numerical *constants* in precisely the same positions in order for closure to occur. 'F_1, $\bar{F}_1$' is a pair of opposites, as is 'G_{21}, $\bar{G}_{21}$'. 'F_1, $\bar{F}_2$' is *not* a pair of opposites, nor is 'G_{21}, $\bar{G}_{12}$'. Examples are given in Figure 7.

```
∀x C K Fx Gx Fx                      ∀x ∀y C Fxy A Gxy Fxy
 1                                    1  2
∀̸x C̸ K̸ F̄x Ḡx Fx                      ∀̸x ∀̸y C̸ F̄xy A̸ Gxy Fxy
        1  1  1                                12     12  12
 1,     F̄1, Ḡ1, F1  [closed]           1,  2,   F̄12,   G12 F12  [closed]
```

FIGURE 7

Since two of the quantifier cancellation rules introduce, to the branch chart, dummies rather than numerical constants, further steps will be necessary in many cases in order to determine whether or not a branch is closed. Some branches may be open after cancellation because their basic formulas contain occurrences of dummy constants, as 'Fab' does in Figure 6 in §5.2. The object of *consolidation* is to close as many branches as possible *by letting constants replace dummies in the branch chart.* The replacement moves which may be made in order to close a branch will be made more explicit in the following sections. The examples given in Figure 8 involving such replacements will be permissible. If a consolidation closes

```
C ∀x Fx ∀x Fx                         ∀x C ∃y Fxy ∃z Fxz
   a       1                           1      2       a
C̸ ∀̸x F̄x ∀̸x Fx                         ∀̸x C̸ ∃̄y F̄xy ∄z Fxz
      a      1                                    12      1a
   a, F̄a,  1, F1   [a = 1 ]            1,      2, F̄12  a, F1a   [a = 2 ]
                    [closed]                                     [closed]
```

FIGURE 8

every branch by permissible replacements, then the original formula is shown to be a truth of logic. If branches are left open, nothing is settled, for a proof may be forthcoming through cancellation of duplicate subformulas as described below in §5.4.4.

The replacement of dummy constants by numerical constants may be carried out within two restrictions, the Uniformity Requirement and the Order Requirement.

5.3.1 The Uniformity Requirement. If a dummy occurs in the branch chart to the left of a split marker, that dummy is said to *govern* the split created by that marker. The replacement of a dummy by a numerical constant must be uniform through every branch governed by that dummy. Here in the example in Figure 9, the occurrence of 'a' to the left of the split marker invokes the uniformity restriction, requiring that 'a' be replaced by the same constant in each of the two branches. Therefore the replacement of 'a' by '1' in the upper branch and by '2' in the lower branch is forbidden, and there is no closure.

A certain shortening and simplification of the cancellation procedure is possible if we modify the requirement that the first *unused* constant be introduced at each point where the cancellation rules demand a constant: *If a constant is introduced in the left-hand side of a split marker, it does not count*

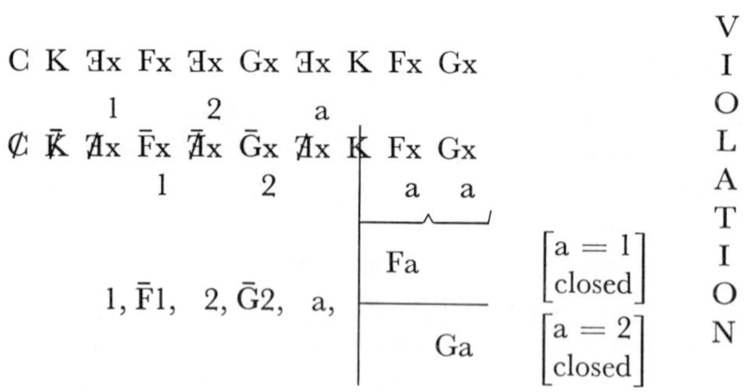

FIGURE 9

as "used" in choosing constants to be introduced in the right-hand side of the same split marker. Since the two cancelled quantifiers are in opposite halves of a split, the results of their cancellations will never appear in the same branch, and no harm will result if the same constant is introduced in the two distinct branches. In the example in Figure 10, the dummy 'a' governs the split marker, invoking the uniformity restriction. Since the two universal quantifiers occur in opposite halves of the same split marker, it is permissible to replace both with the constant '1'.

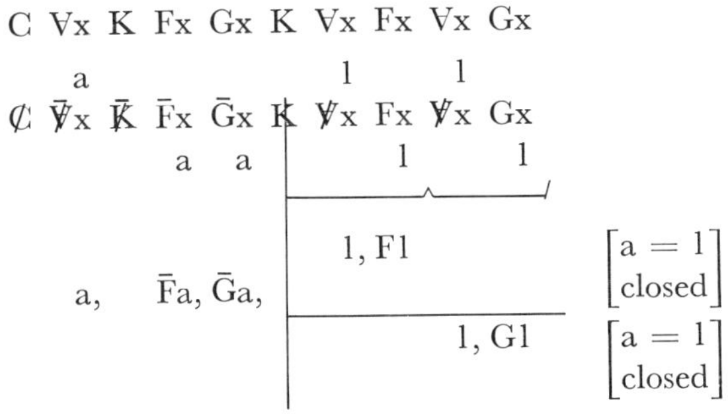

FIGURE 10

The point of the Uniformity Requirement is easy enough to see. The fact that some individual has a given property, F, and that some individual has a given property, G, is not sufficient to guarantee that some individual has both properties, F and G. Let 'Fx' represent "x is a freshman," and 'Gx' represent "x is a giraffe." There are some freshmen, '∃x Fx', and there are some giraffes, '∃x Gx', but it does not follow that something is both a freshman and a giraffe—'∃x K Fx Gx' is not implied by the conjunction of '∃x Fx' and '∃x Gx'.

5.3.2 Order Requirement. This restriction is rather opaque and complicated to state, but it can be made simple and automatic in practice. For each cancellation we construct a quantifier chart (Q-chart) which reflects the order in which quantifiers are cancelled and their overlapping scopes.

5.3.3 Q-Chart Construction. When the first quantifier in a formula is cancelled, its dummy or constant index is written in the Q-chart. When the *n*th quantifier is cancelled, its index is written in the Q-chart under the index of the $(n-1)$th quantifier if the *n*th quantifier is within the scope of the $(n-1)$th; if the *n*th quantifier is not within the scope of the $(n-1)$th, its index initiates a new column; if the *n*th quantifier is not within the scope of the $(n-1)$th, but both are within the scope of the

$(n-m)$th, both are written under the index of the $(n-m)$th quantifier, but neither under the other. If any constants occur in the original uncancelled formula, they are written at the top of the entire Q-chart, so as to head every column.

If replacements of dummy constants by numerical constants are permissible under the Order Requirement, it will be possible to trace a *path* through the Q-chart in conformity with the following principles:

(1) The path strikes each dummy or numerical constant precisely once.
(2) The path strikes a higher entry in any column before a lower entry in the same column.
(3) The path strikes a fixed constant before it strikes any dummy which is to be replaced by that constant.

In the first example in Figure 11, no path can be traced which meets the three conditions and which allows the dummy 'a' to be replaced by the

	$\exists x\ \forall y\ Fxy$	$K\ \exists x\ Fx\ \forall x\ Gx$	$\forall x\ C\ \exists y\ Fxy\ \exists z\ Fzx$
	a 1	a 1	1 2 a
	$\not\exists x\ \not\forall y\ Fxy$	$\not K\ \not\exists x\ Fx\ \not\forall x\ Gx$	$\not\forall x\ \not C\ \bar{\not\exists} y\ \bar{F}xy\ \not\exists z\ Fzx$
	al	a 1	12 al
Q-charts	a / 1	a 1	1 / 2 a

FIGURE 11

constant '1'. In the second example, a path can be traced allowing 'a' to be replaced by '1', and in the third example, paths can be traced to allow the replacement of 'a' by either of '1' or '2'.

5.3.4 Q-Closure Conditions. In addition to replacing dummies by numerical constants which occur in the branch chart, we must in some circumstances introduce new constants into a cancellation, where dummies occur in both members of a potential closure pair and no constant is present which might replace them:

$$C\ \forall x\ Fx\ \exists x\ Fx$$

a b

$$\not C\ \bar{\not\forall} x\ \bar{F}x\ \not\exists x\ Fx$$

a b

a, $\bar{F}$a, b, Fb $\left[\begin{array}{c} a = b = 1 \\ \text{closed} \end{array}\right]$

This special situation reflects the sense of the universal quantifier: "For all x (if there are any): —." The universal quantifier of modern logic carries with it no commitment to the existence of any individuals at all, while the existential quantifier, under the usual interpretations, does have such import: "There is at least one x such that:—." As others have noted, '∀x Fx' implies '∃x Fx' only on the (tacit) assumption that the universe is non-empty. The importance of this assumption will be made clear in later chapters, but the assumption is reflected in our permission to create an individual constant and insert it in a branch chart where no constant appeared originally, as in the example above.

We can now state succinctly the closure conditions for quantificational formulas.

Q-closure I. *Dummy constants in the branch chart may be replaced by numerical constants from the branch chart within the restrictions of the Uniformity Requirement and the Order Requirement.*

Q-closure II. *When two dummies appear in a pair of opposing formulas and no constant is present to replace them, a new numerical constant may be introduced into the cancellation to replace the two dummies.*

Exercises

Demonstrate by cancellation that the first formula of each pair implies the second.

1. ∀x Fx ∃x Fx	2. ∃x K Fx Gx K ∃x Fx ∃x Gx	3. A ∀x Fx ∀x Gx ∀x A Fx Gx
4. ∀x C Fx Gx C ∀x Fx ∀x Gx	5. ∀x C Fx K Gx Hx ∀x C Fx Hx	6. ∀x C A Fx Gx Hx ∀x C Fx Hx
7. ∀x C Fx Gx C ∃x Fx ∃x Gx	8. ∀x C Fx Gx $C\,\exists x\,\bar{G}x\,\exists x\,\bar{F}x$	

Demonstrate by cancellation the equivalence of each pair of formulas.

9. ∃x Fx $\bar{\forall}x\,\bar{F}x$	10. ∀x Fx $\bar{\exists}x\,\bar{F}x$	11. C ∃x Fx p ∀x C Fx p
12. C ∀x Fx p ∃x C Fx p	13. C p ∃x Fx ∃x C p Fx	14. C p ∀x Fx ∀x C p Fx
15. ∃x A Fx Gx A ∃x Fx ∃x Gx	16. ∀x K Fx Gx K ∀x Fx ∀x Gx	

5.4 Failure of Closure and Duplicates. There are several circumstances under which a branch may fail to close after the procedures of consolidation have been followed. Under some of these circumstances we can be sure that the branch can never be closed, and that the original formula is not provable. Under other circumstances, we cannot be certain whether or not the formula is provable, and further procedures must be followed in an attempt to close the branch.

5.4.1 No Opposites Present.

F1a, G2, Fba, H1

There is no pair of basic subformulas (bsfs) in the branch containing occurrences of the same predicate constant, with precisely one of the pair negated. *This branch can never be closed.*

5.4.2 Opposites, but with Different Constants Opposed.

F1a, G2, $\bar{\text{F}}$2a

An appropriate pair of bsfs occurs in the branch, but each of them contains a different *numerical constant* in a given position. *This branch can never be closed.*

5.4.3 Closure Prevented by the Uniformity Requirement. In the example in Figure 12, an appropriate pair of bsfs occurs in each branch, but closure

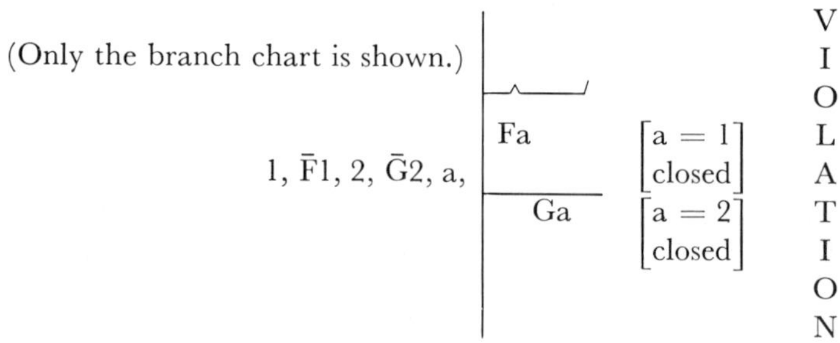

FIGURE 12

is prevented by the Uniformity Requirement. *This cancellation can never be fully closed.* One or the other of its branches will always remain open.

5.4.4 *Closure Prevented by the Order Requirement.* In Figure 13, the branch does not close. The replacement 'a = 1' would violate the Order Requirement. The branch, however, contains the dummy 'a'.

∃x C Fx ∀y Fy

a 1 Q-chart

∄x ⊄ F̄x ∀̸y Fy a

a 1 1

a, F̄a, F1

FIGURE 13

If closure fails, the occurrences of dummies in the branch chart may be taken to be abbreviations of the formulas introduced by the quantifiers whose cancellation resulted in the introduction of those dummies. In the present example, the dummy 'a' was introduced when the existential quantifier was cancelled, and this quantifier introduced the entire formula. In other cases, the quantifier which introduces a dummy may range over only part of the original formula. Such formulas, occurring in the branch chart in the abbreviated form of dummies, may then be recancelled and closure attempted again. In such a case, the formulas are called *duplicates*. While duplicates are never written out in unabbreviated form, we can show, for the example above, the sense of the duplicate abbreviation 'a' as follows:

$$[\exists x \, C \, Fx \, \forall y \, Fy], \quad \bar{F}a, \quad F1.$$

Open branches may be expanded by such duplicates as they contain in the hope that they can be closed on a subsequent consolidation by further cancellation of the duplicates. If an open branch contains no duplicates—no dummies occurring in the branch to abbreviate recurrences of the formulas they introduced—then there is no hope of further cancellation and consolidation, and we know that the branch can never be closed.

A duplicate is attacked by cancelling it, constructing its branch chart, and adding the results to the original branch. It is not necessary actually to go through the process of cancelling the duplicate; the main part of this job has already been done in the original cancellation. The branch chart for a duplicate will always be some portion of the original branch chart with some new dummies and fixed constants. The relevant portion of the branch chart will be that underneath the occurrence of the duplicate whose quantifier is flagged by a dummy in the original formula. The Q-chart gives a quick way of telling which quantifiers must be re-cancelled when a duplicate is attacked. All entries which occur under the relevant dummy, in the same column of the Q-chart, must be re-cancelled. In the present example, the relevant portion of the branch chart is the entire branch chart, since the quantifier which introduced the dummy 'a' was the first operator in the entire original formula. In the next example, this will not be the case.

In introducing new dummies and constants through the cancellation of a duplicate, it is convenient to use the old ones with superscripts. The superscript is always the dummy that abbreviated the duplicate being attacked. Attacking the duplicate 'a' of the present problem, we obtain the expanded branch chart and Q-chart given in Figure 14.

The branch may be closed by the replacement '$a^a = 1$', or '$a = 1^a$'. Either replacement satisfies the Order Requirement.

$$\text{a}, \bar{\text{F}}\text{a}, \text{F1}, \text{a}^{\text{a}}, \bar{\text{F}}\text{a}^{\text{a}}, \text{F1}^{\text{a}}$$

Q-chart	
a	a^{a}
1	1^{a}

FIGURE 14

If closure still fails after a duplicate has been cancelled and the results added to the branch, further duplicates may be cancelled. First, any additional dummies (duplicates) from the original cancellation are attacked, then duplicates which are abbreviated by dummies which are introduced by the cancellation of those duplicates, and so on.

The duplicates abbreviated by dummies in the original branch chart are called *first-generation duplicates*. Duplicates produced by attacking first-generation duplicates are called second-generation duplicates, and so on. The formula abbreviated by 'a^{a}' in the last version of the branch chart will be a second-generation duplicate. In order to ensure that no avenues of closure are overlooked, we require that no duplicate of the $(n+1)$th generation may be attacked until all duplicates of the nth generation have been attacked.

The cancellation of duplicates will not, of course, allow closure of every branch where closure is prevented by the Order Requirement. We can, however, set aside certain formulas which can always be closed in this way, now that we have used one of them to generate a simple example. *Formulas that contain no predicates of degree greater than one are exempted from the Order Requirement.* If the Order Requirement is all that prevents closure of the branch chart when the original formula contains only predicates of degree one (one-place predicates), then the cancellation of duplicates will always allow closure, and the Order Requirement can therefore be ignored.[6] If closure fails for one of the reasons considered in §§5.4.1, 5.4.2 or 5.4.3, however, there is no such way out, and the branches cannot be closed.

In the example given in Figure 15, the predicates are all of degree two, and the Order Requirement prevents closure of one branch. The middle branch does not close. The replacement '$\text{a} = \text{c} = 2$' would violate the Order Requirement. The branch, however, contains three dummies, 'a', 'b' and 'c'. Before attacking the first duplicate, 'a', we may pause

[6] The reason for this is easy enough to see. The Order Requirement can prevent closure only when quantifiers overlap in scope, as reflected in there being more than one entry in a vertical column of the Q-chart. Where all predicates are of degree one, overlapping quantifiers can be untangled in an equivalent formula where no quantifiers overlap and where no column in the resulting Q-chart has more than one entry. The example just considered, for example, '∃x C Fx ∀y Fy', is equivalent to 'C ∀x Fx ∀y Fy'. The Q-chart of this formula would be simply: a 1, and the Order Requirement would not prevent the replacement of 'a' by '1'.

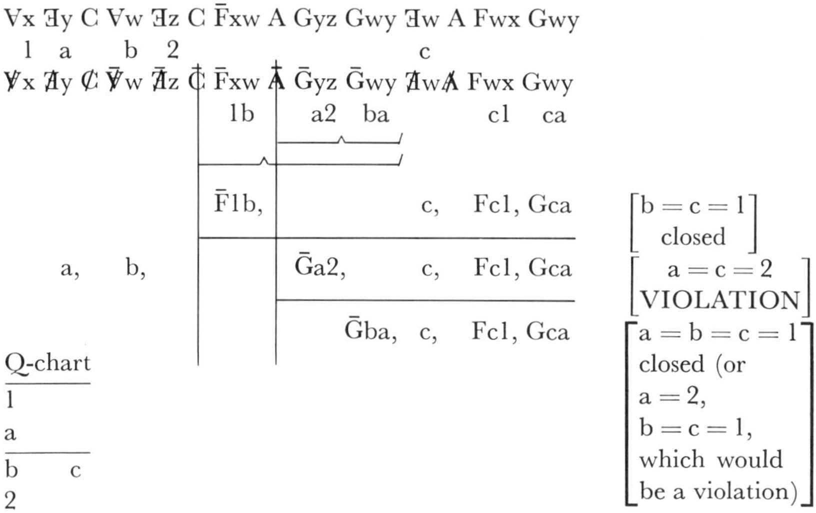

FIGURE 15

to consider precisely what each duplicate in the middle branch abbreviates, and what other dummies and constants will be introduced upon such cancellation.

The duplicate 'a' is introduced by the cancellation of the first existential quantifier in the formula. It is an abbreviation for the entire formula over which that quantifier ranges at the point where it is cancelled. (Since the first universal quantifier has already been cancelled at this point, the variable 'x' will already have been replaced by '1'.) Specifically, the dummy 'a' abbreviates an occurrence in the branch chart of the following formula:

∃y C ∀w ∃z C F̄1w A Gyz Gwy ∃w A Fw1 Gwy.

The dummies and constants which must be involved in the cancellation of the duplicate are all those which occur below 'a' in a vertical column of the Q-chart: 'b', '2' and 'c'. This is the dummy that we will shortly attack.

The duplicate 'b' is introduced by the cancellation of '∀̄w' in the antecedent of the main conditional of the formula. The dummy 'b' abbreviates an occurrence of the entire wff over which this quantifier ranges, specifically:

∀̄w ∃z C F̄1w A Gaz Gwa.

The duplicate 'c' is introduced by the cancellation of '∃w' in the consequent of the main conditional. It abbreviates an occurrence of the formula:

∃w A Fw1 Gwa.

Again, it is never necessary actually to cancel such formulas, since this has already been done. One must simply pick out the portion of the original formula which the dummy abbreviates, and rewrite the portion of the branch chart which resulted from the cancellation of that formula, with appropriate replacement of dummies and constants by superscripted dummies and constants.

Let us attack the duplicate 'a' of the unclosed branch of the present problem. Its branch chart is given in Figure 16 (omitting branches of the original which are already closed, we show only the middle branch plus the result of cancelling the duplicate abbreviated by 'a' in that branch). The full cancellation of this formula is shown in Figure 17.

<table>
<tr><td rowspan="3">$[a], b, \bar{G}a2, c, Fc1, Gc\partial, a^a, b^a,$</td><td rowspan="3">$\bar{F}1b^a,$</td><td colspan="2">$c^a, Fc^a1, Gc^aa^a$</td></tr>
<tr><td>$\bar{G}a^a2^a,$</td><td>c^a, Fc^a1, Gc^aa^a</td></tr>
<tr><td colspan="2">$\bar{G}b^aa^a, c^a, Fc^a1, Gc^aa^a$</td></tr>
</table>

Q-chart

<table>
<tr><td colspan="4">1</td></tr>
<tr><td colspan="2">a</td><td colspan="2">a^a</td></tr>
<tr><td>b</td><td>c</td><td>b^a</td><td>c^a</td></tr>
<tr><td>2</td><td></td><td></td><td></td></tr>
<tr><td></td><td></td><td>2^a</td><td></td></tr>
</table>

All branches: $\begin{bmatrix} c^a = a = 1;\ a^a = 2 \\ \text{closed} \end{bmatrix}$

FIGURE 16

All these branches are closed on the pair $\bar{G}a2$, Gc^aa^a, when the indicated replacements are made. Since all branches of the cancellation are now closed, the original formula is shown to be provable.

If the cancellation of duplicates does not immediately yield closure, we cannot be certain that the original formula is not provable. Each time a duplicate is cancelled, it generates new duplicates, and it is possible for a cancellation to cycle through duplicates indefinitely without closure. The following formula, for example, should not cancel to closed branches:

$$C\ \forall x\ \exists y\ Fyx\ \exists x\ \forall y\ Fxy.$$

That this formula should not hold for every two-place predicate can be made clear if we interpret 'Fxy' as "x is the father of y:" "If everybody has a father, then someone is everybody's father." The formula does not cancel to closed branches, and closure is prevented by the Order Requirement. The conditions described in §§5.4.1, 5.4.2 and 5.4.3 do not occur, so we cannot be sure on those grounds that the formula will not eventually close.

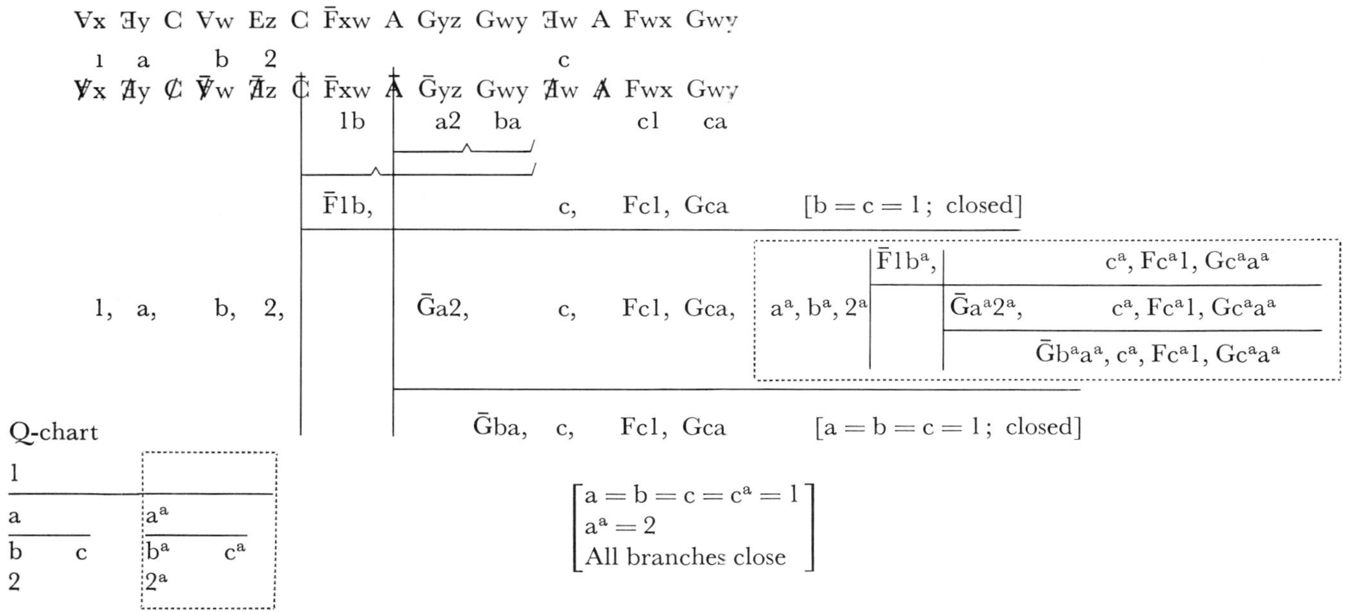

The replacement of 'a' by '2', which would have closed the center branch, was prevented by the Order Requirement. The cancellation of the duplicate abbreviated by the occurrence of 'a' in the branch chart introduces the elements in the branch chart enclosed in the broken line, and the superscripted entries in the Q-chart.

FIGURE 17

Successive generations of duplicates will be generated, and closure must be attempted each time. It can be demonstrated easily enough that a permissible closure pair will never occur no matter how many duplicates are generated, i.e., that a permissible path can never be traced through the Q-chart. But such a demonstration must be carried out outside the cancellation system itself.

5.5 The cancellation procedure just described will be the basic technique employed for detecting provable formulas for the remainder of the book. It is, once mastered, straightforward and mechanical. It is rarely necessary to resort to the cancellation of duplicates. In general, a quick operation with paper and pencil will give a definite answer as to whether or not a given formula is provable. A few practice problems should enable the student to master the basic procedures of the method before dealing with the extension of the method to modal logic. A group of such problems is found at the end of this section.

In Figure 18 are two examples of cancellations of quantified formulas, with the Q-charts and replacements shown.

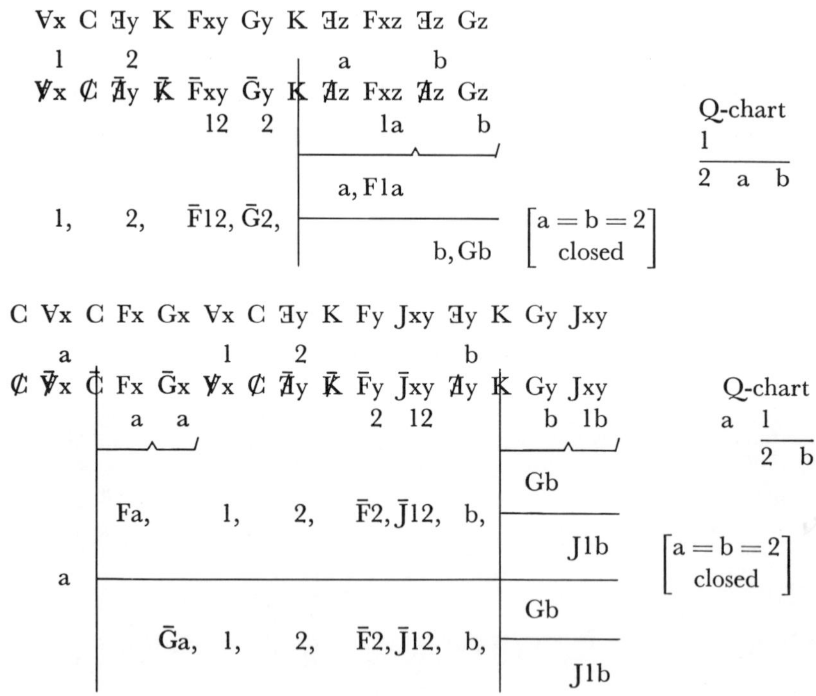

FIGURE 18

Exercises

Demonstrate by cancellation that the following formulas are tautologies.

1. $C \forall x\, C\, Fx\, \forall y\, C\, Gy\, Jxy\, C\, \forall x\, \bar{J}xx\, \bar{\exists} x\, K\, Fx\, Gx$
2. $C \forall x\, Jxx\, \forall x\, \exists y\, Jyx$
3. $C \forall x\, \forall y\, C\, Fxy\, Jxy\, C\, \exists x\, Fxx\, \exists x\, \exists y\, Jxy$

BIBLIOGRAPHY—CHAPTER II

[II.1] Binkley, R. W. and R. L. Clark. "A Cancellation Algorithm for Elementary Logic," *Theoria* **33** (1967), pp. 79–97.

[II.2] Clark, R. L. and Paul Welsh. *Introduction to Logic.* (Van Nostrand Reinhold, New York, 1962)

[II.3] Kanger, Stig. *Provability in Logic.* (Almqvist and Wikrell, Stockholm, 1957)

[II.4] Lewis, Clarence, I. *A Survey of Symbolic Logic.* (University of California Press, Berkeley, 1918)

[II.5] Lewis, Clarence, I. and Cooper H. Langford. *Symbolic Logic.* (Century, New York, 1938)

Chapter III

THE MONADIC MODALITIES

1 INTRODUCTION

Contemporary research in modal logic had its beginning with C. I. Lewis, who published as early as 1912 certain criticisms of the analysis of implication given by Russell and Whitehead in their systems of logic based on truth-functional connectives. Although the truth-functional material conditional is useful in evaluating elementary argument forms in order to determine their validity, and in testing for implication and equivalence in many contexts, it has definite limitations for analysis of "if-then" connections. These limitations are reflected partly in what are called the "paradoxes" of material implication.

The so-called paradoxes of material implication are not paradoxes in the literal sense of that term. They are paradoxical only if we give a strong "if-then" reading to the material implication formula '$C\alpha\beta$'. If 'C' is interpreted only as an abbreviation for expressions containing 'A' and '—' or expressions containing 'K' and '—', as it has been by some logicians, then there are no paradoxes. There are, however, a number of tautologies of truth-functional logic which become paradoxical if 'C' is given the strong "if-then" reading. The first of these we call the *antecedent paradox of material implication*: '$C\bar{p}Cpq$'; a false statement materially implies every statement. The corresponding *consequent paradox of material implication* is '$CqCpq$'; a true statement is implied by every statement. If we think of "is implied by" in the strong sense, as "is a consequence of," the absurdity becomes clear. So long as it is false that my car is red, then anything at all, say, "The Moon is made of green cheese," is a consequence of "My car is red." So long as it is true that my car is blue, then "My car is blue" is a consequence of anything at all, say, "Napoleon lost at Waterloo." One need not suggest that material implication is without use in order to argue that a stronger sense of implication might be very useful indeed. This is precisely the line that Lewis and other modal logicians have taken. The systems of modal logic to be studied here and in later chapters are all stronger than material implication, at least in the sense that they are not subject to the paradoxes of material implication.

Other paradoxical forms that hold for material implication but not for stronger senses of implication can be derived from the antecedent and consequent paradoxes. Of any two statements, either the first materially implies the second or the second materially implies the first: 'ACpqCqp'. Either "Napoleon lost at Waterloo" implies "The Moon is made of green cheese," or else "The Moon is made of green cheese" implies "Napoleon lost at Waterloo." Of any two statements, one materially implies either the other or its negation: 'ACpqCp$\bar{q}$', —that is, one or the other of 'Cpq' or 'Cp$\bar{q}$' is true for every 'p' and 'q'. "Today is Friday" implies either "Swiss cheese is made from goat's milk" or "Swiss cheese is not made from goat's milk." Any two false statements (and any two true statements) imply each other—i.e., are materially equivalent:

'CK$\bar{p}\bar{q}$KCpqCqp' and 'CKpqKCpqCqp'.

The paradoxes, which Lewis felt condemned material implication as a purported analysis of the if-then relation, are not theorems of systems of modal logic if we replace 'C' in the paradox formulas with the strongest if-then operator of a given modal system.

1.1 The Lewis Systems. Lewis attempted to give an improved analysis of implication by introducing into the Russell-Whitehead systems of formal logic a new logical operator, representing "It is possible that—." Two additional operators were defined, using the first: "It is necessary that—" and a strong if-then operator which Lewis called "strict implication." The operators are shown in Chapter I in Lewis' form, which is intended for use with the notation of *Principia Mathematica*. They are shown in Table 6 in the Polish form to be used in this text.

In Polish notation, we do not use a special operator for strict implication.

TABLE 6

Possibility	Mα	It is possible that α.
Necessity	Lα	It is necessary that α.
Strict Implication	LC$\alpha\beta$	If α, then β; or 'α' strictly implies 'β'.

1.2 Definitions. The operators 'M' and 'L' are interdefinable:

$$L\alpha =_{df} \bar{M}\bar{\alpha}.$$

In elementary logic, the material implication formula '$C\alpha\beta$' is equivalent by definition to the negation of the conjunction '$K\alpha\bar{\beta}$'. 'α' materially implies 'β' if and only if it is not the case that 'α' is true and 'β' false. The strict implication formula '$LC\alpha\beta$' is equivalent to the *impossibility* of the conjunction '$K\alpha\bar{\beta}$'. 'α' strictly implies 'β' if and only if it is impossible that 'α' is true and 'β' false:

$$C\alpha\beta =_{df} \bar{K}\alpha\bar{\beta}$$
$$LC\alpha\beta =_{df} \bar{M}K\alpha\bar{\beta}$$

As was the case with truth-functional operators and quantifiers, definitions involving modal operators have no place in the formal system itself, since they are built into the cancellation rules. When the cancellation rules for modal operators are introduced in §2, all of the pairs of formulas that are given here as equivalent by definition will cancel in such a way as to produce identical branch charts.

A second relation between statements, expressed by combining a modal operator with a truth-functional operator, is consistency. Two statements, 'α' and 'β', are said to be *consistent* if their conjunction is possible:

$$MK\alpha\beta.$$

1·3 Reading Modal Formulas. One of the difficulties with reading formulas containing modal operators arises from the need to make clear the scope of a modal operator: whether the operator ranges over a whole complex formula or some subformula within the larger formula. A common source of confusion is the statement form "If p, then necessarily q." This may be construed as either 'CpLq' or 'LCpq', although the latter formula is better read as "Necessarily if p then q," or, shifting from use to mention, "'p' strictly implies 'q'." In general, the operators may be read either as sentence adverbs modifying the statements over which they range ("Necessarily the sky is blue," "Possibly we will go to the movies tonight") or as expressions introducing a subject clause ("It is necessary that whales are mammals," "It is possible that someone will fly to the moon"). When we combine quantification with modality in Chapter V, we must be somewhat more careful, because we can then make a formal distinction between "It is possible that someone is a Martian" and "There is someone who is possibly a Martian". This distinction should be observed verbally as well.

In giving verbal interpretations of the formulas which are theorems of the systems of modal logic in this chapter, we will be somewhat free, in order to give the most general statement of each theorem. The formula 'LCpp', for example, may be read in several ways:

> It is necessary that if p then p.
> Necessarily, if p then p.
> 'p' strictly implies 'p'.

But since we are studying the theorems of the various formal systems primarily to understand the properties of the systems, we will often give a more general sort of reading to formulas like 'LCpp': "A statement strictly implies itself."

We say that an occurrence of a basic formula *falls under* a given occurrence of a modal operator if a wff in which that basic formula occurs is preceded by that modal operator. It may be that a given basic formula falls under no modal operators, as happens with 'p' in the wff 'CpMq'. On the other hand, an atomic formula may fall under several occurrences of modal operators, as is the case with 'p' in 'MAqMp'. Here 'p' falls under both occurrences of 'M', while 'q' falls under only the first occurrence of 'M'.

1.4 Strict Implication, Validity and Equivalence. We say that a wff, α, *strictly implies* another, β, if 'LCαβ' cancels to closed branches. An argument is *strictly valid* if the conjunction of its premises strictly implies its conclusion. Two formulas are *strictly equivalent* if they strictly imply each other. Strict implication, strict validity, and strict equivalence are then defined in the same way that *material* implication, validity and equivalence were defined at II.3.5, except that the tests are carried out with the strict implication complex 'LC' instead of the material implication operator 'C'.

Exercises

Symbolize the following:

1. Necessarily, if both p and q then p.
2. If p, then it is necessary that both q and r.
3. If it is necessary that both p and q, then it is possible that either r or s.
4. It is necessary that either p or q, but it is impossible that r.
5. If 'p' strictly implies 'q', then either 'p' is impossible or 'p' is consistent with 'q'.

2 MODAL CANCELLATION

We develop now a succession of unquantified modal logics within the framework of cancellation methods. The sequence of systems to be developed are cancellation equivalents (in terms of formulas provable within them) to the systems **Mn**, **M**, **M′** and **M″** of G. H. von Wright. The last two mentioned systems are equivalent, in turn, to the systems **S4** and **S5** of C. I. Lewis, and the system **M** is equivalent to the system **T** of Feys and Gödel. The nomenclature of the formal systems to be studied here is relatively simple: We will use the name of the most familiar deductive version of a given formal system, prefixed with the letter '**C**', for cancellation. Thus, the cancellation version of **Mn** will be called '**CMn**', and the remaining systems '**CM**', '**CS4**' and '**CS5**'.

2.1 Vocabulary and Formation. The vocabulary of the modal cancellation systems is the vocabulary of Chapter II, with the addition of the modal operators 'M' and 'L'. Quantifiers are omitted here but will be added to modal cancellation in Chapter V.

Since it will be desirable to differentiate formulas which have been cancelled or partially cancelled from those which have not, we amend the formation rules of Polish notation to distinguish initial (uncancelled) well-formed formulas within the overall class of well-formed formulas.

2.1.1 IWFF. An *initial well-formed formula* (iwff) of the modal cancellation systems is one of the following:

(i) an atomic formula ('p', 'q', 'r', etc.);
(ii) the result of writing 'K', 'A', or 'C' to the left of a pair of iwffs ('Kpq', 'ApKqr', etc.);
(iii) the result of writing 'L' or 'M' to the left of an iwff ('LCpq', 'Mp', 'MMLKpq', etc.);
(iv) the result of writing the negation bar over the left-most sign of an iwff ('$\bar{\text{p}}$', '$\bar{\text{M}}$Kpq', '$\bar{\text{L}}$Mp', etc.).

2.1.2 Additional Vocabulary. To the vocabulary of iwffs, we add the following:

A numerically ordered, denumerable stock of *constants* (Arabic numerals), 1, 2, 3, 4, 5, ···
An alphabetically-ordered, denumerable stock of *dummy constants* (lower-case Roman letters from 'a' to 'k', perhaps with primes), a, b, c, ··· j, k, a′, b′, ··· k′, a″, b″, ···
A special modal constant, 0.

These dummies and constants will be introduced in applications of the modal cancellation rules. They are, with the exception of '0', the same dummies and constants that were used in Chapter II for the cancellation of quantifiers. This will not cause any difficulty in combining modality and quantification in Chapter V.

2.1.3 WFF. A *well-formed formula* (wff) is an iwff or a formula resulting from the cancellation or partial cancellation of an iwff. A wff includes any constants or dummies written as a result of application of the modal cancellation rules below.

2.1.4 Iterated Modalities. According to formation rule (iii), the result of prefixing a modal operator to an iwff produces a new iwff. The formation rules allow not just formulas like 'Mp', 'Lp' and 'MKpq', but also cases where there is more than one modal operator before a formula. Cases where an iwff whose left-most sign is a modal operator falls under an additional modal operator are called formulas of *iterated modality*: 'MLp', 'MLLKpq', 'MLMMLp', 'CMLpMMq'. Cases where a truth-functional compound having modal operators in its components falls under one or more additional modal operators are called cases of *nested modality*: 'LCMpMq', 'MKLMpMq', 'MAMpMKpMq'.

2.2 Modal Cancellation Rules. For the modal operators 'M' and 'L' and their negations, we add the following four cancellation rules to the rules for truth-functional cancellations given at II.4.1:

$$\begin{array}{cccc} & \text{d} & & \text{c} \\ M\alpha \rightarrow & \not{M}\,\alpha & \bar{M}\alpha \rightarrow & \bar{\not{M}}\,\bar{\alpha} \\ 0 & 0 & 0 & 0 \\ & \text{d} & & \text{c} \end{array}$$

$$\begin{array}{cccc} & \text{c} & & \text{d} \\ L\alpha \rightarrow & \not{L}\,\alpha & \bar{L}\alpha \rightarrow & \bar{\not{L}}\,\bar{\alpha} \\ 0 & 0 & 0 & 0 \\ & \text{c} & & \text{d} \end{array}$$

'd' represents the first unused dummy in the alphabetically-ordered stock of dummies; 'c' represents the first unused numerical constant. An occurrence of the special constant '0' is assumed (i.e., written in advance) under each basic subformula (bsf) in the formula. When a modal operator is cancelled, an additional dummy or numerical constant is written under each bsf within the scope of that operator (see Figure 1).

$$\begin{array}{lll} LCpq & CLKpqMArs & KMpMq \\ 1 & \quad\ \text{a} \qquad\quad \text{b} & \quad \text{a} \quad \text{b} \\ \not{L}\not{C}\bar{p}q & \not{C}\bar{\not{L}}\bar{\not{K}}\bar{p}\bar{q}\not{M}\not{A}rs & \not{K}\not{M}p\not{M}q \\ \quad 0\,0 & \qquad 0\,0 \qquad 0\,0 & \qquad 0 \quad 0 \\ \quad 1\,1 & \qquad \text{a a} \qquad \text{b b} & \qquad \text{a} \quad \text{b} \end{array}$$

FIGURE 1

In the cancellation of iterated and nested modalities, successive dummies and constants are written in the order of introduction in a column under each bsf within the scope of the cancelled operators, as in Figure 2. We call such columns *modal index columns*; the set of dummy and constant entries we call the *modal index* of the bsf to which it is appended.

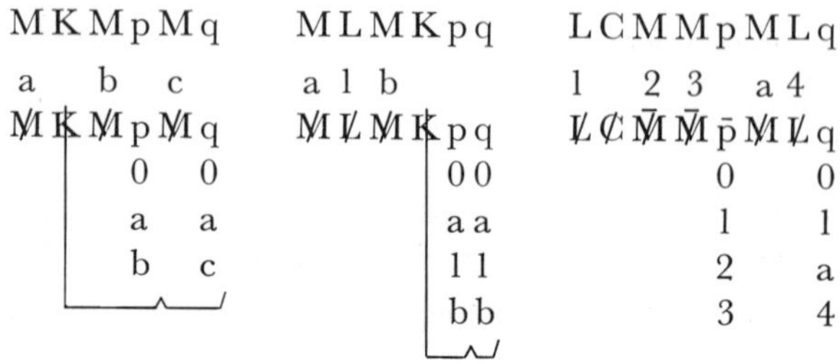

FIGURE 2

As was the case with quantifier cancellation, we allow a constant to be considered unused with respect to the right-hand side of a split marker if it is introduced in the left-hand side of that same split marker.

2.3 The Branch Chart. Basic formulas, with their modal indices, are entered in the branch chart in the same manner as for truth-functional cancellation. In addition, the extra occurrences of dummies written over the cancelled modal operators are entered in the branch chart as if they were basic formulas.

If a dummy occurs in the branch chart to the left of a split marker, that dummy is said to *govern* the two branches created by that split marker.

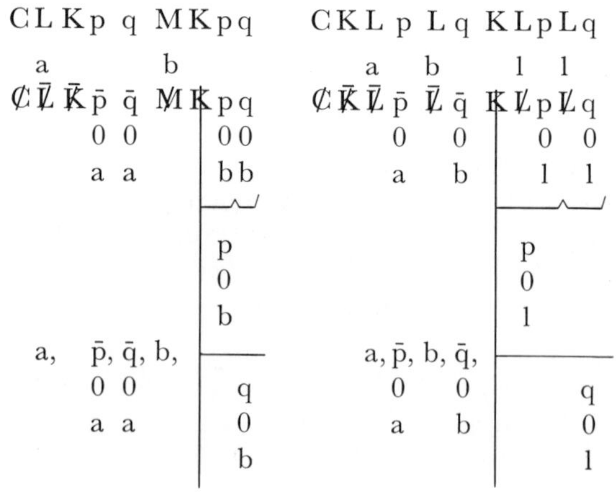

FIGURE 3

In both examples in Figure 3, the dummies 'a' and 'b' govern the split markers. In the second example, both occurrences of 'L' introduce the constant '1', since they occur in opposite sides of the same split marker.

2.4 Closure. We say that an iwff is *provable* if each branch of its branch chart closes. As before, we say that a branch is closed if the sequent of wffs entered in its branch space contains a pair of opposites (see Figure 4). *Opposites* now consist of a wff and its negation *with the same modal index*.

```
L C p p
1
L̸ C̸ p̄ p
    0 0
    1 1
1,  p̄, p     [closed]
    0 0
    1 1
```

FIGURE 4

2.5 Consolidation of the Branch Chart. The various systems of monadic modal logic to be studied in this chapter will differ from one another with respect to the sorts of changes that may be made in the modal indices of bsfs in order to effect closure. With respect to the cancellation procedure itself, the initial indexing of bsfs and the entering of formulas in the branch chart, the formal systems of modal logic will not differ from each other. The basic procedure has now been established; it remains to explore the variations in consolidation and closure that render various iwffs provable or not provable in a given formal system of modal logic.

Exercises

Cancel the following formulas and construct branch charts.

1. CMApqAMpLq
2. CLCpqAMp̄Lq
3. MAMMpLLq
4. LKLApp̄MCqq

3 THE CORE SYSTEM OF MONADIC MODALITY, **CMn**

For the basic system of monadic modal logic, we will use a cancellation version of the system **Mn** of G. H. von Wright. An axiomatic (deductive) presentation of this same system and the remaining systems to be discussed in this chapter is given in Appendix C.

The system **Mn** studies formulas in which each (occurrence of an) atomic formula falls under the same number of (occurrences of) modal operators. Such formulas are called *Mn-formulas*. In an *M*1-formula, for example, each atomic formula falls under precisely one modal operator, as in 'CLpMp', and 'CMKpqKMpMq'. In an *M*2-formula, each atomic formula falls under precisely two modal operators, as in

LCLpMp, CLMpMMq, LLCpq,

and so forth. In an *M*0-formula, each atomic formula falls under *no* modal operators. The formulas studied in the cancellation systems of Chapter **II** were *M*0-formulas.

All of the systems of modal logic to be studied here contain truth-functional logic in the sense that all truth-functional tautologies are theorems of the systems of modal logic. The cancellation rules for modal cancellation include the cancellation rules for truth-functional operators as given in Chapter II, in addition to the rules for modal cancellation given in §2 above.

3.1 CMn Closure. The manipulation of modal index columns which is permissible for the system **CMn** in order to consolidate the branch chart is minimal, and is given in the following two closure conditions.

> **CMn**-*closure* I. *Dummy constants in a modal index column may be replaced by numerical constants from the branch chart within the restrictions of the Uniformity Requirement* (cf. II.5.3.1) *and the Order Requirement.*[1]

> **CMn**-*closure* II. *When two dummies appear in a pair of opposing formulas and no constant is present to replace them, a new numerical constant may be introduced into the cancellation to replace the two dummies.*

Examples of applications of the **CMn** closure conditions will be given in the following section.

3.2 Theorems and Laws of Modal Logic. The following sequence of theorems reflects what have been called *Laws of Modal Logic*. These formulas will be theorems, also, of all subsequent systems of monadic modal logic studied. For the most part, they reflect traditionally accepted principles governing possibility and necessity that predate contemporary work in the field. All the theorems in the first sequence are *M*1-formulas. Cancellation proofs are shown for some of these theorems; the proof of others is left to the student.

[1] In modal cancellation without quantifiers, Q-Charts may be omitted except in cases where the original formula contains iterated or nested modal operators.

T-1 CLpMp

This formula provides the link between necessity and possibility: *If a statement is necessary, then it is also possible.* We may regard this as the distinguishing formula of the system **Mn.**

$$
\begin{array}{cc}
\mathrm{C\,L\,p} & \mathrm{M\,p} \\
\mathrm{a} & \mathrm{b} \\
\not{\mathrm{C}}\,\not{\bar{\mathrm{L}}}\,\bar{\mathrm{p}} & \not{\mathrm{M}}\,\mathrm{p} \\
0 & 0 \\
\mathrm{a} & \mathrm{b} \\
\mathrm{a},\bar{\mathrm{p}}, & \mathrm{b},\mathrm{p} \\
0 & 0 \\
\mathrm{a} & \mathrm{b}
\end{array}
\qquad
\left[\begin{array}{c}
\mathrm{a}=\mathrm{b}=1 \\
\textbf{CMn II} \\
\text{closed}
\end{array}\right]
$$

Both **T-1** itself and its cancellation proof are strongly similar to the quantificational formula '$\mathrm{C}\forall\mathrm{x}\,\mathrm{Fx}\,\exists\mathrm{x}\,\mathrm{Fx}$':

$$
\begin{array}{cccc}
\mathrm{C}\forall\mathrm{x} & \mathrm{Fx} & \exists\mathrm{x} & \mathrm{Fx} \\
\mathrm{a} & & \mathrm{b} & \\
\not{\mathrm{C}}\,\not{\bar{\forall}}\mathrm{x} & \bar{\mathrm{F}}\mathrm{x} & \not{\exists}\mathrm{x} & \mathrm{Fx} \\
& \mathrm{a} & & \mathrm{b} \\
\mathrm{a}, & \bar{\mathrm{F}}\mathrm{a}, & \mathrm{b}, & \mathrm{Fb}
\end{array}
\qquad
\left[\begin{array}{c}
\mathrm{a}=\mathrm{b}=1 \\
\mathrm{Q\,II} \\
\text{closed}
\end{array}\right]
$$

The relationship between the stronger and weaker of the two modal operators is very similar to the relationship between the universal and existential quantifiers. Both the closure conditions and the semantic assumptions to be discussed in later chapters which justify the closure conditions are quite similar in both cases. This relationship between strong and weak operators is common to all the modal and quasi-modal concepts discussed in the Introduction: What is necessary is at least possible; what is obligatory is at least permissible; what is known is at least believed; what is always the case is at least sometimes the case.

With respect to the alethic modal notions of necessity and possibility, our intuitions demand another theorem: 'CLpp', "What is necessary is true," and the equivalent formula 'CpMp', "What is true is at least possible." These formulas will be considered in a later section. They are not *Mn*-formulas, and are not therefore theorems of the system **CMn.**

A formula which is equivalent to **T-1** by virtue of the interdefinability of 'M' and 'L', and which is perhaps more interesting than **T-1**, is the following:

T-2 AMpM$\bar{p}$

Of a statement and its negation, one must be possible.

Again, we can generalize to other modal notions, so long as the strong and weak operators are interdefinable: one or the other of a given pair of opposites must be permissible, must be sometimes the case. With respect to epistemic modality, where "know" and "believe" seem not to be interdefinable, **T-1** holds, but **T-2** does not seem plausible (remember that **T-1** and **T-2** are equivalent only on the assumption that the two operators are interdefinable): one could fail to believe either of any given pair of opposites. At this writing, I do not believe that there are diamonds on Mars, nor do I believe that there are no diamonds on Mars. The apparent epistemic analogue of **T-2**, "Either it is believed that p or it is believed that not-p," does not always hold.

T-3 CMApqAMpMq

If the alternation of two statements is possible, then one or the other of the statements is possible.

The converse of **T-3** is also a theorem:

T-4 CAMpMqMApq

If at least one of two statements is possible, then their alternation is possible.

The following two theorems are equivalent to **T-3** and **T-4**:

T-5 CKLpLqLKpq

T-6 CLKpqKLpLq

If two statements are necessary, then their conjunction is necessary, and *if the conjunction of two statements is necessary, then each of the statements taken individually is necessary.*

Stated as biconditionals, the pairs **T-3**, **T-4** and **T-5**, **T-6** read as follows: *The alternation of two statements is possible if and only if at least one of the statements is possible; the conjunction of two statements is necessary if and only if both statements are necessary.*

Cancellation proofs of **T-5** and **T-6** appear in Figure 5.

Note in the cancellation of **T-6** that the constant '1' is introduced in both sides of the split marker.

These pairs of theorems are analogous to theorems of elementary logic whereby quantifiers are "distributed" through truth-functions. The existential quantifier distributes through an alternation, as does the possibility operator; the universal quantifier distributes through a conjunction, as does the necessity operator (cf. II.5, Problems 15 and 16, p. 49).

As is also the case with quantifiers, modal operators do not distribute through all truth-functions in both directions. Possibility distributes through a conjunction:

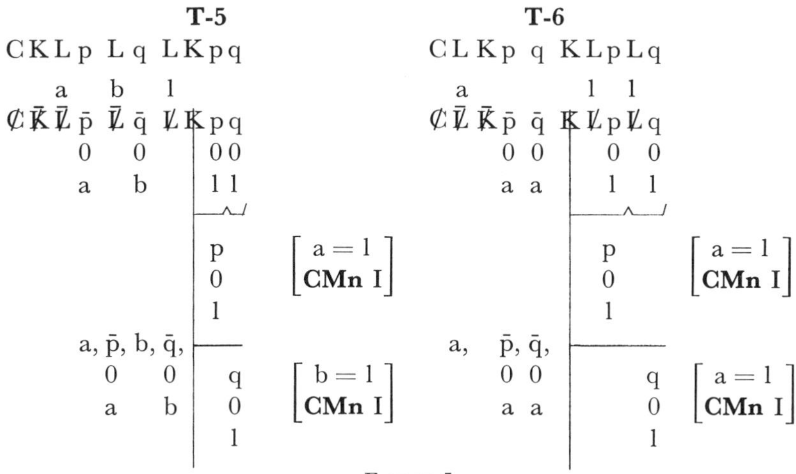

FIGURE 5

T-7 CMKpqKMpMq.

If two statements are consistent, then each of them is possible; or *if the conjunction of two statements is possible, then each of them is possible.*

The converse of **T-7** is not a theorem: 'CKMpMqMKpq' does not cancel to closed branches (see Figure 6):

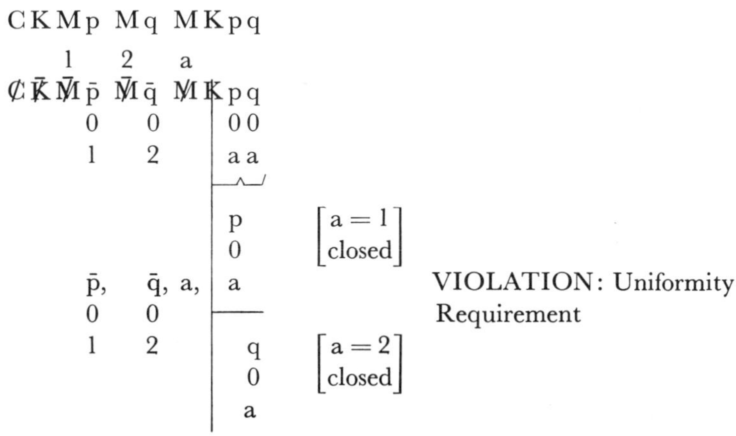

FIGURE 6

Not every two possible statements are consistent with each other. Suppose, for example, that 'p' is a statement which is contingently true or contingently false. That is, suppose that each of 'p' and 'p̄' is possible: 'KMpMp̄'. It could never be the case that the conjunction of 'p' and 'p̄' is possible: 'MKpp̄'. Let 'p' represent some obviously contingent statement, say, "Susie is five feet tall." Presumably it is not necessary that Susie be five feet tall, or that she not be. She just *happens* to be. But this is simply to say that both of "Susie is five feet tall" and "Susie is not five feet tall" are possible, no matter which happens to be true. It does not follow, of course, that Susie might be both five feet tall and not five feet tall at the same time.

On other modal and quasi-modal interpretations, **T-7** seems to be acceptable and its converse unacceptable. For example, if it is permissible to do both of two things, then it must be permissible to do each of them; but not every two actions which are separately permissible are jointly so—it is permissible to drink and permissible to drive, but not, let's say, to do both at once.

T-8 CALpLqLApq

T-8 is obviously the equivalent for necessity of **T-7**: *If at least one of two statements is necessary, then their alternation is necessary.*

Again, the converse does not hold: If the alternation of two statements is necessary, it does not follow that one or the other of the two statements taken by itself is necessary. Consider again the case where 'q' is just 'p̄': it surely does not follow from the logical truth 'LApp̄' that one or the other of 'p' or 'p̄' is necessary.

3.2.1 Generalizations About Cancellation. Theorems **T-1** and **T-3** through **T-8** have been shown as material implication formulas. All of these theorems, and indeed any theorem of the systems of modal logic in this chapter which are conditionals, can be shown to hold also for strict implication. **T-1**, for example, 'CLpMp', can be strengthened to 'LCLpMp', as shown in Figure 7.

CLp Mp		LCLp Mp	Q-chart
a b		1 a b	2
C̸L̸ p̄ M̸p		L̸C̸L̸ p̄ M̸p	1
0 0		0 0	a b
a b		1 1	
		a b	
a, p̄, b, p	[a = b = 1	a, p̄, b, p	[a = b = 2
0 0	**CMn II**	0 0	**CMn II**
a b	closed]	1 1	closed]
		a b	

FIGURE 7

It is easy enough to demonstrate the following general truth about the cancellation systems of modal logic: *If some wff* α *is cancellation provable, then* Lα *is also cancellation provable.* Consider a wff α which cancels to closed branches. No matter how complex α may be, in every branch of its cancellation there will be a pair of opposites, i.e., a pair of bsfs whose indices are identical and which comprise a pair of opposites. The addition of the operator 'L' to the left of the entire formula would alter the index of every bsf in the entire cancellation in precisely the same way, by adding an occurrence of the numerical constant '1' directly under the occurrence of '0' at the top of the modal index column. Every numerical constant n in the cancellation of α is then replaced by the constant $n+1$ in the cancellation of Lα, and if the cancellation of α results in closed branches, the change in numerical constants will have no effect on closure. Suppose, for example, that α is 'A M K p q $\bar{\text{M}}$ K p q':

```
A M K p q M̄ K p q          L A M K p q M̄ K p q

  a       1                 1   a       2
A̸ M̸ K̸ p q M̸̄ K̸̄ p̄ q̄          L̸ A̸ M̸ K̸ p q M̸̄ K̸̄ p̄ q̄
     | 0 0     0 0                   | 0 0     0 0
     | a a     1 1                   | 1 1     1 1
     |__⌣_|                          | a a     2 2
                                     |__⌣_|
```

Without carrying out the branch chart, the effect of the operator 'L' to the left of the formula is clear: the constant '1' is added at the top of each index column, and where '1' appeared in the cancellation of α, '2' appears in the cancellation of Lα.

This general truth about **CMn** and the subsequent systems of modal logic reflects what is called the "Rule of Tautology" in deductive presentations of these systems (cf. Appendix C):

The L-*Rule of Tautology. If* α *is provable, then so is* Lα.

A related and weaker general truth about **CMn** and the subsequent systems is the following:

The M-*Rule of Tautology. If* α *is provable, then so is* Mα.

The L-Rule of Tautology should not be misconstrued as justifying the inference from some formula 'p' to 'Lp'. 'CpLp' is not a theorem of any of the systems of modal logic considered here. It is only when 'p' is *provable*—i.e., when 'p' is itself a theorem of the formal system—that we can say that 'Lp' is also provable.

A second pair of general truths may be shown to hold for the cancellation systems of modal logic. These are also rules of inference in deductive presentations of the same formal systems.

The L-*Rule of Extensionality. If* α *and* β *materially imply each other, then* L α *and* L β *materially imply each other.*

As with the L-Rule of Tautology, the L-Rule of Extensionality lends itself to misunderstanding. This rule is not a claim for the theoremhood of 'CKCpqCqpKCLpLqCLqLp'. What the L-Rule of Extensionality does claim is that if 'KCpqCqp' is *provable*, then so is

KCLpLqCLqLp.

That this claim is justified with respect to the cancellation systems can easily be shown. It is sufficient, as a matter of fact, to show that if a formula of the form 'Cαβ' cancels to closed branches, then so does 'CLαLβ'. The argument is similar to that given above for the L-Rule of Tautology. Suppose that 'Cαβ' cancels to closed branches. Then the modal indices of the bsfs in 'α' and 'β' are of such a nature that replacements of dummies by constants will effect closure on pairs of opposites. The cancellation of 'CLαLβ' will be like that of 'Cαβ' except that a dummy will occur immediately under '0' in the indices of all bsfs in 'α', and a constant will appear in the same position in the indices of all bsfs in 'β'. All dummies and constants resulting from the cancellation of 'Cαβ' will be moved up, as before, but will occur in the same configuration. Suppose, for example, that 'α' is 'LKpq' and 'β' is 'KLpLq'. 'Cαβ' then cancels as in **T-6**, above, producing the branch chart shown for **T-6**. 'CLαLβ' is given in Figure 8.

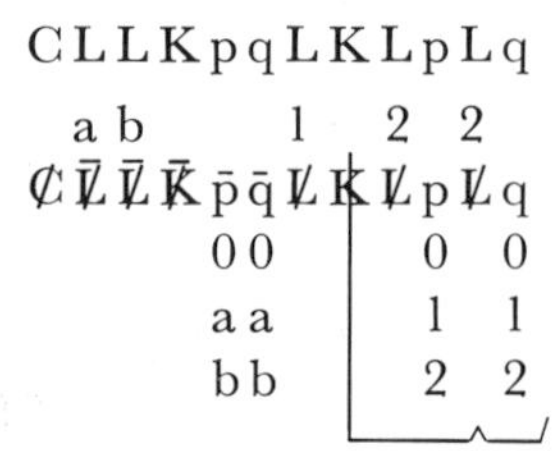

FIGURE 8

Compare this cancellation with the cancellation of **T-6**. The branches will close on the replacements 'a = 1, b = 2'.

The M-*Rule of Extensionality. If* α *and* β *materially imply each other, then* M α *and* M β *materially imply each other.*

3.2.2 Theorems Regarding Strict Implication and Consistency. The remaining theorems of **CMn** that we will discuss concern the properties of strict implication and consistency.

T-9 CLCpqCLpLq

Necessary statements strictly imply only necessary statements; or *if one statement strictly implies another, then if the first is necessary, so is the second.*

This theorem brings up once more the strong analogy between modal operators and quantifiers. A familiar theorem of elementary logic is 'CVxCFxGxCVxFxVxGx' (cf. II. 4. 4, Problem 4). The occurrences of dummies and constants in the cancellation of the quantified formula parallel precisely the occurrences of dummies and constants in the cancellation of **T-9**, as shown in Figure 9.

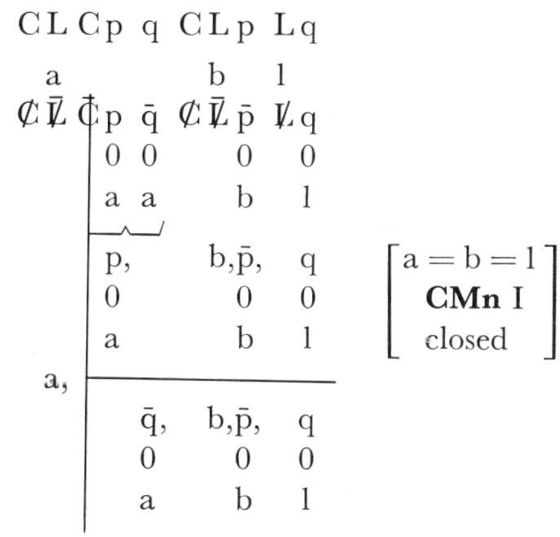

FIGURE 9

T-10 CLCpqCMpMq

Possible statements strictly imply only possible statements.

T-9 and **T-10** are obviously closely related. If we think of 'Mp', 'p' and 'Lp' as representing three degrees of modal strength that a statement may have, we can generalize from **T-9** and **T-10** in the following way: If one statement strictly implies another, then the second must have at least the same modal strength as the first.

Here we may anticipate somewhat the result of mixing modalities in a single formula. If we interpret the occurrences of 'M' and 'L' in the consequents of **T-9** and **T-10** as deontic, we have the following: If one statement strictly implies another and the first is obligatory, then so is the second; if the first is permissible, then so is the second.

T-11 CKLpLCKpqrLCqr

"*A necessary premise may be omitted.*" *If the conjunction of two statements strictly implies a third, and one of the two implying statements taken by itself is necessary, then the remaining statement taken by itself strictly implies the third.*

T-12 CKLpLqMKpq

This theorem is obviously closely related to **T-5**, above. If we think of the question of whether or not two statements are consistent as the question of whether or not they can both be true at the same time, then it is obvious that *any two necessary statements must be consistent.* Necessary statements are always true. If two statements are independently necessary, then both of them are always true and hence consistent with each other.

T-13 CKLpMqMKpq

A necessary statement is consistent with every possible statement.

On the temporal interpretation, where what is necessary is always true and what is possible is sometimes true, this theorem becomes quite clear: If 'p' is always true and 'q' is sometimes true, then sometimes 'p' and 'q' are true at the same time. In a similar sense, if 'p' is known and 'q' believed, then both are at least believed; if 'p' is obligatory and 'q' permissible, then both are permissible.

T-14 CKMpLCpqMKpq

A possible statement is consistent with every statement it strictly implies.

This theorem is not just a restatement of **T-10**, although it is obviously related to **T-10**. An equivalent and perhaps more interesting form of **T-14** is the following:

T-15 CMpAMKpqMKp$\bar{q}$.

A possible statement is consistent with one or the other of any pair of opposites.

One of the familiar properties of material implication is exportation: if two statements taken together materially imply a third, as in 'CKpqr', then the first materially implies that the second materially implies the third, 'CpCqr', and conversely. Exportation is not in general a feature of if-then relations that are stronger than material implication. The following, however, is a theorem for strict implication:

T-16 CLCKpqrLCpCqr.

If two statements taken together strictly imply a third, then the first strictly implies that the second materially *implies the third.*

The formula 'CLCKpqrLCp*L*Cqr' is not a theorem of the present systems of monadic modal logic. (It is not an *Mn*-formula, and fails as a theorem of **CMn** for that reason alone. But it will not be a theorem of subsequent systems either.)

There are certain key properties that strict implication has in common with material implication. For example, strict implication is transitive: *If a statement strictly implies a second and that one a third, then the first strictly implies the third*:

T-17 CKLCpqLCqrLCpr.

Strict implication is also reflexive. *Every statement strictly implies itself*:

T-18 LCpp.

The paradoxes of material implication mentioned in §1 above have analogues in strict implication. While it is not the case that a false statement strictly implies every statement or that a true statement is strictly implied by every statement, the following two theorems show that there are paradoxical properties which strict implication inherits from its ancestor, material implication.

T-19 C$\bar{\text{M}}$pLCpq
T-20 CL qLCpq

An impossible statement strictly implies every statement, and a necessary statement is strictly implied by every statement.

These are, respectively, the *antecedent paradox of strict implication* and the *consequent paradox of strict implication* and will be referred to as such later. Note in the cancellation of **T-19** (Figure 10) that closure is accomplished without any use of the element 'q', and in the cancellation of **T-20** (Figure 10) that closure is effected without 'p'.

C$\bar{\text{M}}$p	LCp q		CLq	LCp q	
a	1		a	1	
₵M̸̄p	L̸₵p̄ q		₵L̸̄q̄	L̸₵p̄ q	
0	0 0		0	0 0	
a	1 1		a	1 1	
a, p,	p̄, q	[a = 1, **CMn I**]	a, q̄,	p̄, q	[a = 1, **CMn I**]
0	0 0		0	0 0	
a	1 1		a	1 1	

FIGURE 10

In the Introduction we discussed only one implication relation that was not in the alethic modes. This was moral implication or obligation, in the sense that one act or state of affairs morally implies another: breaking a window obligates one (morally implies one) to have it repaired. If we interpret moral implication as an analogue of strict implication, we have a rather bothersome paradoxical result in **T-19**: If an act is morally forbidden, then committing that act obligates one to do everything (or better yet, anything one likes)!

Strict implication has in common with material implication the following properties which are sometimes taken to be paradoxical.

T-21 $LCKp\bar{p}q$

T-22 $LCpAq\bar{q}$

A contradiction strictly implies every statement, and a tautology is strictly implied by every statement.

If we think of strict implication as a relational property between statements and the operator 'L' ranging over an atomic formula as expressing a monadic property of statements, then we can draw a connection between strict implication and monadic necessity.

T-23 $CLpLC\bar{p}p$

T-24 $CLC\bar{p}pLp$

A statement is necessary if (**T-24**) *and only if* (**T-23**) *it is strictly implied by its own negation.*

An outgrowth of the paradoxical properties of strict implication at **T-19** and **T-20**, above, are the following two theorems.

T-25 $C\bar{M}pKLCpqLCp\bar{q}$

An impossible statement strictly implies both members of every pair of opposites.

T-26 $CKLCpqLCp\bar{q}\bar{M}p$

If a statement strictly implies both members of a pair of opposites, then it is impossible (the converse of **T-25**).

T-26 reflects the basis for the deductive technique called "indirect proof" or "*reductio ad absurdum.*" One way to show that '$\bar{p}$' must be the case is to show the equivalent result that 'p' is impossible, and one way to show that 'p' is impossible is to show that it (materially or strictly) implies a contradiction.

3.2.3 Mn-formulas where $n > 1$. Theorems **T-1** through **T-26** are all *M*1-formulas, formulas where each bsf falls under precisely one modal operator. The system **CMn** includes in its scope any formula where all subformulas fall under the same number of modal operators. The following theorem, for example, is an *M*2-formula (its cancellation appears in Figure 11).

T-27 $CLLpMMp$

The following is an *M*3-formula, and also a theorem of **CMn**.

T-28 $CLLMpMMMp$

Exercise

Perform cancellation proofs of **T-3, T-15** and **T-28**. Show all replacements needed for closure.

C L L p M M p

a b c d

$\not{C}\,\bar{\not{L}}\,\bar{\not{L}}\,\bar{p}\ \not{M}\,\not{M}\,p$

0 0

a c

b d

a, b, $\bar{p}$, c, d, p

0 0

a c

b d

$\left[\begin{array}{l} a = c = 1 \\ b = d = 2 \\ \textbf{CMn II} \end{array}\right]$

FIGURE 11

4 THE SYSTEM **CM**

The system **CM** studies monadic modal expressions with no limitation placed upon the number of modal operators under which an atomic subformula falls. A deductive presentation of an equivalent system, von Wright's **M**, is given in Appendix C.

The system **CM** contains truth-functional logic and the system **CMn** in the sense that all theorems of the former systems are theorems of **CM**. The cancellation rules for the system **CM** are those for the system **CMn**.

4.1 CM Closure Condition. In addition to the closure conditions for the system **CMn** given in §3.1 above, the system **CM** has the following closure condition.

CM-*closure* I. *A dummy constant may be deleted uniformly in the indices of bsfs in any branch.*[2]

Examples of applications of the **CM** closure condition will be given in the following section.

4.2 Theorems of CM and Laws of Modal Logic. We will begin by studying those theorems of the system **CM** which are $Mn + (Mn + 1)$-formulas: cases in which bsfs within a larger formula fall under different numbers of modal operators, so long as the difference in number of modal operators is no greater than 1. Theorems **T-29** through **T-36**, following, are all $M0 + M1$-formulas.

[2]An equivalent closure condition for **CM** is the following: A numerical constant may be repeated immediately below its original occurrence in any modal index column. It is always assumed that replacements of dummies by numerical constants is carried out beginning at the top of a modal index column.

T-29 CpMp

$$\begin{array}{ccccc}
 & \mathrm{Cp} & \mathrm{Mp} & & \\
 & & \mathrm{a} & & \\
 & \not{\mathrm{C}}\bar{\mathrm{p}} & \not{\mathrm{M}}\mathrm{p} & & \\
 & 0 & 0 & & \\
 & & \mathrm{a} & & \\
\bar{\mathrm{p}}, & \mathrm{a}, & \mathrm{p} & & [\mathbf{CM}\ \mathrm{I}]^{3} \\
0 & & 0 & & \\
 & & \not{\mathrm{a}} & &
\end{array}$$

A true statement is also possible. This theorem, sometimes called the *Principle of Possibility*, is a key theorem of any acceptable system of alethic modal logic, and may be regarded as the distinguishing formula of the system **CM**. The following two theorems are equivalent to **T-29**, given the interdefinability of 'M' and 'L' which is built into the cancellation rules.

T-30 C$\bar{\mathrm{M}}$p$\bar{\mathrm{p}}$

T-31 CLpp

An impossible statement is also false; a necessary statement is also true.

Note, however, that alethic modality and simple deontic modality part company at this point. Interpreting 'M' and 'L' as "It is permissible that—" and "It is obligatory that—" yields the following unhappy results for **T-29** and **T-31**: "If p, then it is permissible that p" and "If it is obligatory that p, then p." These are simply not true of a world where people don't always do what is obligatory and often do things which are not permissible, in both the legal and moral senses of those terms.

On the simple epistemic interpretation, it is **T-29** and **T-31** which begin sorting out the several senses of "know." The analogy between alethic modality and epistemic notions breaks down here. Although knowing entails believing, as necessity entails possibility, the relation becomes more complex when we interpose the question of the *truth* of a statement. A necessary statement must be true. In one sense of "know," a statement which is *known* to be true must be true. There is a weaker sense of "know," which has to do only with one's epistemic attitude toward a statement without bearing any commitment to whether or not the statement is true. In this sense, it is perfectly consistent to say "I *knew* that this was the case, but it turned out not to be"—"I had a feeling of certainty, but I turned

[3] The equivalent closure condition given in footnote 1 would have the following effect on the branch chart: the branch '$\begin{array}{ccc}\bar{\mathrm{p}}, & \mathrm{a}, & \mathrm{p} \\ 0 & & 0 \\ & & \mathrm{a}\end{array}$' is changed to '$\begin{array}{ccc}\bar{\mathrm{p}}, & \mathrm{a}, & \mathrm{p} \\ 0 & & 0 \\ 0 & & \mathrm{a}\end{array}$' by the closure condition, and 'a' is replaced by '0' for closure.

out to be mistaken." The parallel between possibility and belief is completely broken at this point. A statement which is true must be at least possible, but it is not the case that a statement which is true must be believed by anyone. As noted in the discussion of **T-2** above, the definition relation that we assume to hold between possibility and necessity does not hold between belief and knowledge.

The following two theorems concerning consistency and strict implication are obvious instances of **T-29** and **T-31**.

T-32 CKpqMKpq
T-33 CLCpq Cpq

If two statements are true, then they are consistent, and if one statement strictly implies another, then it also materially implies the other. Truth entails consistency; strict implication entails material implication.

The converse does not hold true, of course, although if a material implication formula is *provable*, then so is the corresponding strict implication formula (see §3.1.1 above).

T-34 CCpqCpMq

If one statement materially implies a second, it also materially implies that the second is possible.

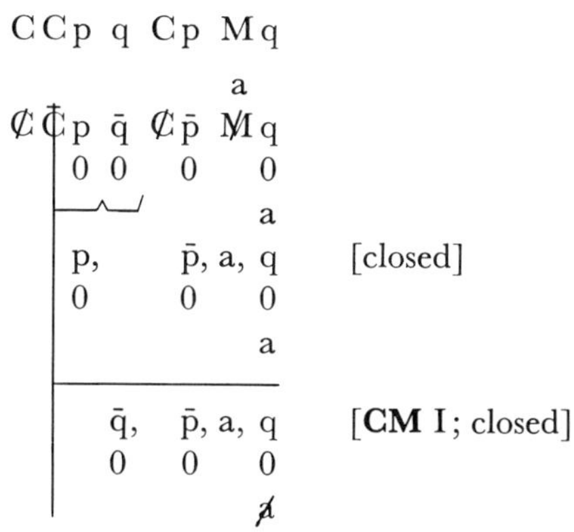

Note that **T-34** also is not acceptable on the deontic interpretation. Robbing a bank may imply doing something dishonest; robbing a bank does not imply that it is permissible to do something dishonest. On the temporal interpretation, this theorem is acceptable, and on the epistemic interpretation, it is unacceptable for the same reasons that **T-29** was unacceptable.

T-35 CCpqCLpq

This is a close analogue to **T-34**. Again, this theorem fails on other interpretations, but is perfectly reasonable with respect to alethic modality.

T-36 C K L p q M K p q

A necessary statement is consistent with any true statement.

4.3 Mn + (Mn + 1)-formulas where n > 0. T-29 through **T-36** may be strengthened to strict implication formulas by prefixing the operator ‘L’ to them, making them $M1 + M2$-formulas. Theorems **T-34** and **T-35** may also be shown to hold for strict implication internally:

T-37 C L C p q L C p M q
T-38 C L C p q L C L p q

These last two theorems are $M1 + M2$-formulas. If they are prefixed with the operator ‘L’, they yield $M2 + M3$-formulas, which are also theorems of the system **CM**:

T-39 L C L C p q L C p M q
T-40 L C L C p q L C L p q

```
L C L C p q L C L p q
1   a         2   b
L̸ C̸ L̸̄ C̸̄ p q̄ L̸ C̸ L̸̄ p̄ q
        |0 0         0 0      [a = 2 CMn I]
        |1 1         1 1      [CM I       ]
        |a a         2 2
        └─^─/        b̸
```

The construction of the branch chart is obvious.

The following four theorems of **CM** relate $M2$-formulas of iterated modality to $M1$-formulas.

T-41 C L M p M p

A statement that is necessarily possible is possible. (cf. **T-31**)

T-42 C L p M L p

A necessary statement is possibly necessary. (cf. **T-29**)

T-43 C L L p L p

A statement that is necessarily necessary is (simply) necessary.

T-44 C M p M M p

A possible statement is possibly possible.

The formulas which are the converses of **T-41** through **T-44** are not theorems of **CM**. The converse of each of these formulas has been the subject of some controversy and will be taken up in a later section. We may regard **T-41** through **T-44** as part of the standard—i.e., uncontroversial—system of modal logic, along with the earlier theorems of the system **CM**. The various extensions of the system **CM** to be considered in

§6 below are not generally agreed upon, nor is there a traditional core of Laws of Modal Logic covering complicated cases of iteration and nesting of modal operators.

5 DETERMINATE VALUE TABLES FOR $M0+M1$-EXPRESSIONS

Before going on to consider extensions of the system **CM**, we will do well to take stock of the modal apparatus already in our grasp, and to look more carefully into the relationships among necessity, possibility, strict implication and consistency.

The statements 'Mp', 'p', '$\bar{\mathrm{L}}$p' and '$\bar{\mathrm{p}}$' are indeterminate in the following sense: each of them is consistent with another $M0$- or $M1$-expression containing an occurrence of 'p' and no other statement letter, and is also consistent with the negation of that expression. 'Mp', for example, is consistent with both of 'p' and '$\bar{\mathrm{p}}$', and with both of 'Lp' and '$\bar{\mathrm{L}}$p'.

In this section we will describe *determinate* modal expressions and will use such expressions to explicate further the notions of strict implication and consistency.

We can list exhaustively all $M0$- and $M1$-expressions for the statement 'p': 'Mp', 'p', 'Lp', 'M$\bar{\mathrm{p}}$', '$\bar{\mathrm{p}}$', 'L$\bar{\mathrm{p}}$', '$\bar{\mathrm{M}}$p', '$\bar{\mathrm{L}}$p', '$\bar{\mathrm{M}}\bar{\mathrm{p}}$', '$\bar{\mathrm{L}}\bar{\mathrm{p}}$'. Given the interdefinability of 'M' and 'L', we may eliminate from this list one of each pair of equivalent expressions, choosing in such a way that each expression remaining either contains no negation or contains a negation bar only over its left-most sign. From the equivalent pair of formulas 'Mp' and '$\bar{\mathrm{L}}\bar{\mathrm{p}}$', for example, we choose 'Mp'; from the equivalent pair 'M$\bar{\mathrm{p}}$' and '$\bar{\mathrm{L}}$p', we choose '$\bar{\mathrm{L}}$p'. There are six remaining formulas: 'Mp', 'p', 'Lp', and their negations, '$\bar{\mathrm{M}}$p', '$\bar{\mathrm{p}}$' and '$\bar{\mathrm{L}}$p'.

By a *determinate combination* of $M0$- and $M1$-expressions we mean a conjunction of one or more of these six expressions for a given statement letter which either entails or entails the negation of every other $M0$- and $M1$-expression containing precisely one occurrence of that statement letter and no other statement letter. This will be a conjunction which includes or entails one of 'p' or '$\bar{\mathrm{p}}$', and one of 'Mp' or '$\bar{\mathrm{M}}$p', and one of 'Lp' or '$\bar{\mathrm{L}}$p'. By constructing conjunctions which contain precisely one of each of these three pairs of opposites, we arrive at the following eight formulas:

(1)	KKMppLp	*(1)	KK$\bar{\mathrm{M}}$ppLp
(2)	KKMpp$\bar{\mathrm{L}}$p	*(6)	KK$\bar{\mathrm{M}}$pp$\bar{\mathrm{L}}$p
*(3)	KKMp$\bar{\mathrm{p}}$Lp	*(7)	KK$\bar{\mathrm{M}}$p$\bar{\mathrm{p}}$Lp
(4)	KKMp$\bar{\mathrm{p}}\bar{\mathrm{L}}$p	(8)	KK$\bar{\mathrm{M}}$p$\bar{\mathrm{p}}\bar{\mathrm{L}}$p

The starred conjunctions, (3), (5), (6) and (7), are not consistent. They may be shown to be contradictory by showing that their negations are theorems of **CM.** To show that (3) 'K K M p p̄ L p' is inconsistent, for example, we cancel 'K̄ K M p p̄ L p'. (This should hardly be necessary to convince us, however. **T-29**, 'C L p p', rules out the case where 'L p' and 'p̄' are both true.)

$$\begin{array}{lllll}
\bar{K}KM & p & \bar{p} & L & p \\
 & 1 & & a & \\
\not{\bar{K}}\not{\bar{K}}\not{\bar{M}} & \bar{p} & \not{\bar{p}} & \not{\bar{L}} & \bar{p} \\
 & 0 & 0 & & 0 \\
 & 1 & & & a \\
 & \bar{p}, & p, & a, & \bar{p} \\
 & 0 & 0 & & 0 \\
 & 1 & & & \not{a}
\end{array}
\qquad
\begin{bmatrix} \textbf{CM I} \\ \text{closed} \end{bmatrix}$$

There remain four non-contradictory determinate combinations for a single statement letter, those at (1), (2), (4) and (8) above. The alternation of these four determinate combinations is a theorem of **CM.** The four determinate combinations are mutually exclusive—not more than one of them can be true for a given statement—and jointly exhaustive—at least one of them must be true for a given statement.

In truth-functional logic, the familiar truth-table (see Table 7) lists exhaustively the truth combinations or "logical states of affairs" that can obtain for a given statement or group of statements.

TABLE 7

p
T
F

p	q
T	T
F	T
T	F
F	F

p	q	r
T	T	T
F	T	T
T	F	T
F	F	T
T	T	F
F	T	F
T	F	F
F	F	F

In Table 8, we construct a table for modal logic which lists exhaustively the logical states of affairs for a given statement by assigning truth values to each of 'M p', 'p' and 'L p' in such a way that precisely one of the determinate combinations (1), (2), (4) and (8) is true on a given table line. We call such a table a *Determinate Value Table* (DVT). Because it exhausts the consistent determinate combinations containing *M*0- and *M*1-expressions for a given atomic formula, we call this particular table a DVT of degree one. A table which exhausts the consistent determinate combinations containing *M*0-, *M*1- and *M*2-expressions for a given atomic formula is a DVT of degree two. Such a table will have additional columns at the left-hand side for 'M M p', 'M L p', 'L M p' and 'L L p'.

TABLE 8

			(1)	(2)	(4)	(8)
Mp	p	Lp	KKMppLp	$KKMpp\bar{L}p$	$KKMp\bar{p}\bar{L}p$	$KK\bar{M}p\bar{p}\bar{L}p$
T	T	T	T	F	F	F
T	T	F	F	T	F	F
T	F	F	F	F	T	F
F	F	F	F	F	F	T

A DVT of degree one for two statements, 'p' and 'q', will have sixteen lines. We say that a complex formula is *fully tabulated* with respect to a given formal system if the values T or F can be assigned to that formula on every line of the DVT; i.e., if either that formula or its negation is implied by the unique determinate combination for each line of the table.

All *M*0-formulas are fully tabulated.

The *M*1-formulas 'M A p q' and 'L K p q', for example, are fully tabulated. In Table 9 'M A p q' is true on all lines except line 16, and is false on line 16; 'L K p q' is false on all lines except line 1, and true on line 1.

The *M*1-formulas 'M K p q' and 'L C p q'—consistency and strict implication—are *not* fully tabulated. These formulas are shown in Table 9 alongside related truth-functional formulas. Related theorem numbers are shown next to each entry in the table. The appeal to previously claimed theorems is not necessary in order to establish whether a given formula should be assigned 'T' or 'F' or no value at all on a given line. The test for tabulation can be carried out as follows: set up the unique determinate combination which is true on a given line of the DVT, and determine whether it implies the formula at the head of the column. If so, write 'T'. If it does not, then determine whether it implies the negation of the formula at the head of the column. If so, write 'F'. If the unique determinate combination for a given table line implies neither the formula at the head of the column nor the negation of that formula, then no entry should appear under that formula on that line of the DVT.

The property of implication and entailment relations that is called

TABLE 9. A DETERMINATE VALUE TABLE OF DEGREE ONE, WITH VALUES CALCULATED FOR CERTAIN *M*0- AND *M*1-EXPRESSIONS

	Mp	p	Lp	Mq	q	Lq	MApq	LKpq	Cpq	LCpq	Kpq	MKpq
1	T	T	T	T	T	T	T (**T-4**)	T (**T-5**)	T	T (**T-20**)	T	T (**T-12**)
2	T	T	F	T	T	T	T (**T-4**)	F (**T-6**)	T	T (**T-20**)	T	T (**T-36**)
3	T	F	F	T	T	T	T (**T-4**)	F (**T-6**)	T	T (**T-20**)	F	T (**T-13**)
4	F	F	F	T	T	T	T (**T-4**)	F (**T-6**)	T	T (**T-20**)	F	F (**T-7**)
5	T	T	T	T	T	F	T (**T-4**)	F (**T-6**)	T	F (**T-9**)	T	T (**T-32**)
6	T	T	F	T	T	F	T (**T-4**)	F (**T-6**)	T	—	T	T (**T-32**)
7	T	F	F	T	T	F	T (**T-4**)	F (**T-6**)	T	—	F	—
8	F	F	F	T	T	F	T (**T-4**)	F (**T-6**)	T	T (**T-19**)	F	F (**T-7**)
9	T	T	T	T	F	F	T (**T-4**)	F (**T-6**)	F	F (**T-9**)	F	T (**T-13**)
10	T	T	F	T	F	F	T (**T-4**)	F (**T-6**)	F	F (**T-33**)	F	—
11	T	F	F	T	F	F	T (**T-4**)	F (**T-6**)	T	—	F	—
12	F	F	F	T	F	F	T (**T-4**)	F (**T-6**)	T	T (**T-19**)	F	F (**T-7**)
13	T	T	T	F	F	F	T (**T-4**)	F (**T-6**)	F	F (**T-33**)	F	F (**T-7**)
14	T	T	F	F	F	F	T (**T-4**)	F (**T-6**)	F	F (**T-33**)	F	F (**T-7**)
15	T	F	F	F	F	F	T (**T-4**)	F (**T-6**)	T	F (**T-10**)	F	F (**T-7**)
16	F	F	F	F	F	F	F (**T-3**)	F (**T-6**)	T	T (**T-19**)	F	F (**T-7**)

" independence " can be clarified in terms of the DVT as follows: an if-then relation is *independent* of the truth-value and modal status of its components if (i) it is not fully tabulated, and (ii) the only value that can be determined for it is 'F'. Any case where an if-then relation can be assigned the value 'T' may be regarded as a paradoxical case—one where the relation holds not because of any connection between the components but merely because of the truth-value or modal status of the components as laid out in the coordinates of the table.

A comparison of the DVT entries under 'C p q' and 'L C p q' indicates the differences between material and strict implication. On lines 5 and 15 of the DVT, the material implication formula is paradoxically true while the strict implication formula is false. Strict implication, as we have observed, demands that the consequent have at least the same modal strength as the antecedent. On line 15, for example, the possibility of the antecedent 'p' and the impossibility of the consequent 'q' is sufficient to falsify the strict implication formula.

On lines 6, 7 and 11, the material implication formula is paradoxically true, while no value can be calculated for the strict implication formula. Note the coordinates of these three lines of the DVT: on line 6, 'p' and 'q' are both contingently true neither is necessary; on line 7, 'p' is contingently false, 'q' is contingently true; on line 11, both 'p' and 'q' are contingently false. The only line of the DVT where both 'p' and 'q' are contingent and a value can be calculated for 'L C p q' is line 10. Here again, the strict implication formula is false because its antecedent is true and its consequent false—the antecedent has greater modal strength than the consequent.

Exercises

On each of the sixteen lines of the DVT, determine whether each of the following formulas is true or false.

1. A M p M q
2. K M p M q
3. A L p M K p q
4. A M q M K p q

6 COLLAPSING ITERATED AND NESTED MODALITIES

It has been suggested that formulas of iterated and nested modality such as 'M L M L K L M q L L M p', though monstrous and mind-boggling, are no more so than some of the wffs of truth-functional logic, like

C C p C q C r s C C C p r s K p C r C s p.

Neither of these formulas can be read in an intelligible manner, and neither is likely to occur in an application of the formal resources to an analytic task.[4] Nevertheless, such formulas are well-formed and some sense must be made of them.

The truth-functional formula can be made somewhat more intelligible by constructing an equivalent formula that is more readable, perhaps with alternations and conjunctions in place of the nested conditionals. For the modal formula, however, there is no formula that is both equivalent in **CM** and significantly more compact. A formula containing an occurrence of 'L' implies another that is the same except for the omission of 'L' (cf. **T-31**), and every formula implies another that is the same except for the addition of an occurrence of 'M' (cf. **T-29**). But in **CM** there is no way in which iterations of modal operators can be reduced to less complex but equivalent forms.

The Lewis systems **S4** and **S5** are systems in which iterations of modal operators can be "collapsed" to less complex and equivalent forms. We will present the equivalent cancellation systems **CS4** and **CS5**, but with the note that here we depart from the area of modal logic where there is general agreement as to the acceptability of the theorems of the systems presented. The various considerations involved in deciding whether or not **CS4** and **CS5** provide acceptable formal systems of alethic modal logic are rather complex. We will consider some of the difficulties with both systems in this chapter, but an attempt at evaluating the two formal systems cannot be made until after the systems have been interpreted in Chapter IV.

6.1 The System CS4. In the system **CS4**, iterated modalities that involve repetitions of the same modal operator have equivalent formulas that omit the repetition. 'L L p' is equivalent in **CS4** to 'L p', and 'M M p' is equivalent in **CS4** to 'M p'.

CS4 is an extension of the system **CM**. That is, all the theorems of **CM** are also theorems of **CS4**. The cancellation rules for the system **CS4** are those for the system **CM**.

6.1.1 **CS4** *Closure Condition.* In addition to the closure conditions for the system **CM**, the system **CS4** has the following closure condition.

[4] Surely there can be no argument here. Not even the most hardened professional logician could claim that either of the following made any sense at all:

Possibly necessarily possibly necessarily both necessarily possibly q and necessarily possibly p.

If if p, then if q then if r then s, then if if if p, then r then s then both p and if r then if s then p.

CS4-*closure* I. *If two constants (excluding zero) occur consecutively in a cancellation index, it is permissible to delete the upper constant before making replacements of dummies by constants. (This rule may be applied more than once in a given index.)*

Examples of applications of the **CS4** closure condition will be given in the following section.

6.1.2 Theorems of **CS4**. The key theorems of the system **CS4** are $M1 + M2$-formulas which, taken together with certain theorems of the system **CM**, yield the equivalences mentioned in §6.1.

T-45 C M M p M p
If a statement is possibly possible, then it is possible.

T-45 is the distinguishing formula of the system **CS4**. Its cancellation is given in Figure 12. Taken together with **T-44**, 'C M p M M p', **T-45**

```
CMMp Mp
 1 2   a
C̸M̸̄M̸̄p̄ M̸p
    0   0
    1   a
    2
    p̄, a, p      ⎡CS4 I;   CMn I;⎤
    0   0        ⎣a = 2    closed⎦
    1̸   a
    2
```

FIGURE 12

yields the material equivalence of 'M p' and 'M M p'. Since both of these theorems can be strengthened to strict implication formulas, we have also the strict equivalence in **CS4** of 'M p' and 'M M p'.

T-46 C L p L L p
A statement that is necessary is necessarily necessary.

Taken together with **T-43** above, **T-46** yields the material and strict equivalence of 'L p' and 'L L p'.

In general terms, the **CS4** equivalences allow collapsing of iterated modalities to less complex formulas, so long as the iterated modalities involve occurrences of the same modal operator. 'M M M M p' is equivalent in **CS4** to 'M M M p', and this in turn is equivalent to 'M M p', and this, finally, is equivalent to 'M p'. *Any uninterrupted string of occurrences of* *the same modal operator in a formula may be reduced to one occurrence of that modal operator*. **T-47** and **T-48**, for example, show that a string of five occurrences of the operator 'L' may be reduced to one occurrence. (Note that **T-47** is a theorem of both **CM** and **CS4**, and that **T-48** is a theorem only of **CS4**.)

T-47 CLLLLLpLp
T-48 CLpLLLLLp

The cancellation of **T-48** is similar to that of **T-45**, with the application of the **CS4** closure condition repeated four times.

Strings of occurrences of the same modal operator which are surrounded by occurrences of other operators may also be reduced. Each of the stuttering recurrences of 'M' and 'L' in 'MMLLLMMMp' may be reduced to one occurrence: the formula is equivalent in **CS4** to just 'MLMp', as the following two theorems show.

T-49 CMMLLLMMMpMLMp
T-50 CMLMpMMLLLMMMp

The formula 'MLMp', however, is not further reducible in **CS4**.

6.2 The System CS5. In the system **CS5**, iterated modalities may be collapsed completely to *M*1-formulas. The **CS4** equivalences hold in **CS5**, and, in addition, 'LMp' is equivalent in **CS5** to 'Mp', and 'MLp' is equivalent in **CS5** to 'Lp'. Not just iterated, but also nested modalities can be collapsed in **CS5**. *For every formula containing iterated or nested modal operators there is an M*1*-formula which is its equivalent in* **CS5**.

CS5 is an extension of **CS4**. All theorems of **CS4** (and therefore all theorems of **CMn** and **CM**) are theorems of **CS5**. The cancellation rules are the same as for the preceding systems.

6.2.2 **CS5** *Closure Condition.* The **CS5** closure condition includes, or, for all practical purposes, *bypasses* the closure conditions for **CM** and **CS4**. It is never necessary in **CS5** to use any closure conditions other than **CMn** I and **CMn** II, which govern the replacement of dummies by constants, and the **CS5** condition which follows.

CS5-*Closure* I. *It is permissible to delete all entries in a modal index column except the bottom entry.*

Examples will be given in the following section.

6.2.3 Theorems of **CS5**. The most important theorems of **CS5** are $M1 + M2$-formulas which, conjoined with theorems of **CM**, yield the remaining equivalences mentioned in §6.1. The distinguishing formula of **CS5** is now given.

T-51 CMLpLp

If a statement is possibly necessary, then it is necessary (see Figure 13). Taken together with **T-42** of **CM**, 'CLpMLp', **T-51** yields the material and strict equivalence of 'Lp' and 'MLp'.

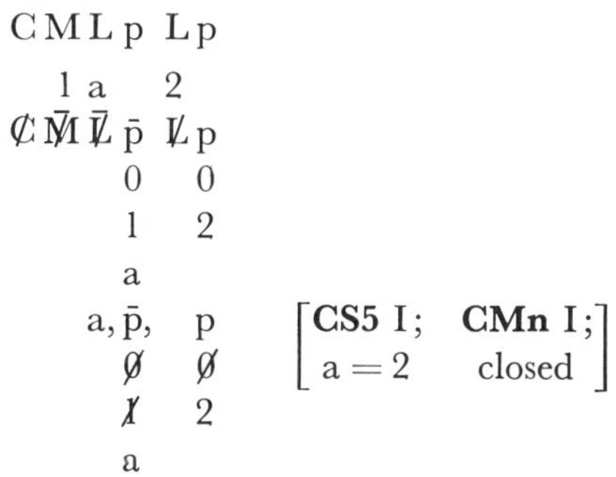

FIGURE 13

T-52 CMpLMp

A possible statement is necessarily possible.

Taken together with **T-41**, 'CLMpMp', **T-52** yields the material and strict equivalence of 'Mp' and 'LMp'.

The equivalences of **CS4** are also, of course, theorems of **CS5**. The four equivalences taken together allow the collapsing of all iterated and nested modalities. In **CS5**, *any uninterrupted string of modal operators occurring in a formula may be reduced to one occurrence of the last modal operator in that string.* This amounts to ignoring all but the right-most modal operator in any uninterrupted string of modal operators. A jumble of operators like that in 'MLLMMMLp', for example, is just a redundant way of writing 'Lp' in **CS5**. As may easily enough be demonstrated by cancellation, the following two formulas are theorems of **CS5**.

T-53 CMLLMMLpLp
T-54 CLpMLLMMLp

In §6, we introduced a horrible example of a formula that was not collapsible in **CM**: 'MLMLKLMqLLMp'. In **CS5**, this formula is equivalent to simply 'KMqMp'.

T-55 CMLMLKLMqLLMpKMqMp
T-56 CKMqMpMLMLKLMqLLMp

As may be apparent, there is inherent in the cancellation proof technique the machinery for finding the simplest *M*1-formula which is equivalent in **CS5** to any given formula containing iterated and nested modalities. It is simply a matter of cancelling the more complicated formula, applying the **CS5** closure condition (i.e., striking all but the bottom-most entry in each modal index column), and then finding the *M*1-formula whose cancellation would yield such a result. This latter part of the procedure, the reconstruction of uncancelled formulas from the debris of the branch chart, will be described in Appendix D. For now, we will let the matter rest with an illustration. In Figure 14 is a cancellation of our horrible

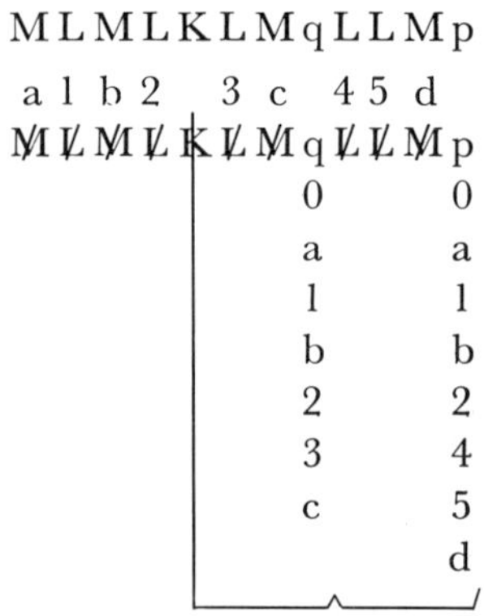

FIGURE 14

example. The **CS5** closure condition permits striking all entries in the modal index columns except the bottom entries, leaving 'q' and 'p' with dummies in the bottom positions. Now observe the cancellation of 'KMqMp':

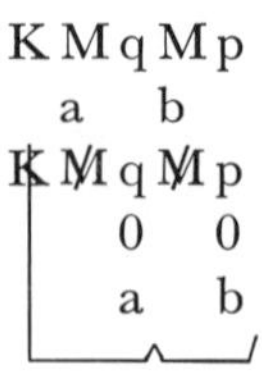

The resulting configuration in the branch chart is the same. The procedure becomes somewhat more complicated when indices end in occurrences of the same dummy or constant, but it is essentially mechanical and simple.

6.3 The Effect of Collapsing Iterations. Before considering whether or not the **CS4** and **CS5** equivalences are intuitively acceptable, or whether or not they are acceptable on other grounds, we can get a sense of the effect of these equivalences by comparing the three systems **CM**, **CS4** and **CS5** on a Determinate Value Table.

A Determinate Value Table which includes among its coordinates entries for *M*0-, *M*1- and *M*2-formulas is called a DVT of degree two. The number of consistent determinate combinations of *M*0-, *M*1- and *M*2-formulas will vary according to which formal system is being tabulated. In the system **CMn**, for example, a determinate combination which contains as conjuncts ‘L L p’ and ‘$\bar{\text{L}}$ p’ is consistent, while such a determinate combination is not consistent in any of the subsequent formal systems. For the system **CM**, there are twelve consistent determinate combinations of *M*0-, *M*1- and *M*2-formulas, and hence twelve lines on the DVT of degree two for the system **CM**. We will make no attempt to tabulate formulas on the DVT of degree two for **CMn** as we did in §5 above with the DVT of degree one. Such a table would contain 144 lines, and its sheer length would obviate any clarification of the formal systems it might present. We can, however, compare **CM**, **CS4** and **CS5** using the DVT of degree two for one statement in the system **CM**, as shown in Table 10.

The differences among the three systems can be seen by comparing the number of consistent determinate combinations. In **CMn** there are twelve; in **CS4**, ten; in **CS5**, four. The determinate combination for line 2 of the table, for example, is at odds with **T-46** of **CS4**, ‘C L p L L p’, and hence could not appear on the DVT of degree two for **CS4**. Given our earlier observations about **CS5**, we can state as a general result that a DVT of *any* degree for **CS5** will contain only four lines—the same number as a DVT of degree one.

Exercises

1. Perform cancellation proofs in **CS4** and **CS5** for **T-47** and **T-49**, and in **CS5** for **T-53**.
2. Show that the following pairs of formulas are equivalent in **CS5**
   (a) M L p and L L p
   (b) M M p and L M p

TABLE 10.* A DETERMINATE VALUE TABLE OF DEGREE TWO FOR THE SYSTEM **CM**

	MMp	Mp	LMp	p	MLp	Lp	LLp	Consistent in **CS4**	Consistent in **CS5**
1	T	T	T	T	T	T	T	Yes	Yes
2	T	T	T	T	T	T	F	No	No
3	T	T	T	T	T	F	F	Yes	No
4	T	T	T	F	T	F	F	Yes	No
5	T	T	F	T	T	F	F	Yes	No
6	T	T	F	F	T	F	F	Yes	No
7	T	T	T	T	F	F	F	Yes	Yes
8	T	T	T	F	F	F	F	Yes	Yes
9	T	T	F	T	F	F	F	Yes	No
10	T	T	F	F	F	F	F	Yes	No
11	T	F	F	F	F	F	F	No	No
12	F	F	F	F	F	F	F	Yes	Yes

* A DVT of degree two for the system **CS4** would contain only those lines which are consistent in **CS4** (10 lines), and a DVT of degree two for the system **CS5** would contain only those lines which are consistent in **CS5** (four lines; the same number of lines as a DVT of degree one).

7 ARGUMENTS CONCERNING A STANDARD SYSTEM OF MODAL LOGIC

The system **CM** and the Laws of Modal Logic whose counterparts are theorems of **CM** have a clear intuitive appeal, and are generally agreed upon as correct so far as they go. There is no provision in **CM** for dealing with iterated modality, however, and there has been considerable disagreement among logicians concerning the correctness of the systems presented here as **CS4** and **CS5** as acceptable means of collapsing iterations. We will not be able to resolve these disagreements in this chapter, since informal or intuitive verbal readings of formulas like 'MLMMp' are either not helpful or, worse yet, misleading. We can, however, consider some of the informal arguments that have been advanced.

We noted in the Introduction that within alethic modality there are several senses of possibility, necessity, compatibility and entailment. No attempt has been made in the present chapter to distinguish these senses formally; we have assumed that a single sort of possibility and necessity was involved in each of the formulas considered—logical possibility and necessity, or theoretical possibility and necessity, or perhaps some other sense. One should be on guard against mixing the various senses of alethic modality in arguments concerning iterated modal operators. One might, for example, argue for a consistent interpretation of 'K$\bar{\text{M}}$pMMp' (which is inconsistent in both **CS4** and **CS5**) by supposing 'p' to be a statement which is physically impossible with respect to some common sense or theoretical framework (hence '$\bar{\text{M}}$p') and hold at the same time that it is logically possible that the world could be otherwise (hence 'MMp'). Suppose 'p' is "I can walk unaided on the ceiling." This is, in some intuitive sense, physically impossible. Yet one can imagine without internal contradiction that gravity is other than what it is or that I am capable of psychokinesis. This sort of argument is probably sound, but whether it is an argument against the **CS4** theorem 'CMMpMp' is not clear because there are clearly two different senses of possibility involved in it.

Erik Stenius[5] and others have argued that a truth of mathematics, say 'q', is necessary (hence 'Lq'), but contingently so (hence

$$\text{KMLqM}\bar{\text{L}}\text{q}).$$

Such a conjunction would be inconsistent in **CS5**. Stenius' argument seems to be the following: Truths of mathematics are truths of logic with respect to a given system of axioms, rules and definitions, and are thus necessary.

[5] The argument was advanced by Professor Stenius during a lecture at Duke University in the spring of 1963.

But that these axioms, rules and, in particular, these definitions should be the "correct" ones is only contingently the case. The necessity that is claimed for truths of mathematics is derived within a particular formal system, but the contingency of this necessity is a fact *about* that particular formal system. Again, it is not clear that this sort of argument has any bearing against the **CS4** and **CS5** equivalences, where we must assume a single sense of modality.

The problem of mixing various senses of modality in a single formula can be treated formally, and will be in Chapter VI. But verbal attempts to sort out such proliferations seem not to be helpful. The expressions "necessarily necessary," "possibly possible," "possibly possibly necessary" don't seem to have much place in ordinary or even extraordinary discourse. Attempts to argue from analogous notions fall into the same difficulty: "provably provable," "permissibly permissible," and similar expressions seem to be sheer gibberish. We must drop the matter as a verbal consideration and return to it only when we have laid a semantic foundation for interpreting the formal systems.

The systems of monadic modal logic presented in this chapter are taken to be generic: so long as one is dealing consistently with one species of alethic modality, the theorems of **CM** seem to hold. The systems **CS4** and **CS5** seem to be neither more nor less problematic for logical possibility and necessity than for theoretical possibility and necessity or other senses. We are not yet ready to consider the integration of the species of alethic modality, let alone the integration of alethic modality with deontic, epistemic, or temporal notions, but it is clear that distinctions among all of these groups of somewhat analogous notions will have to be observed, and that no single formal system will serve as a reasonable interpretation of them all, unless there are distinct operators in that system for each set of notions, and each operator obeys its own set of rules. Whether or not the distinguishing formula of **CS4** ('C M M p M p') is acceptable is a separate question from whether or not the following are acceptable:

> If it is logically possible that it is theoretically possible that p, then it is theoretically possible that p.
> If it is logically possible that it is theoretically possible that p, then it is logically possible that p.

8 SUMMARY LIST OF THEOREMS

Following is a list of all the numbered theorems discussed in this chapter. Those for which cancellation proofs have been given are marked '*'.

8.1 Theorems of CMn.

***T-1** CLpMp (Distinguishing Formula of **CMn**)
T-2 $\mathrm{AMpM\bar{p}}$
T-3 CMApqAMpMq
T-4 CAMpMqMApq
***T-5** CKLpLqLKpq
***T-6** CLKpqKLpLq
T-7 CMKpqKMpMq
T-8 CALpLqLApq
***T-9** CLCpqCLpLq
T-10 CLCpqCMpMq
T-11 CKLpLCKpqrLCqr
T-12 CKLpLqMKpq
T-13 CKLpMqMKpq
T-14 CKMpLCpqMKpq
T-15 $\mathrm{CMpAMKpqMKp\bar{q}}$
T-16 CLCKpqrLCpCqr
T-17 CKLCpqLCqrLCpr
T-18 LCpp
***T-19** $\mathrm{C\bar{M}pLCpq}$
***T-20** CLqLCpq
T-21 $\mathrm{LCKp\bar{p}q}$
T-22 $\mathrm{LCpAq\bar{q}}$
T-23 $\mathrm{CLpLC\bar{p}p}$
T-24 $\mathrm{CLC\bar{p}pLp}$
T-25 $\mathrm{C\bar{M}pKLCpqLCp\bar{q}}$
T-26 $\mathrm{CKLCpqLCp\bar{q}\bar{M}p}$
***T-27** CLLpMMp
T-28 CLLMpMMMp

8.2 Theorems of CM. All of **T-1** through **T-28** are theorems of **CM.** The following formulas are also theorems of **CM.**

***T-29** CpMp (Distinguishing Formula of **CM**)
T-30 $\mathrm{C\bar{M}p\bar{p}}$
T-31 CLpp
T-32 CKpqMKpq
T-33 CLCpqCpq
***T-34** CCpqCpMq
T-35 CCpqCLpq
T-36 CKLpqMKpq
T-37 CLCpqLCpMq
T-38 CLCpqLCLpq

T-39 LCLCpqLCpMq
***T-40** LCLCpqLCLpq
T-41 CLMpMp
T-42 CLpMLp
T-43 CLLpLp
T-44 CMpMMp

8.3 Theorems of CS4. All of **T-1** through **T-44** are theorems of **CS4.** The following formulas are also theorems of **CS4.**

***T-45** CMMpMp (Distinguishing Formula of **CS4**)
T-46 CLpLLp
T-47 CLLLLLpLp
T-48 CLpLLLLLp
T-49 CMMLLLMMMpMLMp
T-50 CMLMpMMLLLMMMp

8.4 Theorems of CS5. In addition to **T-1** through **T-50**, the following formulas are theorems of **CS5.**

***T-51** CMLpLp (Distinguishing Formula of **CS5**)
T-52 CMpLMp
T-53 CMLLMMLpLp
T-54 CLpMLLMMLp
T-55 CMLMLKLMqLLMpKMqMp
T-56 CKMqMpMLMLKLMqLLMp

Review Exercises for Chapter III

1. Construct cancellation proofs for at least two theorems of each of the systems, **CMn, CM, CS4** and **CS5**, using the closure conditions given in the chapter. (Do not choose theorems that were starred in §8.)

2. If you had to use an eraser more than twice in doing the first problem, construct cancellation proofs of *all* of **T-1** through **T-56**. The procedure, though mechanical, requires the development of some skills. It should be mastered before we go on.

3. At the end of Chapter II, exercises were given involving tests for material implication, material validity and material equivalence. Each test involved cancelling a formula of the form '$C\alpha\beta$'. *Strict* implication, *strict* validity and *strict* equivalence involve the same tests, but carried out with a formula of the form '$LC\alpha\beta$'. Using the system **CM**, show that the arguments in Problems 1–9 following II.3.5 are strictly valid as well as materially valid. Show that the pairs of formulas in Problems 10–24 following II.3.5 are strictly equivalent as well as materially equivalent.

BIBLIOGRAPHY—CHAPTER III

[III.1] Anderson, Alan Ross. "Independent Axiom Schemata for von Wright's *M*," *Journal of Symbolic Logic* **22** (1957), pp. 241–244.

[III. 2] Von Wright, G. H. *An Essay in Modal Logic.* (North Holland, Amsterdam, 1951)

Chapter IV

INTERPRETATION OF MONADIC MODAL LOGIC

1 INTERPRETING A FORMAL SYSTEM

A formal system is *interpreted* by establishing a relationship between the elements of the formal system and a set of objects (called a *model*) in such a way that truth or falsity can be assigned to the simplest formulas of the formal system. Conditions are then laid down for assigning truth or falsity to complex formulas.

The procedure for interpreting a system of logic reflects in an artificial way the relationship between any natural language and the world, at least with respect to the way in which the simpler statements of the language are interpreted. Suppose I were to utter the sentence "Maggie has white fetlocks," and you wished to determine whether or not I had made a true statement. You would first have to ascertain two things: who or what "Maggie" refers to in my use of it, and what the predicate "—has white fetlocks" means. *Interpretation* in this case might consist of establishing that "Maggie" corresponds to my dog, Maggie, and "—has white fetlocks" corresponds to the property of having tufts of white hair on the lower hind leg. Under this interpretation the statement is true. Under another interpretation it might not be true—e.g., if "Maggie" is assigned to my nextdoor neighbor and "—has white fetlocks" to having some antique guns of a certain type.

If a procedure is laid down for determining whether or not simple statements are true, then it can be extended to determine whether or not certain complex statements are true. For example, to determine truth or falsity for "Maggie has white fetlocks and a white tail" under a particular interpretation, we would first determine truth or falsity for each of "Maggie has white fetlocks" and "Maggie has a white tail," and then set down the condition governing the use of "and." Presumably *this* condition would remain the same no matter how we interpret "Maggie," "—has white fetlocks" and "—has a white tail." A reasonable condition governing the use of "and" would be the following: If 'α' and 'β' are both statements, then 'α and β' is true just in case both of 'α' and 'β' are true.

The point in interpreting a formal system is not to determine what formulas are true under some particular interpretation. It is rather to determine what formulas are true under *all* interpretations of that system which meet certain criteria. These criteria consist of conditions which *any* interpretation of that formal system must meet—conditions governing the logical operators of that formal system. To show that a given formula is true under one interpretation is merely to show that it is a contingent formula. To show that a given formula is true under all interpretations of its formal system is to show that that formula is a *truth of logic* with respect to that formal system.

One of the aims of this chapter will be to define "truth of logic" for each of the formal systems considered in Chapter III. This is a prerequisite to discussing certain properties of those formal systems. We will say, for example, that a formal system is *complete* if all truths of logic which can be expressed in its notation are theorems of the formal system, and that a formal system is *sound* if all its theorems are truths of logic. "Theorem" is a syntactical term. It has to do with the evaluation of a wff *within* the formal system. "Truth of logic," as we will use it, is a semantical term. It has to do with the evaluation of a wff of the formal system from outside the formal system—i.e., from the point of view of an interpretation of that formal system.[1]

Beyond giving a means for determining completeness and soundness, interpreting formal systems on certain sorts of models is an important step toward understanding what the choice of a given formal system commits us to. We have already seen in Chapter III a case where several alternative formal systems were presented—**CM**, **CS4** and **CS5**—and where our logical intuitions were of little help in choosing among them. Purely syntactical considerations—those having to do with the formal systems themselves—did not clarify matters much, nor did the unsystematic appeal to verbal examples. Semantical considerations, on the other hand—those having to do with the relationship between the formal system and a model—are carried out systematically in such a way as to interpret every formula in the system rather than random examples. They

[1] The expression "truth of logic" is used here in preference to the alternative expressions "valid formula" or "valid schema." There are several reasons for the choice: first, to reserve the term "valid" for *arguments* whose conclusions follow from their premises; second, to drive home yet one more time the point that "Logical Truth" should not be capitalized—that it can only be made sense of with respect to a given formal system and its model in the present context, or with respect to a given interpretation of logical particles, in the wider sense.

will often be of help in deciding which formal system presents the most defensible rendering of its logical constants, and which is best suited to a particular task of analysis. As we move into even more complex formal systems involving quantifiers and dyadic modal operators, we will have to depend upon semantical considerations to help choose those formal systems which best reflect the modal notions of possibility, necessity and law-likeness.

2 MODEL SETS

Jaakko Hintikka [IV.2] has presented a particularly fertile device for the interpretation of modal calculi which he calls the *model system*. This semantic device makes use in a natural and convenient way of the ancient notion of possible worlds: what is necessary is true in all possible worlds; what is possible is true in some possible world (but perhaps not in this one). Hintikka's device makes it unnecessary to speak of possible worlds and allows us to speak instead about marks on paper, sets of occurrences of formulas which may or may not be construed as abbreviated statements. One model set of formulas is taken to be fixed and may, if you like, be construed as the set of true statements which describe the world. In a model system, additional model sets may be constructed which have different constituent formulas from the fixed set, and which qualify as *alternatives* to the fixed set by virtue of being constructed according to certain conditions. These may be construed as sets of possible and perhaps true statements which describe possible worlds.[2]

A model set is a set of marks which are occurrences of atomic formulas, plus other marks that are contained in the set in accordance with certain conditions. To distinguish marks comprising the model from expressions of the formal systems themselves, we will refer to the former as *sentences* and to the latter as *wffs*.

The language of the formal system is usually called the *object language*, and the language of the semantic model is usually called the *metalanguage*. In the metalanguage, we use the Greek letters 'μ', 'λ' and 'ν' to represent names of model sets. '0' is the name of that particular model set which is fixed (the "real world" set). 'α' and 'β' are used as before, to represent any wffs of the object language. '$[\alpha]$' represents the sentence of the model

[2] The choice of Hintikka's semantic approach is, to some extent, a matter of stylistic preference and to a larger extent a matter of convenience, since the cancellation techniques mesh most easily with this sort of model structure. The technique is similar to that of Saul A. Kripke, whose work is perhaps more historically important (see [IV.4], [IV.5]). Both Hintikka and Kripke owe a debt to Beth's notion of "semantic tableaux," and Beth, in turn, to Carnap's "state descriptions."

which is associated with the wff 'α'. '$\in$' is used to signify the membership of a sentence in a model set; '$\notin$' to signify non-membership.

This metalinguistic vocabulary makes it easier to describe the relationship between the formal system and its model. Suppose we are interpreting a system of truth-functional logic on the fixed model set. The fixed model set, 0, may (but need not) be taken to be a set of marks, each of which represents a true statement about the world. Thus, if α is a wff of the object language, it will be determined to be true just in case $[\alpha]$—the sentence of the model assigned to α—is in the set of sentences which comprise 0. In the vocabulary of the preceding paragraph, α is true just in case $[\alpha] \in 0$. Similarly, the wff β will be determined to be true just in case $[\beta] \in 0$. But the aim of interpreting a formal system, as we have noted, is not to determine whether or not particular atomic formulas are true under particular interpretations. We are interested in the conditions according to which complex formulas are, or are not, determined to be true. The *conditions* on the model set will give means for determining truth or falsity for $\bar{\alpha}$, $K\alpha\beta$, $A\alpha\beta$, and so on, under *any* interpretation. These conditions will hold not just for the fixed model set 0, but for every model set in the model system.

2.1 Truth-Functional Conditions on Model Sets. A model set μ is a set of sentences satisfying the following conditions.

2.1.1 $[\bar{\alpha}] \in \mu$ if and only if $[\alpha] \notin \mu$.
2.1.2 $[K\alpha\beta] \in \mu$ if and only if $[\alpha] \in \mu$ and $[\beta] \in \mu$.
2.1.3 $[A\alpha\beta] \in \mu$ if and only if either $[\alpha] \in \mu$ or $[\beta] \in \mu$.
2.1.4 $[C\alpha\beta] \in \mu$ if and only if $[A\,\bar{\alpha}\beta] \in \mu$.
2.1.5 $[\bar{K}\alpha\beta] \in \mu$ if and only if $[A\bar{\alpha}\bar{\beta}] \in \mu$.
2.1.6 $[\bar{A}\alpha\beta] \in \mu$ if and only if $[K\bar{\alpha}\bar{\beta}] \in \mu$.
2.1.7 $[\bar{C}\alpha\beta] \in \mu$ if and only if $[K\alpha\bar{\beta}] \in \mu$.
2.1.8 $[\bar{\bar{\alpha}}] \in \mu$ if and only if $[\bar{\alpha}] \notin \mu$.

Note that all the conditions after 2.1.3 are dependent upon prior conditions. There is some choice here, in that conditions 2.1.4 through 2.1.8 could be stated without reference to earlier conditions. Condition 2.1.3 could be cast in the same way as the later conditions by defining '$A\alpha\beta$' as '$\bar{K}\bar{\alpha}\bar{\beta}$'. There is nothing more at stake than economy, however, and it is convenient to state the conditions as we have done.

If there is a limited number of sentences in a given model set, assignments can still be made to every atomic wff of the formal system. If there are more statement letters in the formal system than there are atomic sentences in the model set, we simply assign more than one statement letter to a given sentence. [p], [q] and [r] might, under some interpretation, be the same sentence of the model set. We are guaranteed, then, that for every wff α of the formal system, one or the other of [α] or [$\bar{\alpha}$] is a member of a given model set.

2.2 An Example of Interpretation. It is easy enough to show that a truth of statement logic, say 'C K p C p q q', must be interpreted as true with respect to any model set constructed according to the conditions of §2.1. In particular, we can show that [C K p C p q q] must be a member of the fixed model set, 0, under any interpretation—i.e., under any assignment of sentences of the model to wffs of the formal system. It will not be neccessary for our later work to develop a formal procedure for such proofs. The following illustration therefore makes informal appeal to the conditions on model sets in 2.1, and is somewhat longer and less economical than it needs to be.

(1) [C K p C p q q] $\in$ 0 if and only if [A $\bar{\text{K}}$ p C p q q] $\in$ 0. (2.1.4)
(2) [A $\bar{\text{K}}$ p C p q q] $\in$ 0 if and only if either [$\bar{\text{K}}$ p C p q] $\in$ 0 or [q] $\in$ 0. (2.1.3)
(3) [$\bar{\text{K}}$ p C p q] $\in$ 0 if and only if either [$\bar{\text{p}}$] $\in$ 0 or [$\bar{\text{C}}$ p q] $\in$ 0. (2.1.5 and 2.1.3)
(4) [$\bar{\text{C}}$ p q] $\in$ 0 if and only if both [p] $\in$ 0 and [$\bar{\text{q}}$] $\in$ 0.
(5) Summarizing, [C K p C p q q] $\in$ 0 if and only if [q] $\in$ 0, or [$\bar{\text{p}}$] $\in$ 0, or both [p] $\in$ 0 and [$\bar{\text{q}}$] $\in$ 0.
(6) One or the other of [p] or [$\bar{\text{p}}$] and one or the other of [q] or [$\bar{\text{q}}$] is a member of 0. (cf. 2.1 above)
(7) If [$\bar{\text{p}}$] $\in$ 0 or [q] $\in$ 0, then [C K p C p q q] $\in$ 0. ((5) above)
(8) It remains to consider the case where neither [$\bar{\text{p}}$] $\in$ 0 nor [q] $\in$ 0.
(9) If [$\bar{\text{p}}$] $\notin$ 0, then [p] $\in$ 0, and if [q] $\notin$ 0, then [$\bar{\text{q}}$] $\in$ 0. (2.1.1)
(10) If [$\bar{\text{p}}$] $\notin$ 0 and [q] $\notin$ 0, then [p] $\in$ 0 and [$\bar{\text{q}}$] $\in$ 0. ((9))
(11) If [p] $\in$ 0 and [$\bar{\text{q}}$] $\in$ 0, then [C K p C p q q] $\in$ 0. ((5) above)
(12) In any case, then [C K p C p q q] $\in$ 0. ((6), (7), (8), (11) above)

Exercise

Show that a formula of the form 'A α $\bar{\alpha}$' is true with respect to any model set constructed according to the conditions in 2.1. This may be done by showing that [A α $\bar{\alpha}$] is a member of every model set constructed according to these conditions.

3 MODEL SYSTEMS

A *model system* is a *set* of model sets like the one just described. Each model set is constructed according to the conditions of §2.1, and each contains a number of basic sentences, plus sentences compounded from these according to the conditions. The model sets in a model system stand in a certain relationship to each other. The relation is called *H-alternativeness*, but the connotations of the word "alternative" in ordinary English should not mislead us into supposing that the relation has any special properties at the outset. For the present, H-alternativeness may be taken to be a relation between model sets, whose presence or absence is determined solely by the model sets' being constructed according to certain conditions. In the following additional conditions on model systems, '$H\lambda\mu$' will be used as an abbreviation for "(model set) λ is an H-alternative to (model set) μ."[3]

3.1 A model set, μ, in a model system, Ω, satisfies the following conditions in addition to those given in §2.1, above.

3.1.1 $[M\alpha] \in \mu$ if and only if there is in Ω at least one model set λ such that $H\lambda\mu$ and $[\alpha] \in \lambda$.

This condition reflects the intuitive notion that what is possible is "true in some possible world." Consider the special case of the condition where μ is the fixed model set, 0, and α is associated with some simple statement, say, "Barbra Streisand has blonde hair." According to this condition, $[M\alpha] \in 0$—"*It is possible that Barbra Streisand has blonde hair*" *is one of the statements which describe the real world*—if and only if there is in Ω at least one model set λ such that $H\lambda\mu$—*just in case there is a set of statements which obeys all the structural conditions obeyed by the set of statements which describes the real world*—and $[\alpha] \in \lambda$—*and* "*Barbra Streisand has blonde hair*" *is a member of that set of statements.*

3.1.2 $[L\alpha] \in \mu$ if and only if for every model set λ in Ω such that $H\lambda\mu$, $[\alpha] \in \lambda$.

This condition reflects in turn the notion that what is necessary is "true in all possible worlds." The demonstration that [CKpCpqq] must be a member of the model set 0 can be generalized to a demonstration that [CKpCpqq] must be a member of every model set constructed according to the conditions of §2.1. In fact, the sentence of the model assigned to *any* tautology of the formal system must be a member of every model set in the model system: All tautologies are necessary.

3.1.3 Every model set in Ω contains at least one sentence.

[3] 'H' for Hintikka. Other alternativeness relations will be introduced later which will differ somewhat from the H-relation.

This is a "housekeeping" condition, and may be deemed unnecessary, given that we have stipulated that a model set is a set of sentences. The use of the term "set," however, often allows for the case where a given set is empty. The present condition simply makes it explicit, if perhaps redundant, that no model set is empty.

3.1.4 For every model set μ in Ω there is at least one model set in Ω which is H-alternative to μ.

While this condition seems to require that there be more than one model set in the model system, in fact it makes no such requirement. The condition is satisfied if there is precisely one model set of sentences, the "real world" set, so long as we can interpret that set as being alternative to itself. The intention, of course, is to allow for an indefinite number of different model sets, each of which has different component basic sentences, but each of which obeys the same rules governing its logical structure—the conditions of §2.1 and the present section.

3.2 Summary Description of Model System. We can describe the constituents of a model system easily if we adopt the following abbreviation: Let 'H^μ' represent the set of model sets which are H-alternatives to the model set μ. A model system, then, is a set, Ω, whose members are model sets. The model sets that comprise Ω include the fixed model set, 0; a subset of Ω, H^0, whose members are the model sets of Ω that are H-alternatives to 0; and for each model set μ that is a member of H^0, a subset H^μ of Ω, whose members are the model sets that are H-alternatives to μ, and so on. The model system Ω contains at least one member, 0 itself. The subset H^0 then contains at least one member, by 3.1.4, above, and for every model set μ the set H^μ contains at least one member, also by 3.1.4.

Note that there are no conditions imposed on the constituents of model sets other than conditions on logical constants, including the modal operators. There is nothing in the stipulations to prevent combinations of statements like "Schulz is married and Schulz is a bachelor," "Schulz has a red, green, round, square wife who walks on the ceiling from today to yesterday," or even "Schulz shook hands with O'Flannerty while they were thirteen miles apart." Such statements must be interpreted as impossible in one sense or another, but they are not barred from membership in a model set by the stipulations given so far. Obviously, more must be required of a model set than sheer logical consistency for it to be a description of a state of affairs which is a reasonable alternative to the description of the "real" state of affairs. Such stipulations will be considered in Chapter VI when we try to sort out the various species of possibility and necessity semantically, and to use this sorting out as a means of evaluating **CM**, **CS4** and **CS5** in order to choose a standard system. For now, we may take the conditions of §2 and §3 as providing a minimal articulation of possibility and necessity, allowing us to get on with the limited task of interpreting the formal systems of Chapter III as generic systems of alethic modality.

Exercise

0 is the fixed model set in the model system Ω. Appealing to the conditions at 3.1.1, 3.1.2 and 3.1.3 as needed, show that if [L p] ∈ 0, then [M p] ∈ 0.

4 INTERPRETATION OF THE CANCELLATION SYSTEMS

The cancellation systems of monadic modal logic of Chapter III can now be interpreted with respect to the basic model just described. To interpret the formal systems we must lay down rules for associating with each sign in a wff an element of the model. An interpretation may be described in set-theoretical terms as a function or functional relationship between the signs of the formal system and the marks of the model.

4.1 Interpretation as a function. An interpretation J on a model system Ω is a function J(Z, W), where 'Z' varies over signs in wffs of the formal system and 'W' varies over the marks and sets of marks which constitute the model system. The range of the function is the set of values [T, F]. An interpretation assigns sentences and model sets of the model system to wffs and modal constants of the cancellation systems of modal logic in the following way. (We assume replacement of dummy constants by numerical constants according to the closure rules.)

4.1.1 To the special modal constant '0', which occurs in the top position of every modal index column, an interpretation assigns the fixed model set, 0.

4.1.2 If a constant, 'c', occurs in a modal index column immediately below another constant, 'm', and 'm' is assigned model set 'μ', an interpretation assigns to 'c' a model set which is a member of H^{μ} (i.e., a model set which is an H-alternative to 'μ').

4.1.3 To atomic formulas occurring in wffs of the formal system, an interpretation assigns atomic sentences which are members of the model sets of Ω.[4]

[4] That assignments 4.1.1 and 4.1.2 can be made is assured by the argument in §3.2. That assignment 4.1.3 can be made is assured by 3.1.3.

4.1.4 Given the preceding instructions, then, an interpretation will assign to each wff a sentence of the model, in a manner about to be described, and will assign to each constant entry in a modal index column a model set of the model system. The wff ' p ' will be assigned the following

$$\begin{matrix} 0 \\ 1 \\ 2 \end{matrix}$$

items in the model system: to the wff ' p ' we assign [p] in a manner to be described below; to the modal constant ' 0 ' we assign the fixed model set, 0; to the modal constant ' 1 ', occurring immediately below ' 0 ', we assign a model set which is H-alternative to 0, and to the modal constant ' 2 ', occurring immediately below ' 1 ', we assign a model set which is H-alternative to the model set assigned to ' 1 '.

4.2 A *particular* interpretation determines truth values T or F for a wff α of the formal system. If α is a wff, it consists of an iwff containing a series of signs (statement constants and logical constants) to which is appended an associated matrix of constants, written under the atomic formulas within α as the result of prior cancellation steps. The matrix may in some cases consist of only an occurrence of ' 0 ' under each atomic formula. If α contains more than one modal index column (i.e., if α is truth-functionally complex), the cancellation rules assure that all modal index columns within α will be alike.[5]

4.3 Association of Sentences with Complex Wffs. Let $[\alpha]$ be the sentence of the model which is associated with the wff α according to the following rules.

4.3.1 The first sign of α is the left-most sign of α.

4.3.2 The *n*th sign of α is the sign immediately to the right of the $(n - 1)$th sign of α.

4.3.3 Negation bars are considered as if they were to the left of the signs over which they occur. Constants in the modal index column are not counted.

[5] Consider, for example, the case where α is of the form 'A $\beta\gamma$'. The modal index columns of β and γ could differ only if there were modal operators governing one or the other of β or γ without governing the other. But by the formation rules any such operators would have to be to the right of ' A ', and therefore within α and uncancelled.

4.3.4 Example. The ordering of the signs in a wff is fairly simple. Consider the wff ' M K p A $\bar{\text{q}}$ $\bar{\text{M}}$ r '. The signs are ordered as follows:

0 0 0
1 1 1

First sign: ' M '	Sixth sign: ' q '
Second sign: ' K '	Seventh sign: ' ‾ '
Third sign: ' p '	Eighth sign: ' M '
Fourth sign: ' A '	Ninth sign: ' r '
Fifth sign: ' ‾ '	

4.3.5 The marks of [α] are as follows:

(a) If the *n*th sign of α is a logical constant, the *n*th mark of [α] is (an occurrence of) that same logical constant.

(b) If the *n*th sign of α is an atomic formula, the *n*th mark of [α] is (an occurrence of) the atomic sentence associated with that atomic formula by 4.1.3 above.

If α is the wff of §4.3.4 above, and '**P**', '**Q**' and '**R**' are sentences of the model system associated with the wffs 'p', 'q', and 'r', respectively, by 4.1.3, then [α] will be ' M K **P** A $\bar{\mathbf{Q}}\bar{\text{M}}$ **R** '.

4.4 Determination of Truth-Value. A particular interpretation J determines T for α if and only if [α] ∈ λ, where λ is the model set assigned to the modal constant in the bottom position of the modal index columns of α, as determined by 4.1.1 or 4.1.2 above. Otherwise J determines F for α.

4.5 A truth of logic with respect to a model system is a wff which is determined T under every such interpretation on that model system.

4.6 Examples of Interpretation. In accordance with 4.1.1, 4.1.2 and 4.1.3 above, assignments are made to the various entries in the modal index columns of a wff, and a sentence of the model system is associated with the wff itself. In accordance with 4.4, we can then assign a truth-value to the formula. The wff 'p', for example, is assigned the value T just in case

0
1
2

[p]—the sentence of the model associated with 'p'—is a member of [2]—the model set associated with the modal constant '2' in interpreting this formula—and [2] is H-alternative to [1], and [1] is H-alternative to [0], as demanded by 4.1.1 and 4.1.2. For complex formulas, since all

modal indices will be alike (see footnote 5 above), the reading will be similar: 'A p M K q r' is true just in case [A p M K q r] ∈ [4], and H[4][2]

0	00
2	22
4	44

and H[2][0].

5 MODELS FOR **CMn**, **CM**, **CS4** AND **CS5**

The model system just described is the simplest model for interpreting systems of monadic modal logic. By making additions to the conditions on this model, and later by making additions to its vocabulary as well, we will develop model systems for quantified modal logic and dyadic modal logic.

5.1 CMn. A model for the core system of monadic modal logic, **CMn**, is the model described in §3 and §4 above. The closure conditions **CMn** I and **CMn** II are justified by the conditions imposed on the model in §3. **CMn** I, which allows dummy constants to be replaced by numerical constants already in the branch chart, simply reflects the notions "all" and "some" as they occur in 3.1.1 and 3.1.2. **CMn** II, which allows a new numerical constant to be introduced into a branch chart to replace two dummies, where no constant appeared originally, reflects 3.1.4, which stipulates that for every model set in the model system there is a model set which is H-alternative to the first.

5.2 CM. A model for the standard system of modal logic, **CM**, is obtained from the model for **CMn** by adding the following condition.

5.2.1 Every model set in Ω is H-alternative to itself. (The H-relation is totally reflexive.)

Condition 3.1.4 now becomes redundant, since it can be derived from condition 5.2.1. If every model set in the model system is H-alternative to itself, then for every model set in the model system there is at least one model set which is H-alternative to it (if only that set itself).

5.2.2 The **CM** *Closure Condition.* The various manipulations of modal indices permitted by the closure conditions of **CM**, **CS4** and **CS5** are more than means of juggling symbols to get formulas provable that are wanted as theorems of a given system of logic. They reflect very directly the conditions on the model system. The closure condition for **CM** can now be shown to preserve the truth-value of any wff to which it is applied. That is, if α is a wff of **CM**, and α^* is the result of applying **CM** I, the **CM** closure condition, to α, then an interpretation on the **CM** model system determines T for α if and only if it determines T for α^*.

The **CM** closure condition allows the uniform deletion of a dummy in the modal indices of a branch; '$\underset{\substack{0\\1\\a}}{\text{p}}$', for example, may be changed to '$\underset{\substack{0\\1}}{\text{p}}$'. This, as we noted in Chapter III, has the effect of replacing the dummy by the constant immediately above it, and deleting the repetition: '$\underset{\substack{0\\1\\a}}{\text{p}}$' to '$\underset{\substack{0\\1\\1}}{\text{p}}$' to '$\underset{\substack{0\\1}}{\text{p}}$'. It is sufficient, then, to show that '$\underset{\substack{0\\c}}{\text{p}}$' is determined T with respect to the **CM** model system if and only if '$\underset{\substack{0\\c\\c}}{\text{p}}$' is, where the indices are otherwise the same with respect to any entries occurring above or below 'c'. This is assured by 4.1.2 above, in conjunction with 5.2.1, which specifies that any model set assigned to 'c' must be H-alternative to itself.

Consider, for example, the distinguishing formula of **CM**, 'C p M p'. Cancelled, it produces the branch '$\underset{0}{\bar{\text{p}}}$, a, $\underset{\substack{0\\a}}{\text{p}}$': either [p̄] is a member of 0, or else [p] is a member of some model set [a], which is H-alternative to 0. By 5.2.1, 0 is H-alternative to itself, which allows '$\underset{\substack{0\\0}}{\bar{\text{p}}}$, a, $\underset{\substack{0\\a}}{\text{p}}$': either [p̄] is a member of 0, which is H-alternative to 0, or else [p] is a member of some model set [a], which is H-alternative to 0. But now 0 is just such a set, and 'a' can be replaced by '0'.

5.3 CS4. A model for the first system of collapsed modality, **CS4**, is obtained from the model for **CM** by adding the following condition.

5.3.1 For all model sets λ, μ and ν in Ω, if $H\lambda\mu$ and $H\mu\nu$, then $H\lambda\nu$. (The H-relation is transitive.)

5.3.2 The **CS4** *Closure Condition.* We now show that if α is a wff of **CS4**, and α is changed to α^* by an application of the **CS4** closure condition, then an interpretation on the **CS4** model system determines T for α if it determines T for α^*.

The **CS4** closure allows the deletion of the upper of any two constants in a modal index column; '$p_{\substack{0\\1\\2}}$', for example, becomes '$p_{\substack{0\\2}}$'. It is sufficient to show that '$p_{\substack{0\\c\\c'\\c''}}$' is determined T with respect to the **CS4** model system if '$p_{\substack{0\\c\\c''}}$' is, where the indices are otherwise the same. According to 4.1.2 above, [c] is a model set which is H-alternative to 0, [c′] is a model set which is H-alternative to [c], and [c″] is a model set which is H-alternative to [c′]; and the wff is true if and only if [p] ∈ [c″]. By 5.3.1, if [c″] is H-alternative to [c′], it is also H-alternative to [c]. Thus, if assignments are made according to the conditions imposed on the model system, the deletion of the constant 'c′' preserves the truth-value of the original wff.

The **CS4** distinguishing formula, 'CMMpMp', when cancelled, yields the branch '$\bar{p}_{\substack{0\\1\\2}}$, a, $p_{\substack{0\\a}}$'. The **CS4** closure condition allows deletion of the occurrence of '1', altering the reading of the formula as follows: "either [p̄] ∈ [2] and H[2][1] and H[1]0, or else [p] ∈ [a], where [a] is some H-alternative to 0," becomes "either [p̄] ∈ [2] and H[2]0, or else [p] ∈ [a], where [a] is some H-alternative to 0." By 5.3.2, [2] is just such a set, and 'a' can be replaced by '2', closing the branch.

5.4 CS5. A model for the second system of collapsed modality, **CS5**, is obtained from the model for **CS4** by adding the following condition.

5.4.1 For all model sets λ and μ in Ω, if $H\lambda\mu$, then $H\mu\lambda$. (Symmetry of the H-relation.)

5.4.2 The **CS5** *Closure Condition.* If α is a wff of **CS5**, and α is changed to

α^* by the **CS5** closure condition, then an interpretation on the **CS5** model system determines T for α if it determines T for α^*.

The **CS5** closure condition allows the deletion of all entries in a modal index column except the bottom entry: ‘$\underset{\substack{0\\1\\a\\2\\b}}{\mathrm{p}}$’, for example, may be changed to ‘$\underset{\substack{0\\b}}{\mathrm{p}}$’. The occurrence of ‘0’ may be assumed to remain at the top of each modal index column, since this will not affect closure. It is sufficient to show, then, that the bottom entry in a modal index column must be an H-alternative to the fixed model set, 0, if assignments of model sets to modal indices are made according to 4.1.1 and 4.1.2. This is assured by conditions 5.3.1 and 5.4.1 above.

Exercise

Using the conditions on the model system given in §3 and §5.2.1, show that if $[\mathrm{p}] \in 0$ then $[\mathrm{M\,p}] \in 0$.

6 VARIANT MODELS

There are many conditions and combinations of conditions which may be placed on the H-relation, and for each combination of conditions, a set of object language closure conditions may be devised. Binkley and Clark have considered various combinations of the following conditions, and there are yet others that will be considered.

Ex (Existence)	For every model set μ in Ω, there is at least one model set λ in Ω such that $H\lambda\mu$. (cf. 3.1.4)
Rf (Reflexivity)	For every model set μ in Ω, $H\mu\mu$. (cf. 5.2.1)
Tr (Transitivity)	For any model sets λ, μ and ν in Ω, if $H\lambda\mu$ and $H\mu\nu$, then $H\lambda\nu$. (cf. 5.3.1)
Sm (Symmetry)	For any model sets λ and μ in Ω, if $H\lambda\mu$, then $H\mu\lambda$. (cf. 5.4.1)
Qe (Quasi-equivalence)	For any model sets λ, μ and ν in Ω, if $H\lambda\mu$ and $H\nu\mu$, then $H\lambda\nu$. (Two sets alternative to the same set are alternative to each other.)

Thirty-two different combinations of these properties may be laid out, but these are reduced to fifteen by logical interconnections among the properties themselves. *Qe* and *Rf* taken together, for example, entail both *Sm* and *Tr*, so there could be no model system in which the H-relation had the properties *Qe* and *Rf* without also having *Sm* and *Tr*.

The remaining fifteen combinations are set out in Table 11 with '+' indicating the presence of the property listed at the head of each column. Systems already mentioned in whose models the H-relation has a given combination of properties are listed at the right-hand side.

TABLE 11 PROPERTIES OF THE H-ALTERNATIVENESS RELATION

	Ex	*Rf*	*Tr*	*Sm*	*Qe*	Formal System
1	+	+	+	+	+	**CS5**
2	+	+	+			**CS4**
3	+	+		+		Brouwer
4	+	+				**CM**
5	+		+		+	
6	+		+			
7	+			+		
8	+				+	
9	+					**CMn**
10			+	+	+	
11			+		+	
12			+			
13				+		
14					+	
15						

6.1 Generation of Variant Models. With just the five properties of the H-relation considered in Table 11, fifteen distinct systems of monadic logic can be generated. Additional properties—for example, irreflexivity, asymmetry and intransitivity, as distinct from the absence of reflexivity, symmetry and transitivity—would allow us to generate yet more systems, although the properties of some of the systems generated might become rather bizarre. In the following section we will describe a way to simplify the determination of which properties of the H-relation render which wffs truths of logic with respect to a given model system. But from the sequence of formal systems of Chapter II it is already apparent what some of the effects of the various properties are.

Ex is the property of the H-relation in the model system which allows, in the formal system, the introduction of a constant to replace dummies when no constant is present from the cancellation of operators, as in the cancellation of 'C L p M p'. The formal systems whose models are described by lines 10 through 15 of Table 11 will lack this link between the strong and weak modal operators.

Rf is the property of the H-relation which makes the "link between modality and truth" reflected in the object language theorems 'C p M p' and 'C L p p'. As noted in Chapter III, obligation and permission, and on some interpretations knowledge and belief, if interpreted as modal operators, do not demand this link. Model systems which lack *Rf* as a property of the H-relation would not allow the connection between the strong operator (necessity, obligation, knowledge) and truth, or between truth and the weak operator (possibility, permission, belief). The systems described on lines 5 through 9, which possess *Ex* but not *Rf*, and which possess one or more of the properties which allow collapsing of iterations, will maintain the link between the strong and weak operators, as in 'C L p M p', and lack the links between modality and truth. Such systems may be interesting when interpreted as deontic or epistemic modes. In system 6, for example, the **CS4** distinguishing formula 'C M M p M p' is a truth of logic, but its converse is not: 'C M p M M p'. In Chapter VI, a limited version of reflexivity will be used to develop a model system that makes 'C M p M M p' a truth of logic, but not 'C p M p'.

Tr is the property of the H-relation that produces the first collapsing of iterated modalities. If the H-relation has only the property of reflexivity, each model set has its own cluster of alternative model sets, but nothing is clear about the relation between a model set and the alternatives *to* its alternatives. With the assumption of transitivity, the relation is passed along in a linear way, so that alternatives to alternatives to a given model set become alternatives to that model set. 'C M M p M p' and 'C L p L L p' become truths of logic with respect to the model system when transitivity is assumed.

Sm makes the H-relation "reversible:" If one model set is alternative to a second, then the second is also alternative to the first. Coupled with *Rf* and *Tr*, symmetry allows us to assume that all model sets in a string of model sets that are related by the H-relation are H-alternatives to each other. For this reason, it is possible to ignore all entries in a modal index column except the last entry when the assumptions *Rf*, *Tr* and *Sm* are all taken in the model. The last entry represents the model set where the basic formula is "placed." All the other entries in a modal index column represent model sets, *all of which stand in the same relation to each other.* Iterated modalities all collapse under these circumstances. The alternativeness relation obtains directly between two model sets or not at all; it is never "mediated" by a third model set. 'C M L p L p' and 'C M p L M p' become truths of logic with respect to the model system under this combination of assumptions.

While some general directions are clear with respect to the interpretation of modal systems in terms of obligation and permission, knowledge and belief, and so on, it is clear that additional machinery for manipulating the model structure is needed. In the following section we will develop such machinery, and in Chapter VI, after having developed means for introducing quantifiers and predicates into the object language along with modal operators, we will take up once more the problem of finding systems of modal logic that do justice to deontic and epistemic notions.

7 A QUANTIFICATIONAL ANALOGUE FOR MONADIC MODAL LOGIC

As may be apparent from the frequent use of expressions like "Every model set, " and "Some model set" in describing the model systems for monadic modal logic, the entire semantic structure just described can be accommodated in the notation of quantification theory, with the addition of only a few special items of vocabulary. The cancellation procedures described in Chapter II for quantified (non-modal) formulas provide a convenient way of manipulating the model structure. Our interest here is not the development of a single formal system of modal logic, or even the presentation of classical systems like **M**, **S4** and **S5** for their own sake, but rather the development of techniques for constructing a variety of formal systems to meet a number of different analytic situations, and the development of methods for evaluating and interpreting these systems semantically. It thus becomes desirable to make the semantic structure as easy to manipulate as possible.

Object language formulas can be "translated" into metalinguistic formulas which do not contain modal operators, but which reflect instead the model system described in the preceding sections. The result of thus translating a wff of modal logic will be called the *Quantificational Analogue* of that formula, and the formal system of quantificational logic in which the quantificational analogues of all and only the theorems of a given system of monadic modal logic are provable will be called the Quantificational Analogue of that formal system. The abbreviation QA: followed by either a wff or the name of a formal system of modal logic, will designate respectively the Quantificational Analogue translation of the wff or the Quantificational Analogue of that formal system. The Quantificational Analogue of a given formal system will match, feature for feature, the model structure developed in the preceding sections. Formulas of the Quantificational Analogue will be kept typographically distinct from formulas of the object language by the use of boldface except in cancellation proofs.

7.1 New Vocabulary. To the vocabulary of quantified first-order logic (Chapter II, § 2 and §5) we add the following.[6]

λ, μ, ν	variables ranging over model sets, or "world variables."
0	model set constant; the name of the fixed model set, or "real world."
H——	A predicate of degree 2 (H-alternativeness), taking model set variables or constants as arguments. '**Hλμ**' is read "(Model set) λ is H-alternative to (model set) μ," with the assumption (henceforth tacit) that all model sets under discussion are in the model system Ω.
[—]—	A predicate of degree 2, taking an iwff of the object language for the first argument and a model set variable or constant for the second. '**[α]μ**' is read "[α] is a member of (model set) **μ**," or "'α' is true in world μ," or more simply "α in μ."

7.2 Mapping Rules. The following rules govern the translation of iwffs of modal logic into the Quantificational Analogue. The first step in translation is to enclose the iwff in parentheses and write '0' outside the parentheses. The mapping rules may then be regarded as rules for the removal of parentheses. In translating the iwff 'MLKp$\bar{\text{A}}$qr', for example, we first write '(MLKp$\bar{\text{A}}$qr)**0**', and apply the following rules in order to remove the parentheses.

7.2.1 '(Mα)**μ**' is translated to '**∃λKHλμ** (α)**λ**'.

Compare this translation rule with the condition on the model system at 3.1.1 above. In presenting each translation or mapping rule, we will follow through the example of the preceding paragraph, translating it step by step into the Quantificational Analogue.

[6] The vocabulary presented here contains some rather suspicious-looking items. That variables ranging over model sets and the square brackets are quite respectable abbreviations can be made clear easily enough, if we introduce a new predicate constant of degree one, 'W', taking as argument an individual variable. 'Wx' is interpreted "x is a model set (world)." '**H**– –' will also then take individual variables as arguments. '**∀μ**—' is then an abbreviation for '**∀xCWx**—', and '**∃μ**—' an abbreviation for '**∃xKWx**—'. '**[α]μ**' can be interpreted as an abbreviation for '**[α] ∈ μ**'.

Example

'(M L K p Ā q r)**0**' is translated to '**∃λ K Hλ0** (L K p Ā q r)**λ**'.

7.2.2 '(L α)**μ**' is translated to '**∀λ C H λμ** (α)**λ**'.

This rule reflects precisely the condition on the model system at 3.1.2 above.

Example

'**∃λ K Hλ0** (L K p Ā q r)**λ**' is translated to

'**∃λ K Hλ0 ∀ν C Hνλ** (K p Ā q r)**ν**'

7.2.3 Where 'T' represents one of the truth-functional operators 'K', 'A' or 'C', '(T α β)**μ**' is translated to '**T** (α)**μ** (β)**μ**'.

Compare with the truth-functional conditions on model sets in §2.1 above. '(C α β)**μ**' becomes '**C** (α)**μ** (β)**μ**' and so on.

Example

'**∃λ K Hλ0 ∀ν C Hνλ** (K p Ā q r)**ν**' is translated to

'**∃λ K Hλ0 ∀ν C Hνλ K** (p)**ν** (Ā q r)**ν**'.

7.2.4 Negation is brought outside the parentheses and affects the left-most sign of the next translation step. '($\bar{\alpha}$)**μ**' is translated to '$\overline{(\alpha)}$**μ**'.

Example

'**∃λ K Hλ0 ∀ν C Hνλ K** (p)**ν** (Ā q r)**ν**' is translated, by 7.2.4 and 7.2.3, to '**∃λ K Hλ0 ∀ν C Hνλ K** (p)**ν Ā** (q)**ν** (r)**ν**'.

7.2.5 Where 'α' is an atomic formula, '(α)**μ**' is translated to '**[α]μ**'.

The replacement of parentheses around atomic formulas by square brackets is the last step in translation into the Quantificational Analogue. Where possible, we will adopt the following notational convenience: Square brackets may be dropped, and statement letters changed to capital letters.

Example

The result of the preceding step is translated to

'**∃λ K Hλ0 ∀ν C Hνλ K [p]ν Ā [q]ν [r]ν**',

which in turn is shortened to '**∃λ K Hλ0 ∀ν C Hνλ K Pν $\bar{\mathbf{A}}$ Qν Rν**'.

The final result of the translation steps in the example is then

QA: M L K p A q r.

Following are a few familiar formulas shown with their Quantificational Analogue translations:

'C L p M p' is translated to '**C ∀λ C Hλ0 Pλ ∃λ K Hλ0 Pλ**'.
'C p M p' is translated to '**C P0 ∃λ K Hλ0 Pλ**'.
'C M K p q K M p M q', to
'**C ∃λ K Hλ0 K Pλ Qλ K ∃λ K Hλ0 Pλ ∃λ K Hλ0 Qλ**'.
'C M M p M p', to '**C ∃λ K Hλ0 ∃μ K Hμλ Pμ ∃ λ K Hλ0 Pλ**'.

7.3 Antecedent Assumptions and Closure Conditions. The language of the model system has now become the language of quantification theory with the addition of the special vocabulary at 7.1. The Quantificational Analogue is a formalized metalanguage. Theorems of the object language systems of modal logic will be theorems of the Quantificational Analogues of those systems if we now provide a means of expressing and using in the Quantificational Analogue the various properties of the H-relation that distinguish the models for the various systems of modal logic.

The properties of the H-relation discussed in §6 above, and in the semantic conditions on the model system at 3.1, 5.2, 5.3 and 5.4 above, appear as follows in QA.

Ex	(Existence)	**∀λ ∃μ Hμλ**
Rf	(Reflexivity)	**∀λ Hλλ**
Tr	(Transitivity)	**∀λ ∀μ ∀ν C K Hλμ Hμν Hλν**
Sm	(Symmetry)	**∀λ ∀μ C Hλμ Hμλ**
Qe	(Quasi-equivalence)	**∀λ ∀μ ∀ν C K Hλμ Hνμ Hλν**

The formulas above are used with the Quantificational Analogues of modal wffs. In general, the procedure for determining what sort of model system is needed to make a given object language formula a truth of logic is as follows: To determine the property of the H-relation needed to make α a truth of logic, construct in QA the conditional having **QA: α** as consequent and one of the conditions above, or a conjunction of the conditions above, as antecedent, and cancel the QA formula. If the QA formula cancels to closed branches, then α is a truth of logic with respect to a model system in which the H-relation has the properties attributed to it in the antecedent. We say, then, that **QA: α** is provable given the *antecedent assumption* of those conditions.

The distinguishing formula of **CM**, for example, is 'CpMp'. At several points, we have discussed the way in which the reflexivity of the H-relation makes this formula a truth of logic with respect to the model system. To show it is in QA, in Figure 1 we construct the conditional having *Rf* as antecedent and **QA: CpMp** as consequent. If this conditional cancels to closed branches, we say that **QA: CpMp** is provable given the antecedent assumption of *Rf*.

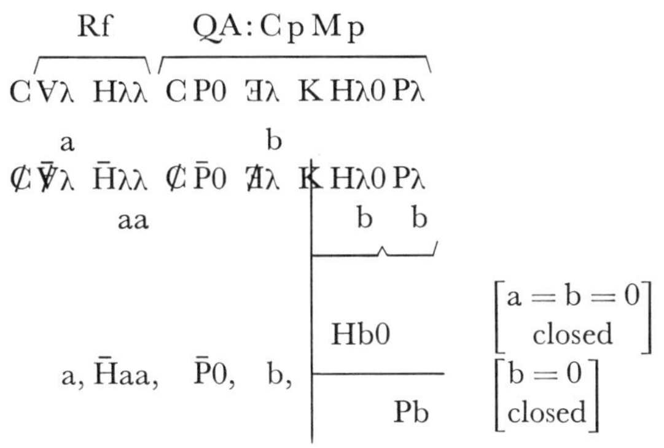

FIGURE 1

Exercise

Demonstrate that **QA: AMpMp** cancels to closed branches in **QA: CMn**, in the following three steps.

(1) Translate 'AMpMp̄' into QA, using the rules in 7.2.
(2) Construct the conditional having *Ex* as antecedent and **QA: AMpMp** as consequent.
(3) Determine whether or not this conditional cancels to closed branches under the rules for quantified elementary logic given in Chapter II.

8 THE SUCCESSION OF FORMAL SYSTEMS IN QA

The Quantificational Analogue allows a great deal of exploration of variant systems of modal logic. For the present, we will limit our explorations to the sequence of formal systems developed in Chapter III. In later chapters, we will use QA to manipulate several sorts of variant formal systems.

8.1 A System Prior to CMn. In §6.1 we mentioned the effect of withholding the property *Ex* from the H-alternativeness relation in the model system. This effect can now be spelled out, using QA. If no properties are attributed to H-alternativeness, its occurrence in QA lacks significance.

The object language system of modal logic which would correspond to such a semantic model would be one in which the modal operators occurred trivially. The system is that corresponding to line 15 of Table 11 (p. 111). All truth-functional tautologies are provable in such a system, of course, as well as all formulas resulting from the uniform substitution of modal formulas for statement letters in truth-functional tautologies, and the results of prefixing 'M' or 'L' to such formulas, as the following examples will show. However, no formulas are provable which hinge upon a relationship between 'M' and 'L', or on any logical properties of 'M' or 'L' (see Figure 2). The second example in Figure 2 is the **CMn** distinguishing formula. Both the object language cancellation and the QA cancellation fail to close without, respectively, **CMn II** and the antecedent assumption of *Ex*, which is the semantic counterpart of **CMn II.**

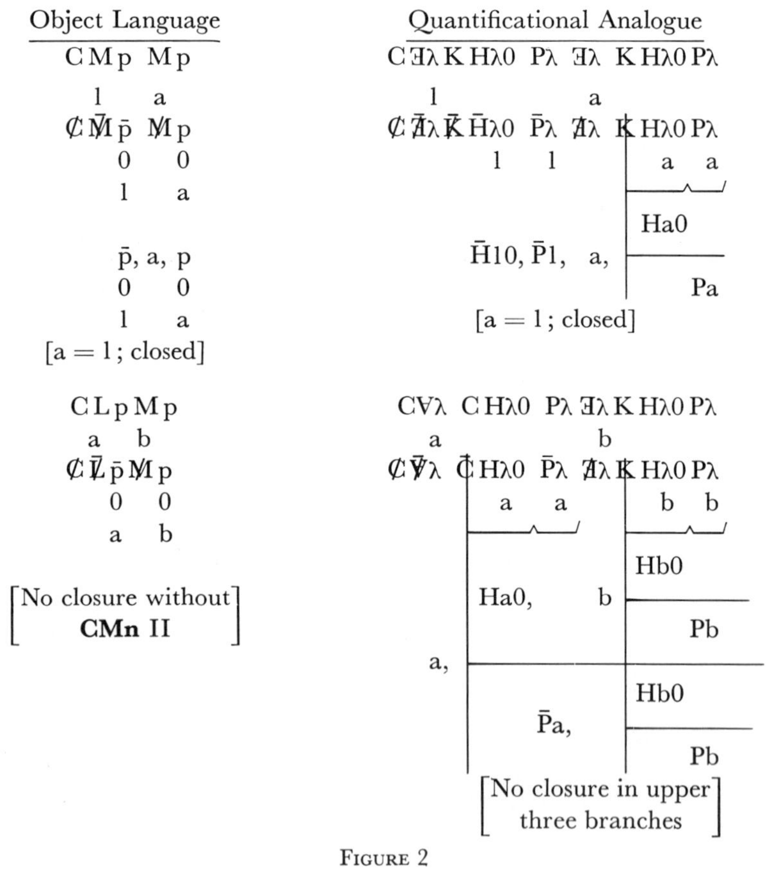

FIGURE 2

8.2 QA: CMn. The Quantificational Analogues of all theorems of **CMn** cancel to closed branches if the formula *Ex* is taken as an antecedent assumption. The cancellation of **QA: C L p M p**, which failed to close in the example in Figure 2, may be closed with the antecedent assumption. (See Figure 3.)

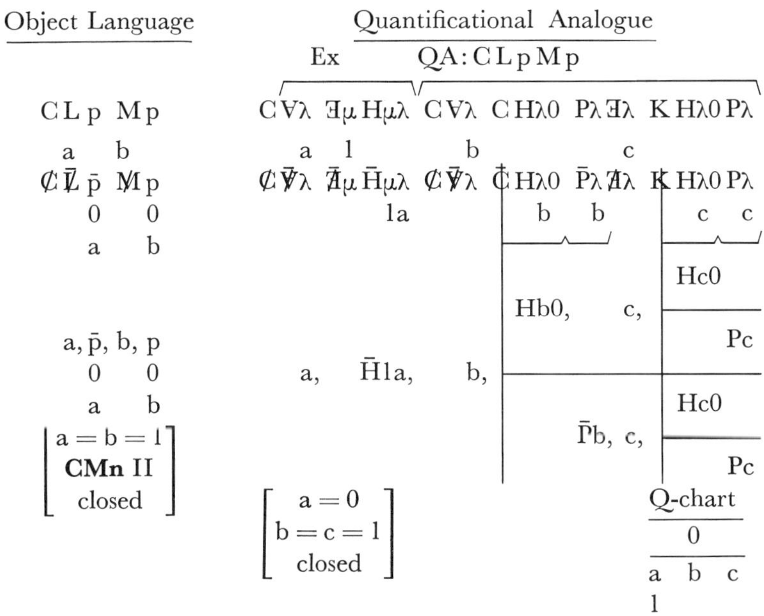

FIGURE 3

8.3 QA: CM. In the Quantificational Analogue of the system **CM**, both *Ex* and *Rf* may be taken as antecedent assumptions. Since *Rf* entails *Ex*, *Rf* alone will suffice. In Figure 1 above, we showed that **QA: C p M p** cancelled to closed branches with the antecedent assumption of *Rf*. It is worth noting, however, that we need not assume that every model set is H-alternative to itself in order to close this cancellation. It would suffice to assume that the fixed model set is H-alternative to itself: **H00**.

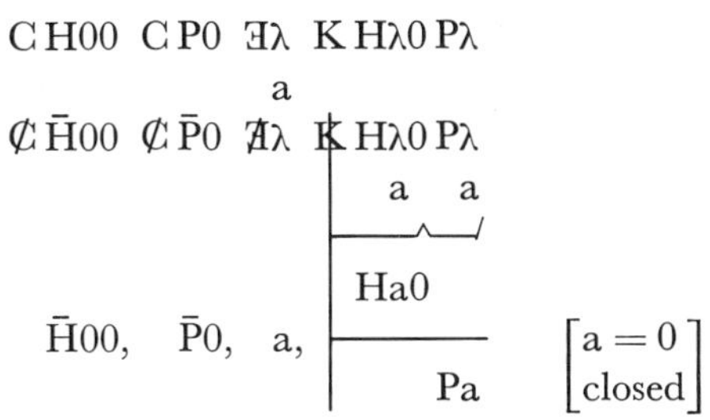

The result of using the assumption **H00** instead of the stronger *Rf* would not be **QA: CM**, however, since the Quantificational Analogues of formulas containing iterated modal operators would not cancel to closed branches. For example, none of

QA: C L p L M p, QA: C L L p L L M p, QA: C M L M p M L M M p

will close with only the antecedent assumption of H00, but all will close with the antecedent assumption of '**∀λHλλ**', which is *Rf*.

8.4 QA: CS4. The antecedent assumptions *Ex*, *Rf* and *Tr* may be taken for **QA: CS4**. Again, *Ex* is redundant, since it is entailed by *Rf*. The distinguishing formula of **CS4** is 'C M M p M p', which appears as follows in the Quantificational Analogue:

C ∃λ K Hλ0 ∃μ K Hμλ Pμ ∃λ K Hλ0 Pλ.

The antecedent assumption *Tr* alone is sufficient to close the cancellation of this formula, but since all theorems of **CMn** and **CM** are also theorems of **CS4**, the conditions *Ex* and *Rf* cannot be omitted from **QA: CS4**. It is clear, however, that any formal system for which *Tr* may be taken as an antecedent assumption in QA—any formal system for which we assume the transitivity of the H-relation semantically—will have 'C M M p M p' as a theorem. Seven of the fifteen systems outlined in Table 11 above have this property.

8.5 QA: CS5. All of the antecedent assumptions described in §6 may be taken for **QA: CS5**—*Ex*, *Rf*, *Tr*, *Sm* and *Qe*. It is sufficient to take *Rf*, *Tr* and *Sm*, since these in combination entail *Ex* and *Qe*. The distinguishing formula of **CS5** is 'C M L p L p'. In QA:

C ∃λ K Hλ0 ∀μ C Hμλ Pμ ∀λ C Hλ0 Pλ.

As was the situation with **CS4**, not all of the permissible antecedent assumptions are needed to close the cancellation of the distinguishing formula of **CS5**, although *Rf*, *Tr* and *Sm* are needed to obtain the full system **CS5**. The conjunction of *Sm* and *Qe* taken as an antecedent assumption will close the cancellation of **QA: C M L p L p** above.

Exercises

1. Show that the claim made in the last sentence of the preceding paragraph is true by performing the cancellation.

2. Determine which of the following antecedent assumptions will make **QA: Cp Mp** cancel to closed branches: **∀λ Hλ0, ∀λ ∃μ Hλμ, ∀λ ∀μ Hλμ.**

9 SYSTEM-BUILDING WITH QA

In the fifteen formal systems outlined in §6, we already have more systems of modal logic available than we have interpretations for. When we add quantifiers to the object language in the following chapter, and later add dyadic operators for entailment and compatibility as well, the fifteen systems will become many times fifteen. With each addition of a symbol to the object language there is some choice as to precisely which conditions on the model system—or antecedent assumptions in QA—provide the best interpretation of that symbol. While there are limitations imposed by the logical interrelations of the assumptions that we have considered, there is obviously great latitude with respect to what may be considered a "truth of modal logic," since this notion is always referred to some particular model system. There is rarely a clear-cut *logical* reason for preferring one of the many model systems over another. All of them are equally respectable from a purely formal point of view. The choice must be made on other grounds. Often the model structure itself will help to clarify what is involved in such a choice, but ultimately the model which is most acceptable will depend upon how one reads the modal operators and what one is trying to do with the formal system. In Chapter VI we will argue on semantic grounds that the **CS5** model system is the most acceptable one for *alethic* modality (possibility and necessity) so long as the formal system in question deals with only one species of alethic modality (logical, theoretical, etc.). We have argued already that a model system containing *Rf* is *not* suitable for deontic modality because *Rf* provides the link between modality and truth that is undesirable in deontic systems.

In some cases, one can get a purchase on the desirable features of the "best" formal system for a given purpose from the object language, as was the case with our intuitions concerning the link between modality and truth in deontic logic. In other cases, such matters become clear only when one looks at the model structure, as will be the case with the choice among **CM**, **CS4** and **CS5** interpreted as alethic systems. Now that both the formal systems and their models are mechanized by the use of cancellation in both the object language and the Quantificational Analogue, we will be able to work back and forth between the two with some facility.

The Quantificational Analogue provides a way of manipulating the model structure in such a way as to determine the object language effects of the various metalinguistic assumptions concerning the H-relation and other features of the model system. In §8.4 above, for example, we found that any model system having *Tr* would make 'CMMpMp' a truth of logic. Conversely, the Quantificational Analogue provides a way to determine what sort of model structure is needed to make some particular "desirable" formula a truth of logic, and to determine at the same time what other features a formal system containing the desirable formula must have. Suppose there is a particular object language formula which is clearly called for on a particular reading of the modal operators, and that we wish to know what sort of model system is needed to make this formula a truth of logic. The first step is to translate the formula into QA and cancel it. It remains, then, to find an antecedent assumption that will provide the elements needed to close the cancellation. If, for example, we are reading 'M' and 'L' in such a way that 'CpLMp' is clearly desirable as a theorem of the object language, we first translate the formula into QA and cancel (see Figure 4). The lower branch closes if a = 0, but the

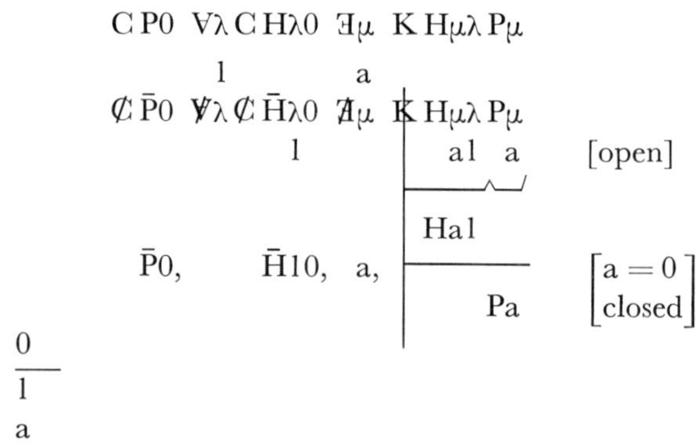

FIGURE 4

upper branch will not close. Since the uniformity requirement is in effect in this branch chart, the occurrence of 'a' in the upper branch must also be replaced by '0'. The upper branch is then '$\mathbf{\bar{P}0, \bar{H}10, a, H01}$'. One thing that would allow the branch to close is permission to reverse '1' and '0' as they occur following 'H'. But this is just *Sm*, the symmetry of the H-relation: '$\mathbf{\forall\lambda\,\forall\mu\,C\,H\lambda\mu\,H\mu\lambda}$'. Without re-cancelling the original formula, we can put '$\mathbf{C\,\forall\lambda\,\forall\mu\,C\,H\lambda\mu\,H\mu\lambda} \cdots$' in front of it to make sure. Since there will be no quantifiers ranging over both the antecedent assumption and the consequent QA formula above, there is no harm in throwing the alphabet out of order for the moment by choosing the first dummies after the ones used above.

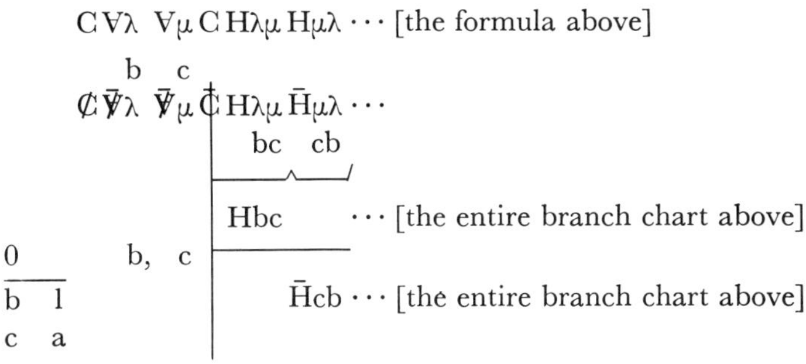

Since the lower branch of the original cancellation closed on '$\bar{\text{P}}$0, P0', we need only worry about the upper branch in each of its two occurrences in the new cancellation. The two open branches are as follows:

b, c, Hbc, $\bar{\text{P}}$0, $\bar{\text{H}}$10, a, H01
b, c, $\bar{\text{H}}$cb, $\bar{\text{P}}$0, $\bar{\text{H}}$10, a, H01

Both branches will close if b = 1 and c = 0. This replacement is permissible. The entire cancellation, put together as if we knew in advance what the antecedent assumption would be, is given in Figure 5 (dummies are kept in the order used above). We know, then, that any system of modal logic that has *Sm* as a feature of its semantic model will have 'C p L M p' There are five such systems in Table 11 of §6.

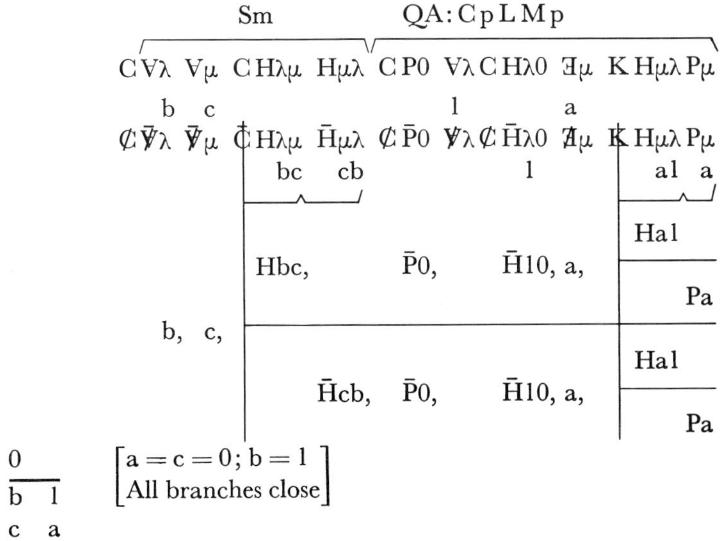

FIGURE 5

The basic machinery for constructing systems of modal logic is now before us: the object language, with its cancellation proof techniques; the model structure, Hintikka's model system; and finally a means of manipulating the model structure by cancellation, the Quantificational Analogue. While additional items of vocabulary will be introduced in the remaining chapters, the machinery will remain the same. It will be necessary to add quantifiers and new modal operators to the object language in order to develop formal systems of the flexibility needed for analytic tasks. Such additions will involve new relations in the metalanguage also, but they will henceforth be additions to an already established machinery, and will involve no completely new techniques beyond those already presented.

Exercises

1. Determine how many of the permissible antecedent assumptions for **CS5** are needed to close the cancellation of **QA: CMLpLp**.

2. Determine by using QA which of the fifteen systems of modal logic described in Table 11, p. 111, will have the following as theorems: 'CpMp', 'CMpMMp', 'CMMpMp', 'CMLpLp', 'CLpMLp', 'CLpMp'. (If all the preceding exercises have been done, much of the work for this one is already at hand.)

BIBLIOGRAPHY—CHAPTER IV

[IV.1] Carnap, Rudolph. *Meaning and Necessity*. (University of Chicago Press, Chicago, 1947)

[IV.2] Hintikka, Jaakko. "The Modes of Modality," *Acta Philosophica Fennica* **16** (1963), pp. 65—81.

[IV.3] Kanger, Stig. *Provability in Logic*. (Almqvist and Wiksell, Stockholm, 1957)

[IV.4] Kripke, Saul. "Semantical Considerations in Modal Logic," *Acta Philosophica Fennica* **16** (1963), pp. 83–94.

[IV.5] ———. "A Completeness Theorem in Modal Logic," *Journal of Symbolic Logic* **24** (1959), pp. 1–14.

[IV.6] ———. "Semantical Analysis of Modal Logic," *Journal of Symbolic Logic* **24** (1959), p. 323.

[IV.7] Pollock, John L. "Logical Validity in Modal Logic," *The Monist* **51** (1967), pp. 128–135.

[IV.8] Schilpp, Paul A. (ed.) *The Philosophy of Rudolf Carnap*. (Open Court, La Salle, Ill., 1963)

[IV.9] von Wright, G. H. "Interpretations of Modal Logic," *Mind* **61** (1952); reprinted in *Logical Studies*. (Routledge and Kegan Paul, London, 1957) pp. 75–88.

Chapter V

QUANTIFIED MONADIC MODAL LOGIC

1 INTRODUCTION

Although we have dealt with quantifier cancellation in Chapter II and have used it in manipulating the Quantificational Analogue for modal logic in Chapter IV, we have not yet introduced quantifiers into the object language and integrated them with modal operators.

In the following sections we will make the necessary additions to the vocabulary and formation rules for the treatment of quantifiers with modal operators. Then a very simple quantificational extension of the system **CMn** will be presented, with appropriate additions to the model system and the Quantificational Analogue. Having laid the foundation, we will then move to a full discussion of the integration of quantifiers with modal operators and consider a range of variant systems of quantified modal logic.

1.1 Vocabulary. The full vocabulary of the cancellation systems of quantified modal logic is as follows.

1.1.1 *Logical Constants.*

A	Alternation
K	Conjunction
C	(material) Conditionality
—	Negation
M	Possibility
L	Necessity
Ǝ	Existential Quantifier
V	Universal Quantifier

1.1.2 A denumerable stock of individual *variables*: x, y, z, x′, y′, ··· x″, ...

1.1.3 A numerically ordered, denumerable stock of *constants* (Arabic numerals).

1.1.4 An alphabetically ordered, denumerable stock of *dummy constants* (lower-case Roman letters, perhaps with primes).

1.1.5 A denumerable stock of predicate constants of degree 0, 1, 2, ···, *n*. (F, G, J, F′, G′, J′, F″, ···)

1.2 Syntactical Variables. 'α', 'β' and 'γ' will be used, as before, to represent well-formed formulas.

'Φ' and 'Ψ' are syntactical variables representing well-formed contexts for individual variables or constants. 'Φ2', for example, will represent a well-formed formula containing one or more occurrences of '2'; perhaps a wff as simple as 'F2', perhaps one as complex as 'Vy C G2y ∃z K Gz2 F2', perhaps one even more complex.

1.3 Formation. As before, we will distinguish between initial well-formed formulas (iwffs) and well-formed formulas (wffs).

1.3.1 IWFFS. An initial well-formed formula of quantified modal logic is one of the following:

(i) A predicate constant of degree *n* followed by *n* individual constants. (A predicate of degree zero we say is a statement.)

(ii) The formula which results from prefixing '∃' or 'V', followed by an individual variable, to an iwff.

(iii) A predicate constant of degree *n* followed by *n* individual variables, *provided* all such variables are to be bound by quantifiers according to (ii).

(iv) The formula which results from prefixing one of 'A', 'K', or 'C' to a pair of iwffs.

(v) The formula which results from prefixing 'M' or 'L' to an iwff.

(vi) The formula which results from placing the negation bar '‾' over the left-most sign of an iwff.

1.3.2 The *index matrix* of a cancelled or partially cancelled formula consists of constants and dummy constants written below the predicate constants in that formula according to the cancellation rules.

1.3.3 The *modal index column* is the left-most column of the index matrix under a predicate constant. It contains dummies and constants written as the result of cancellation of modal operators. The special modal constant '0' occurs in the topmost position of every modal index column.

1.3.4 A *well-formed formula* (wff) is an iwff or a formula resulting from the cancellation or partial cancellation of an iwff. A wff includes any index matrices written below predicate constants within it as the result of applications of the cancellation rules.

1.4 Nomenclature of Formal Systems. In designating the many variant formal systems of quantified modal logic that can be developed, we will need to be systematic. The name of each formal system will begin with the letter '**C**', to indicate that it is a cancellation system. The next letter or group of signs will indicate the basic modal system: '**Mn**', '**M**', '**S4**', '**S5**', or one of the unnamed formal systems of IV.6, indicated by the number assigned to that system on the table IV.6—'**5**', '**6**', and so on. Finally, a letter or group of signs will indicate the quantificational variant: '**Q**' for a system which includes quantifier cancellation without any special closure conditions for quantifiers, '**Q1**', '**Q2**', and so on for systems which include various closure conditions to be introduced below. The cancellation system which includes the **CMn** closure conditions plus quantifier cancellation, but with no special closure rules for quantifiers, for example, will be **CMnQ**; the system which adds to this one all the **CS5** closure conditions will be **CS5Q**, and so on.

1.5 Reading Quantified Modal Formulas. Verbal readings of quantified modal formulas contain no new ingredients. Difficulty arises only when we are inconsistent in reading an operator or when the order of operators is reversed. 'M∃x Fx' may be read "It is possible that something F's," or "It is possible that there is an x such that Fx," or even "Possibly something F's." It should be kept distinct from '∃x M Fx'— "Something possibly F's," or "There is an x such that possibly Fx," or "There is something which possibly F's." The two formulas 'M∃x Fx' and '∃x M Fx' are formally distinct, and are not equivalent in most systems of quantified modal logic to be discussed here. At very informal levels of discussion, the distinction is clear, and it can be emphasized by giving the "There is —" reading to the quantifier.

Where truth-functional compounds fall within quantified modal formulas, the readings become more subtle and lend themselves to ambiguity. Consider the following, where 'Fx' represents "x is a student" and 'Gx' represents "x will pass this course."

M∃x K Fx Gx	It is possible that some student will pass this course.
∃x M K Fx Gx	There is something which is possibly a student who will pass this course.
∃x K Fx M Gx	Some student will possibly pass this course.
∃x K M Fx Gx	Something which is possibly a student will pass this course.

$M\forall x\,C\,Fx\,GX$	It is possible that all students will pass this course.
$\forall x\,M\,C\,Fx\,Gx$	For everything, it is possible that if it is a student it will pass this course.
$\forall x\,C\,Fx\,M\,Gx$	All students will possibly pass this course.
$\forall x\,C\,M\,Fx\,Gx$	Everything which is possibly a student will pass this course.

There are as many different permutations of 'L' with these same formulas. One's logical intuitions are of little help in deciding implication and equivalence relations among the formulas above. '$\exists x\,K\,Fx\,M\,Gx$' seems to imply '$\exists x\,M\,K\,Fx\,Gx$' on the verbal interpretation above: "Some student will possibly pass this course" seems to imply "There is something which is possibly a student who will pass this course." But the verbal reading leads us astray here. In order for the implication to hold, it must be independent of any particular verbal reading, and one counterexample is enough to make it incorrect. Consider the case where 'Gx' is just '$\bar{F}x$'. '$\exists x\,K\,Fx\,M\,Gx$' becomes '$\exists x\,K\,Fx\,M\bar{F}x$'—"Something is F but possibly not F (i.e., F but not necessarily F);" and '$Ex\,M\,K\,Fx\,Gx$' becomes '$\exists x\,M\,K\,Fx\,\bar{F}x$'—"Something is possibly both F and not F," which can never be true. Decisions as to which quantified modal formulas are related by implication or equivalence to others must be reserved until we have arrived at an acceptable semantic interpretation for quantified modal logics and decided upon an appropriate model system.

Exercise

For the purpose of this exercise, suppose that there are precisely five possible worlds:

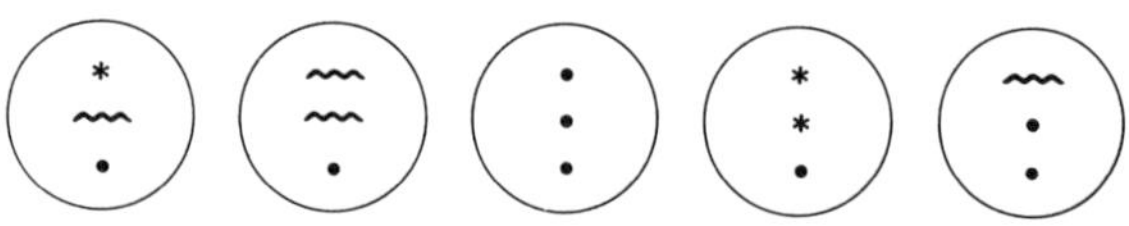

Let the left-most world be the real world. Interpret '$L\alpha$' as "α is true of all (five) possible worlds" and '$M\alpha$' as "α is true of at least one (of the five) possible worlds." Determine which of the following formulas are true. (Vocabulary is given in Chapter I).

1. $L\forall x\,C\,\bar{D}x\,\exists y\,K\,Dy\,Oxy$
2. $\forall x\,C\,Wx\,\forall y\,C\,Wy\,Ixy$
3. $L\forall x\,C\,Wx\,\forall y\,C\,Wy\,Ixy$
4. $M\forall x\,Dx$
5. $M\forall x\,C\,Dx\,\exists y\,K\,\bar{D}y\,Oxy$
6. $\exists x\,K\,Dx\,\exists y\,Oxy$
7. $M\exists x\,K\,Dx\,\exists y\,Oxy$
8. $L\exists x\,KSx\,\forall y\,C\,\bar{I}yx\,Oxy$
9. $M\exists x\,K\,Sx\,\forall y\,C\,Sy\,Iyx$
10. $M\exists x\,\exists y\,K\,K\,K\,Sx\,Sy\,\bar{I}xy$

2 QUANTIFIER CANCELLATION

The quantifier cancellation rules are essentially those introduced in Chapter II. The complicating element is the presence of a modal index column under the bsfs in the formula being cancelled. Dummies and constants resulting from quantifier cancellation must be placed in relation to those resulting from the cancellation of modal operators. We might properly introduce a second set of dummies and a second set of constants to differentiate between those resulting from quantifier cancellation and those resulting from modal cancellation. In practice, however, this is not necessary. The placement of dummies and constants in the index of a basic formula makes the differentiation clear, and the cancellation rules assure that the same dummy or constant cannot be both a modal index and an individual index in a given cancellation proof.

2.1 Cancellation Rules. In addition to the cancellation rules for A, K, C, M, L, $\bar{A}$, $\bar{K}$, $\bar{C}$, $\bar{M}$, $\bar{L}$ and $^{=}$ given in Chapters II and III, the cancellation rules for quantified modal logic include the following rules for the cancellation of quantifiers. In the rules, 'm' is used to represent the bottom entry in the modal index columns under the predicates in a wff when a given rule is to be applied. 'c' and 'd' represent respectively the (numerically) first constant and the (alphabetically) first dummy which has no prior occurrence in the formula. A dummy or constant resulting from the cancellation of a quantifier is written to the right of the modal index column, under each occurrence of the individual variable bound by that quantifier, on a horizontal line with the last entry occurring in the modal index column at the time that quantifier is cancelled.

$$\begin{array}{ccc} & & \mathrm{d} \\ \exists x\, \Phi x & \rightarrow & \not\exists x\, \Phi x \\ 0 & & 0 \\ \vdots & & \vdots \\ \mathrm{m} & & \mathrm{md} \end{array} \qquad \begin{array}{ccc} & & \mathrm{c} \\ \bar{\exists} x\, \Phi x & \rightarrow & \not\bar{\exists} x\, \bar{\Phi} x \\ 0 & & 0 \\ \vdots & & \vdots \\ \mathrm{m} & & \mathrm{mc} \end{array}$$

$$\begin{array}{ccc} & & \mathrm{c} \\ \forall x\, \Phi x & \rightarrow & \not\forall x\, \Phi x \\ 0 & & 0 \\ \vdots & & \vdots \\ \mathrm{m} & & \mathrm{mc} \end{array} \qquad \begin{array}{ccc} & & \mathrm{d} \\ \bar{\forall} x\, \Phi x & \rightarrow & \not\bar{\forall} x\, \bar{\Phi} x \\ 0 & & 0 \\ \vdots & & \vdots \\ \mathrm{m} & & \mathrm{md} \end{array}$$

Examples are given in Figure 1.

$$\begin{array}{l} \exists x\, M\, Fx \\ \\ \;\mathrm{a}\;\;\mathrm{b} \\ \not\exists x \not{M}\, Fx \\ \qquad\quad 0\mathrm{a} \\ \qquad\quad \mathrm{b} \end{array} \qquad \begin{array}{l} M\, \forall x\, Fx \\ \\ \;\mathrm{a}\;\;\; 1 \\ \not{M} \not\forall x\, Fx \\ \qquad\quad 0 \\ \qquad\quad \mathrm{a}1 \end{array} \qquad \begin{array}{l} \forall x\, C\, Fx\, M\, Gx \\ \\ \;1 \qquad\quad \mathrm{a} \\ \not\forall x \not{C}\, \bar{F}x \not{M}\, Gx \\ \qquad\; 01 \quad\; 01 \\ \qquad\qquad\quad \mathrm{a} \end{array} \qquad \begin{array}{l} M\, K\, \exists x\, Fx\, \exists x\, M\, Gx \\ \\ \;\mathrm{a} \qquad \mathrm{b} \qquad \mathrm{c}\;\mathrm{d} \\ \not{M} \not{K} \not\exists x\, Fx \not\exists x \not{M}\, Gx \\ \qquad\quad 0 \qquad\quad 0 \\ \qquad\quad \mathrm{ab} \qquad \mathrm{ac} \\ \qquad\qquad\qquad\;\; \mathrm{d} \end{array}$$

FIGURE 1

2.2 The System CMnQ. All the formal systems of quantified modal logic to be presented here are extensions of the system **CMn.** The first and most simple system, **CMnQ**, will contain no special conditions for the integration of quantifiers and modal operators. The following closure conditions are the Q-closure conditions of II.5.3.4 and the **CMn** closure conditions of III.3.1, reworded to allow for more complicated index matrices. We use the terms "modal dummy" and "modal constant" to indicate dummies and constants which occur in the modal index column under a bsf, and "individual dummy" and "individual constant" to indicate dummies and constants which occur to the right of the modal index column of a bsf.

For the system **CMnQ**, we say that a branch is closed if it contains a pair of opposites with identical index matrices. When dummies occur in index matrices, they may be replaced by constants according to the following conditions.

CMnQ-*closure* I. *Individual dummies in the branch chart may be replaced by individual constants from the branch chart within the restrictions of the Uniformity Requirement and the Order Requirement.*

CMnQ-*closure* II. *When two individual dummies appear in a pair of opposing formulas and no individual constant is present to replace them, a new individual constant may be introduced into the cancellation to replace both dummies.*

CMnQ-*closure* III. *Dummy modal constants in a modal index column may be replaced by modal constants from the branch chart within the restrictions of the Uniformity Requirement.*

CMnQ-*closure* IV. *When two modal dummies appear in a pair of opposing formulas and no modal constant is present to replace them, a new modal constant may be introduced into the cancellation to replace both dummies.*

2.3 Theorems of CMnQ. The system **CMnQ** is a simple, almost trivial extension of **CMn.** Its theorems reflect theorems already cited for formulas containing quantifiers and formulas containing modal operators, but there are no new theorems resulting from the juxtaposition of quantifiers and model operators. The following group of theorems may thus be regarded as basic to any system of quantified modal logic. They hold without any special assumptions or conditions governing quantifiers with the exception of the existence assumptions reflected in the closure conditions **CMnQ** II and **CMnQ** IV: There is at least one possible world (Ex), and there is at least one individual.

To shorten cancellations somewhat, we will omit rewriting bsfs and dummies in the branch charts of cancellations which do not involve split markers. In effect, this simply uses the cancelled formula as its own branch chart.

T-57 C L∀x Fx M ∀x Fx

```
  a b     c  1
Ȼ Ł̄ ∀̸x F̄x M̸ ∀̸x Fx        [      b = 1, CMnQ I      closed]
      0        0          [a = c = 2, CMnQ IV             ]
      ab       c 1
```

If it is necessary that everything F'*s, then it is possible that everything* F'*s.*

This theorem is just **T-1**, with '∀x Fx' substituted for 'p'. **T-1** is 'C L p M p'. A related theorem of elementary logic is 'C ∀x Fx ∃x Fx'. The following five theorems involve permutations of these two theorems, with modal operators and quantifiers occurring in the same formulas, as in **T-57**.

T-58 C ∀x L Fx ∀x M Fx

```
  a b      1 c
Ȼ ∀̸x Ł̄ F̄x ∀̸x M̸ Fx        [      a = 1, CMnQ I      closed]
      0a       01         [b = c = 2, CMnQ IV             ]
      b        c
```

If everything necessarily F'*s, then everything possibly* F'*s.*

T-59 C L ∀x Fx L ∃x Fx

```
  a   b   1  c
Ȼ Ł̄ ∀̸x F̄x Ł ∄ x Fx        [      a = 1, CMnQ III    closed]
      0       0           [b = c = 2, CMnQ II             ]
      ab      1c
```

If it is necessary that everything F'*s, then it is necessary that something* F'*s.*

T-60 C L ∀x Fx M ∃x Fx

If it is necessary that everything F'*s, then it is possible that something does.*

T-61 C ∀x L Fx ∃x M Fx

If everything necessarily F'*s, then something possibly does.*

T-62 C ∀x L Fx ∃x L Fx

If everything necessarily F'*s, then something necessarily* F'*s.*

For each of the theorems of **CMn** given in Chapter III, there are many permutations in **CMnQ**. There is no provision, however, for the mixing of the *order* in which quantifiers and modal operators appear. Formulas like 'C ∀x L Fx M ∃x Fx' are not theorems of **CMnQ**, since there is no provision in this system for jockeying the *positions* of dummies and constants in modal and individual indices. The cancellation of 'C ∀x L Fx M ∃x Gx' produces the branch, 'a, b, F̄ , c, d, F '. The indices in this branch cannot

0a 0
b cd

be made identical by the **CMnQ** closure rules; 'a' and 'd' cannot be brought to corresponding positions. Within this limitation, however, there are many theorems of **CMnQ** that are of interest. The distribution theorems **T-5** and **T-6**, for example, generate the following:

T-63 C K L ∀x Fx L ∀x Gx L ∀x K Fx Gx
T-64 C L ∀x K Fx Gx K L ∀x Fx L ∀x Gx

"*Necessarily everything is* F *and necessarily everything is* G" *is materially* (*and strictly*) *equivalent to* "*Necessarily everything is both* F *and* G."

The order of the modal operators and quantifiers can be *uniformly* reversed:

T-65 C K ∀x L Fx ∀x L Gx ∀x L K Fx Gx
T-66 C ∀x L K Fx Gx K ∀x L Fx ∀x L Gx

but, again, the order of quantifiers and modal operators cannot be mixed within a given formula.

The following two theorems, with their cancellation proofs (Figures 2 and 3) illustrate this point. Both are related to **T-7**, 'C M K p q K M p M q', and the corresponding theorem of elementary logic,

C ∃x K Fx Gx K ∃x Fx ∃x Gx.

T-67 C M ∃x K Fx Gx K M ∃x Fx M ∃x Gx

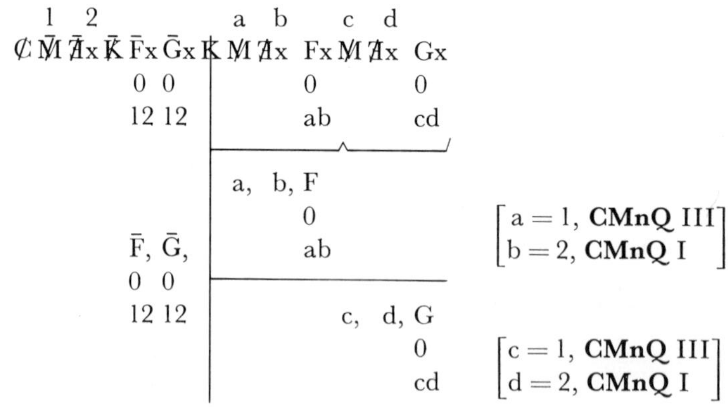

FIGURE 2

T-68 C ∃x M K Fx Gx K ∃x M Fx ∃x M Gx

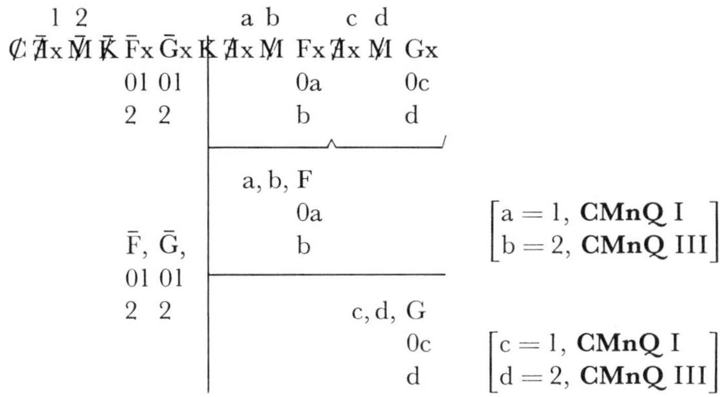

FIGURE 3

"*If it is possible that something is both* F *and* G, *then it is possible that something is* F *and it is possible that something is* G" and "*If there is something which is possibly* F *and* G, *then it is possible that there is something which is* F *and it is possible that there is something which is* G."

Not surprisingly, neither the converse of **T-67** nor the converse of **T-68** is a theorem. "It is possible that something is a freshman and it is possible that something is a giraffe" does not imply "It is possible that something is both a freshman and a giraffe."

One of the reasons that we have avoided the use of a defined operator for strict implication is that the *monadic* modal nature of strict implication makes it behave somewhat differently from material implication. The use of a defined operator as an abbreviation for 'L C—' might prove misleading. Consider **T-9**, 'C L C p q C L p L q', and the related quantificational formula 'C ∀x C Fx Gx C ∀x Fx ∀x Gx'. Quantifiers distribute through a material implication formula, and so does the operator 'L'. The following theorems reflect this feature of the operators.

T-69 C L ∀x C Fx Gx C L ∀x Fx L ∀x Gx
T-70 C ∀x L C Fx Gx ∀x C L Fx L Gx
T-71 C L ∀x C Fx Gx L C ∀x Fx ∀x Gx
T-72 C ∀x L C Fx Gx C ∀x L Fx ∀x L Gx

However, the quantifier does not distribute through a strict implication formula in **CMnQ**. 'C ∀x L C Fx Gx L C ∀x Fx ∀x Gx' is not a theorem of this formal system, and will not be a theorem of most of the systems of quantified modal logic to be studied.

T-73 C L ∀x C Fx Gx C M ∃x Fx M ∃x Gx

If it is necessary that all F's *are* G's, *then if it is possible that something is* F, *it is possible that something is* G.

T-74 C ∀x L C Fx Gx C ∃x M Fx ∃x M Gx

If all F's *are necessarily* G's, *then if something is possibly* F, *something is possibly* G. Both of **T-73** and **T-74** are related to **T-10**.

Verbal intuitions demand the following: If it is necessary that all F's are G's, then it is possible that some F's are G's. But here, as with the quantificational formula 'C ∀x C Fx Gx ∃x K Fx Gx', our intuitions take us beyond what the formal system allows. The universal quantifier still carries no commitment to the existence of anything. In quantification theory, the formula above is not provable without the additional assumption that there are some F's: 'C K ∃x Fx ∀x C Fx Gx ∃x K Fx Gx'. The following two theorems, related to **T-14**, indicate a corresponding situation in quantified modal logic.

T-75 C K M ∃x Fx L ∀x C Fx Gx M ∃x K Fx Gx

T-76 C K ∃x M Fx ∀x L C Fx Gx ∃x M K Fx Gx

If it is possible that something F's *and it is necessary that all* F's *are* G's, *then it is possible that some* F's *are* G's; *and if there is something which possibly* F's *and all* F's *are necessarily* G's, *then there is something which is possibly both* F *and* G.

The paradoxes of strict implication recur, of course, in quantified modal logic. The next four theorems are "paradox" theorems, and other related formulas are also theorems.

T-77 $\mathrm{C}\,\overline{\exists}\mathrm{x}\,\mathrm{M}\,\mathrm{Fx}\,\forall\mathrm{x}\,\mathrm{L}\,\mathrm{C}\,\mathrm{Fx}\,\mathrm{Gx}$

If there is nothing that is possibly F, *then all* F's *are necessarily* G's.

T-78 $\mathrm{C}\,\overline{\mathrm{M}}\,\exists\mathrm{x}\,\mathrm{Fx}\,\mathrm{L}\,\forall\mathrm{x}\,\mathrm{C}\,\mathrm{Fx}\,\mathrm{Gx}$

If it is impossible that something is an F, *then it is necessary that all* F's *are* G's.

T-79 C ∀x L Gx ∀x L C Fx Gx

If everything is necessarily G, *then all* F's *are necessarily* G's.

T-80 C L ∀x Gx L ∀x C Fx Gx

If it is necessary that everything is a G, *then it is necessary that all* F's *are* G's.

The theorems **T-79** and **T-80** seem less paradoxical when combined with quantified formulas. This is partly because the if-then structure is lost verbally. But the conditional is still present in the formula, and to the extent that we interpret the conditional as expressing a relation between statements, the formulas remain paradoxical. There is no way in **T-79** and **T-80** to interpret 'Gx' as a consequence of 'Fx'. The theorems **T-77** and **T-78** become less paradoxical if '∀x' and 'L' are replaced by equivalent constructions containing '∃x', 'M' and negation:

T-81 $\mathrm{C}\,\overline{\exists}\mathrm{x}\,\mathrm{M}\,\mathrm{Fx}\,\overline{\exists}\mathrm{x}\,\mathrm{M}\,\mathrm{K}\,\mathrm{Fx}\,\overline{\mathrm{G}}\mathrm{x}$

If there is nothing that is possibly F, *then there is nothing that is possibly both* F *and non*-G.

T-82 $C \bar{M} \exists x Fx \bar{M} \exists x K Fx \bar{G}x$

If it is impossible that something is F, *then it is impossible that something is both* F *and non*-G.

The fact that the so-called "paradoxes" seem perfectly reasonable when they are read in some way other than "if-then" is not without significance. Those who would argue for a stronger sense of implication or entailment than is provided by strict implication have no quarrel about **T-81** and **T-82.** It is only when "There is nothing that is possibly both F and non-G" or "It is impossible that something is both F and non-G" are taken to be *equivalent to* "G is a consequence of F" or "F entails G" that the quarrel arises. Certainly if 'Fx' entails 'Gx', it *follows* that it is impossible for something to be both F and non-G; but this could also follow from other things, e.g., the impossibility of 'Fx' or the necessity of 'Gx'.

Exercise

Perform cancellation proofs of **T-61**, **T-66**, **T-72**, **T-74** and **T-81**.

3. INTERPRETATION OF QUANTIFIED MODAL FORMULAS

There will be a rather large number of systems of quantified modal logic developed from the system **CMnQ**. First, there will be the extensions which are developed by adding the closure conditions for **CM**, **CS4** and **CS5**, as well as the remaining systems in the list of fifteen sketched in Chapter IV. In addition, there will be various combinations of these basic modal systems with conditions designed to integrate the quantifiers and modal operators. Before confronting several dozen distinct formal systems of quantified modal logic in the object language, we will make the appropriate additions to the metalanguage and the Quantificational Analogue, so that the various formal systems can better be compared as they are presented.

In discussing the elements of the model system which correspond to certain elements of the formal system, we will observe the vocabulary given in Table 12. An *atomic formula* of the formal system consists of a predicate constant of degree *n* followed by *n* individual constants. A *basic formula* is an atomic formula or the negation of an atomic formula. An *atomic sentence* of the model consists of a predicate of degree *n* coupled with *n* individual names. A *basic sentence* of the model is an atomic sentence or the negation of an atomic sentence.

TABLE 12

Formal System	*Model System*
wff	sentence
atomic formula	atomic sentence
individual constant	individual name
predicate constant	predicate

An *interpretation* associates with every predicate constant of the object language a predicate which has an occurrence in one or more of the model sets of the model system, and with every individual constant in the object language an individual name which has an occurrence in one or more model sets in the model system. For every predicate constant 'Ψ' of the object language, there is a predicate [Ψ] of the model system, and for every individual constant 'i' of the object language there is an individual name [i] in the model system.

We say that an individual name or predicate *belongs* to a model set if it occurs in at least one basic sentence of that model set. We say that an individual name *falls under* a predicate (or the negation of a predicate) in a given model set if both that individual name and that predicate (or negated predicate) occur in some basic sentence of that model set. *It is not assumed that every individual name belongs to every model set.*

3.1 Conditions on Model Sets. A model set μ in a model system Ω fulfills the following conditions in addition to those given at IV.2, IV.3 and IV.5.1 above, except that IV.2.1.1 is replaced by 3.1.1, following.

3.1.1. If [i] belongs to μ, then $[\Phi i] \in \mu$ if and only if $[\bar{\Phi} i] \notin \mu$.

3.1.2 $[\exists x\, \Phi x] \in \mu$ if and only if $[\Phi(i/x)] \in \mu$ for at least one individual name [i] which belongs to μ. '$\Phi(i/x)$' is the result of replacing every occurrence of 'x' in 'Φx' by an occurrence of 'i'.

3.1.3 $[\forall x\, \Phi x] \in \mu$ if and only if $[\Phi(i/x)] \in \mu$ for every individual name [i] which belongs to μ.

3.1.4 Basic Formulas. The two preceding conditions on quantifiers provide the *First Interpretation* of basic formulas. Initial well-formed formulas containing individual constants will not be considered until a later chapter, and therefore every fragment of the form 'Φx' that enters into the present treatment will have come from a quantified formula, where the quantifier introducing 'x' has already been interpreted. Formulas containing individual constants may be interpreted without further conditions, but they will not be in this chapter. A number of alternative conditions for interpreting basic formulas will be considered in §6 below. What will distinguish the various interpretations will be the requirement on individual names. For example, how are we to interpret "[i] is [F] in μ" when [i] does not belong to μ? No matter what answer seems appropriate, there will be some difficulty with interpreting "[i] is [$\bar{F}$] in μ." In this First Interpretation we will not tackle these problems. The conditions on quantifiers make the minimal requirements with respect to individual names and the relation "belongs," They are sufficient to provide an elementary interpretation of quantified modal formulas.

The condition on the model system at IV.3.1.3 guarantees that every model set in the model system contain at least one basic sentence. Since a basic sentence contains at least one individual name, we are assured by the definition of "belongs," above, that at least one individual name belongs to each model set. There are no unpopulated worlds in this model system. The following condition on the model system is therefore redundant, but it is important enough to state for reference.

3.1.5 At least one individual name belongs to every model set in the model system.

Finally, we require that individual names belong to model sets; that there be no "floating individuals" without a world to exist in.

3.1.6 Every individual name belongs to at least one model set.

3.2 The New Conditions. The conditions on the model system which govern the interpretation of quantifiers are easy enough to grasp if we speak for the moment of "possible worlds," and instead of individual names which belong to model sets, of individuals which exist in various possible worlds. By condition 3.1.2, '∃x Fx'—say, "Something is a frog"—is true with respect to a world if and only if some individual [i]—say, Freddie—which exists in that world, has the property [F]—is a frog—in that world. "Something is a frog" is true for a given world just in case there is at least one individual, perhaps Freddie, which exists in that world and is a frog. By 3.1.3, '∀x Fx'—"Everything is a frog"—is true for a given world just in case every individual which exists in that world is a frog.

Condition 3.1.5 is analogous to the condition *Ex* on model sets. Where *Ex* was needed to permit 'CLpMp', condition 3.1.5 is needed to permit

'C∀xFx∃xFx' (see II.5.3.4). We will refer to 3.1.5 as the *Principle of Singular Existence* (*SEx*).

The condition at 3.1.6 we call the *Principle of Population* (*Pop*).

4 ADDITIONS TO THE QUANTIFICATIONAL ANALOGUE

The Quantificational Analogue will continue to parallel the model system. The additions to be made are precise reflections of those just made to the model system. To the vocabulary of the Quantificational Analogue given at IV.7.1 we make the following addition.

B—— A predicate of degree 2 ("belongs"), taking an individual variable or constant for the first argument and a model set variable or constant for the second. '**Bxμ**' is read "x belongs to (model set) μ" or "(the individual) x exists in (world) μ."

4.1 Mapping Rules. The following rules, taken together with those given at IV.7.2, allow the translation of iwffs of quantified modal logic into the Quantificational Analogue. As before, mapping rules may be regarded as rules for the removal of parentheses.

4.1.1 '(∃xΦx)μ' is translated to '**∃xKBxμ**(Φx)μ'. (cf. 3.1.2 above).
4.1.2 '(∀xΦx)μ' is translated to '**∀xCBxμ**(Φx)μ'. (cf. 3.1.3 above).
4.1.3 Where 'Ψx' contains no logical constant other than negation, '(Ψx)μ' is translated to '[**Ψx**]μ'.

The mapping rule at 4.1.3 reflects the First Interpretation of basic formulas, and will be replaced in later interpretations.[1]

[1]Once again, the elements in the vocabulary of QA can be expanded, with interesting results. For simplicity, in the text we will use the analogue in the form shown above. But since we understand the model system to consist of marks on paper—occurrences of basic and complex sentences arranged in (model) sets—we can treat these marks and sets of marks in a perfectly straightforward way as a set of concrete objects over which the variables of quantification in QA may range. In the footnote at IV.7.1 we showed how to replace model set variables with individual variables in QA: '**Wx**' represents "x is a model set," and the translation rules are adjusted accordingly. The following further additions can be made to the basic QA vocabulary:

Let '**Dx**' represent "x is an occurrence of an individual name".
Let '**Px**' represent "x is an occurrence of a predicate letter".
Let '**Exyz**' represent "y falls under x in z".

'**Bxy**' is then understood to mean "x occurs in a basic sentence of y," and conditions may be introduced containing the combinations 'K Dx Bxy' and 'K Px Bxy'. It must be noted here that not every formula that is well formed in the vocabulary of QA is part of that system which comprises the Quantificational Analogue of a formal system of modal logic,

As we did with unquantified formulas, we will drop the square brackets as a notational convenience. The effect will be to make an object language predicate of degree n into a metalanguage predicate of degree $n + 1$. The fragment '(Fx)**λ**' becomes, by translation rule 4.1.3, '[**Fx**]**λ**', which we shorten to '**Fxλ**'—"x is an F in λ."

In Figure 4 we give an example of a step by step translation of **T-61** into QA. Translation rules from Chapter IV and from this section are cited as they are used.

C ∀x L Fx ∃x M Fx	
(C ∀x L Fx ∃x M Fx)**0**	IV.7.2
C (∀x L Fx)**0** (∃x M Fx)**0**	IV.7.2.3
C ∀x C Bx0(L Fx)**0** (∃x M Fx)**0**	4.1.2
C ∀x C Bx0 ∀λ C Hλ0(Fx)**λ** (∃x M Fx)**0**	IV.7.2.2
C ∀x C Bx0 ∀λ C Hλ0 Fxλ (∃x M Fx)**0**	4.1.3
C ∀x C Bx0 ∀λ C Hλ0 Fxλ ∃x K Bx0 (M Fx)**0**	4.1.1
C ∀x C Bx0 ∀λ C Hλ0 Fxλ ∃x K Bx0 ∃λ K Hλ0 (Fx)**λ**	IV.7.2.1
C ∀x C Bx0 ∀λ C Hλ0 Fxλ ∃x K Bx0 ∃λ K Hλ0 Fxλ	4.1.3

FIGURE 4

4.2 Antecedent Assumptions. There are three antecedent assumptions for the Quantificational Analogue justified by the model system described in §3. One of these, the condition that requires no floating individuals, will not be needed, but we state it for later reference, under the label *Pop*, for "population" (cf. 3.1.6).

Pop **∀x ∃λ Bxλ**

The two remaining conditions are the existence assumptions for model sets and individuals, respectively.

Ex **∀λ ∃μ Hμλ**
SEx **∀λ ∃x Bxλ**

The Quantificational Analogues of all the theorems of **CMnQ** given in §2.3 above may be closed by taking one or both of the existence conditions

---

but only those wffs which are translations of object language formulas, or antecedent assumptions, or conditionals formed from these. The translation rules above become, in basic QA vocabulary, as follows:

(∃x Φx)**m** is translated to '**K Wm ∃x K K Dx Bxm** (Φx)**m**'
'(∀x Φx)**m** is translated to '**K Wm ∀x C K Dx Bxm** (Φx)**m**'
(Fx)**m**' is translated to '**K K Dx Pf Efxm**'

With appropriate means used for handling negated object language predicates, and with appropriate additional antecedent assumptions, it need not be assumed, as we assume in the text, that every predicate letter has an occurrence in every model set.

as antecedent assumptions. **T-61**, which was translated into QA in the preceding section, requires both *SEx* and *Ex* as antecedent assumptions (see Figure 5).

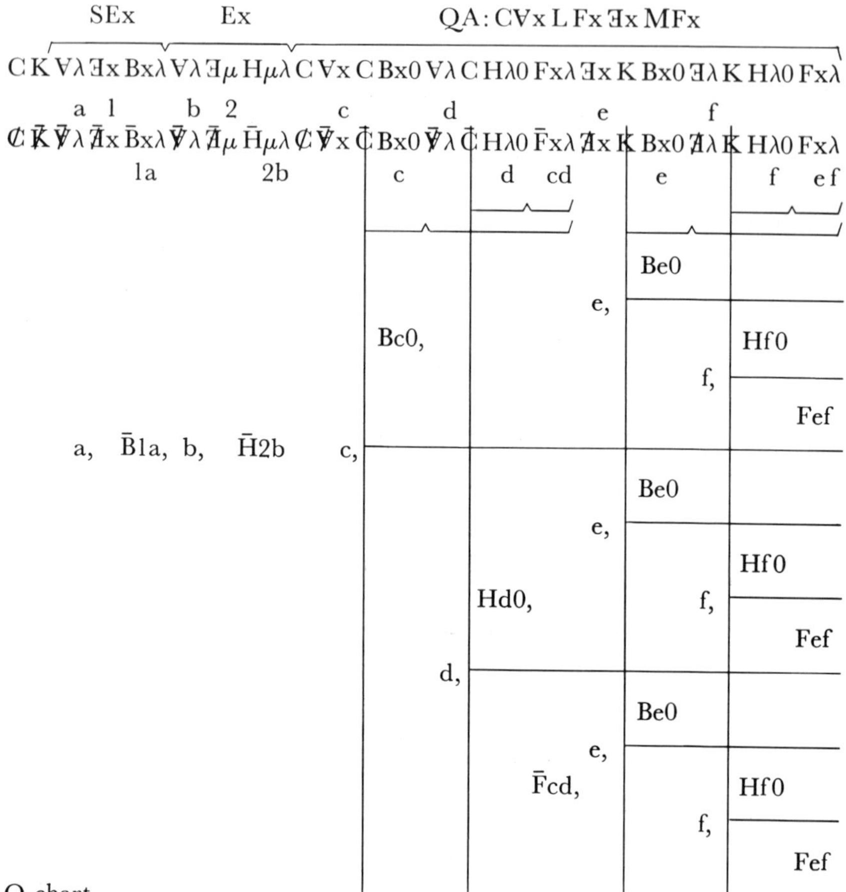

Q-chart

0
a b c e
1 2 d f

[a = b = 0; c = e = 1; d = f = 2 — All branches close]

FIGURE 5

4.3 Systems Beyond CMnQ. The antecedent assumptions in QA for the systems that follow **CMn** are the same as before.

Rf ∀λ Hλλ

Tr ∀λ ∀μ ∀ν C K Hλμ Hμν Hλν

Sm ∀λ ∀μ C Hλμ Hμλ

The system **CMQ** is obtained by adding the antecedent assumption *Rf* to the antecedent assumptions for **CMnQ**; **CS4Q** adds *Tr* to the assumptions for **CMQ**; **CS5Q** adds *Sm* to the assumptions for **CS4Q**. Object language closure conditions will be discussed in §5.4 below.

Exercise

Translate **T-58**, **T-62** and **T-70** into QA and, using the antecedent assumptions *Ex* and *SEx*, show that they cancel to closed branches in QA.

5 INTEGRATION OF QUANTIFIERS WITH MODAL OPERATORS

We have noted that in the system **CMnQ** there are no theorems of the sort 'C ∀x L Fx M ∃x Fx', where quantifiers and modal operators occur in different order before basic formulas. This is also true of the **CM**, **CS4** and **CS5** extensions of **CMnQ**. To allow for mixing the order of quantifiers and modal operators in object language theorems, it is necessary to consider further assumptions concerning the individual names which belong to the various model sets in the model system, or in "world" talk, concerning the individuals which exist in possible worlds. To facilitate the discussion, we will speak of the set of individual names which belong to a model set—the individual names which occur in basic sentences of a model set—as the *population* of that model set.

The choice of a standard quantified modal logic is nearly impossible, and is not our aim here in any case. Given the fifteen basic systems of unquantified modal logic described in IV.6, it is possible to develop something over one hundred distinct and thoroughly respectable systems of quantified modal logic by making various assumptions concerning the populations of alternative model sets and by introducing variant interpretations of quantifiers and atomic formulas. To the extent that formal systems are tools of the analytic trade, there is no reason to suppose that a single super-tool is appropriate to every task that the trade involves. If the population under consideration is the sort of population we cope with in everyday discussions of matters of fact, one group of assumptions might turn out to be preferable over all others. If the population under consideration is the set of natural numbers rather than a set of concrete individuals, quite another group of assumptions will be appropriate, and if it is a specialized and limited set of concrete individuals, yet another group of assumptions might be called for. Again, genera of modality other than the alethic will make demands on the formal system that will differ from those made by possibility and necessity. We have noted this difference already with respect to the interrelations of modal operators in deontic and epistemic logic, and further differences will appear with respect to the integration of quantifiers with modal operators.

5.1 The Barcan Formula and its Converse. In 1946, Ruth Barcan Marcus published a quantified version of the Lewis systems of monadic modal logic [V.13]. In this system, the two formulas '∃x M Fx' and 'M ∃x Fx' are demonstrably equivalent. The so-called Barcan Formula, 'C M ∃x Fx ∃x M Fx', has aroused considerable controversy, and on some interpretations of quantifiers and modal operators is decidedly counter-intuitive: If it is possible that there is something which F's, then there is something which possibly F's. There are several interpretations on which the Barcan Formula is not a desirable theorem of modal logic, one of them being A. N. Prior's temporal interpretation of possibility [V.18, pp. 26ff.]. Consider the possibility that the first woman President will be blue-eyed. According to the Barcan Formula, if it is possible that the first woman President will be blue-eyed, then there is someone who is possibly the first woman President and blue-eyed.

The intuitive difficulty with the Barcan Formula may have to do with the interpretation of the existential quantifier as reflected in the rules we have chosen for translating '∃x Fx' into QA. If that interpretation is modified so that the "existential commitment" of the quantifier is removed or weakened, the Barcan Formula becomes more plausible. We will consider such alternative interpretations in a later section. For the present, however, we will consider the Barcan Formula under the interpretations of the existential quantifier and atomic formulas given above.

There is more at stake in the Barcan Formula than our intuitions or intuitive counterexamples. To accept the equivalence of '∃x M Fx' and 'M ∃x Fx' is to gloss over a distinction of modality which was recognized at least as early as the twelfth century (Abelard) and which is still treated with respect by contemporary logicians [cf. V.19, pp. 8, 25]. This is the distinction between modality *de dicto* and modality *de re*. Modality *de dicto* attaches to entire statements, as in "*It is possible that* someone has been eating my porridge," or, more to the point, "*It is possible that* Sherlock Holmes lived at 221b Baker Street." *De dicto* possibility affects the entire statement, and commits us to nothing more than the possibility of that golden-haired "someone" existing and eating the porridge. Perhaps the porridge hadn't been put in the bowls at all. It commits us to nothing more than the possibility that there was someone named Sherlock Holmes who lived at 221b Baker Street.

Modality *de re*, on the other hand, "intervenes" between the subject and predicate of a statement, as in "There is someone who has possibly been eating my porridge," and "Sherlock Holmes possibly lived at 221b

Baker Street (or possibly it was 227b)." As usually understood, modality *de re* has to do with *actual* individuals who *possibly* did such-and-such, whereas with modality *de dicto* both the existence of the individuals and the reality of what they may have done are left up in the air.

The antecedent and consequent of the Barcan Formula can be viewed as modern reflections of, respectively, possibility *de dicto* (M ∃x Fx) and possibility *de re* (∃x M Fx). The traditional, and seemingly natural, interpretation is that possibility *de re* entails possibility *de dicto*, but not vice versa: "There is someone who has possibly been eating my porridge" entails "It is possible that someone has been eating my porridge," but not the other way around. On this view, the converse of the Barcan Formula should be a theorem of an adequate quantified modal logic where the quantifiers are interpreted conventionally, but the Barcan Formula itself shouldn't. 'C ∃x M Fx M ∃x Fx' should be provable; 'C M ∃x Fx ∃x M Fx' should not. That this view is correct, given only the interpretation of quantifiers in the preceding sections, emerges quite clearly when we translate the antecedent and the consequent of the Barcan Formula into QA:

M ∃x Fx	**∃λ K Hλ0 ∃x K Bxλ Fxλ**
∃x M Fx	**∃x K Bx0 ∃λ K Hλ0 Fxλ**

'M ∃x Fx' does not involve any commitment about an existing individual—an individual name which belongs to the fixed model set—but only about an individual which exists in some possible state of affairs—an individual name which belongs to some H-alternative to the fixed model set. '∃x M Fx', on the other hand, is clearly about an individual which exists—an individual name x, such that **Bx0**.

5.2 Antecedent Assumptions Concerning Singular Existence. Depending upon the applications for which a formal system is intended, there are any number of conditions that might be imposed upon the populations of possible worlds or model sets. If we wish to consider what is possible and necessary only with respect to a fixed set of individuals, it is appropriate to assume that the population of every model set is the same. If, for example, the set of individuals being discussed is the set of natural numbers rather than a set of concrete individuals which come into existence and pass out of existence, the assumption that the population of every model set is uniform is appropriate. There are two different assumptions which have the same effect on the object language and which require uniformity of population. The first we call the *Principle of Uniform Population* (*PUP*): *Every individual belongs to every model set.* In QA, this is '**∀x ∀λ Bxλ**'. A slightly modified form of *PUP* which seems more cautious we call the *Principle of Uniform Possible Individuals* (*PUPI*): *Every individual which belongs* *to a model set belongs to every model set.* In QA, this is '**∀x C ∃λ Bxλ ∀λ Bxλ**'. *PUP* and *PUPI* are equivalent, given *Pop* above. Both the Barcan Formula and its converse are theorems of any formal system for which either *PUP* or *PUPI* is assumed.

Suppose, however, that the formal system is being used for the analysis of a context in which it is appropriate to consider possible states of affairs involving existing individuals plus states of affairs involving existing individuals *and* other individuals that may or may not exist. This might be called for in a temporal interpretation of modality which treats the fixed model set as a set of sentences describing everything that *is* the case or *has been* the case, and H-alternative model sets as sets of statements describing all this *plus* what might be the case in the future. The population of the "real world" would include Socrates, Julius Caesar, Torquemada, George Washington, Ho Chi Minh, you and me, among others. "Future worlds" would have all of these plus additional possible individuals which do not yet exist, but which might exist in some future state of the world. An assumption that describes such an arrangement we call the *Principle of Historical Expansion of Population* (*PHEP*): *Every individual name which belongs to a given model set belongs also to every* H-*alternative to that model set.* In QA, '**∀x ∀λ ∀μ C K Bxλ Hμλ Bxμ**'. The Barcan Formula is not a theorem of **CMn**, **CM** or **CS4** when *PHEP* is taken as an assumption, but it does become a theorem of **CS5** with *PHEP*.

Obviously *PHEP* is not an appropriate assumption for every interpretation of the notion of a possible world or alternative state of affairs. First of all, an individual which exists in actuality exists in every possible world. Suppose we think of constructing a model set different from the real world set intensionally, as "wishful thinking." When you wish for a new car, you are conjuring up a possible world which contains all the individuals that the real world contains, *plus* your new car. So far, so good. But with *PHEP*, you are not allowed to conjure up a possible world which *lacks* any individual that exists in the real world. Acceptance of this condition would forbid your wishing for, or even considering the pleasant possibility of, the non-existence of your worst enemy. The condition also forbids historical musings of the sort "What would have happened if ——— (Caesar, Luther, Churchill, etc.) never existed?"

On the First Interpretation of basic formulas, *PHEP* is the only assumption on individual existence that generates a formal system in which the converse of the Barcan Formula is a theorem and the Barcan Formula itself is not. But in quantified **CS5** with *PHEP*, both the Barcan Formula and its converse are provable.

A principle which is just the reverse of *PHEP* requires the actual existence of all possible individuals. On the somewhat facetious "wishful thinking" interpretation of possible worlds, we might call this condition "Wishing will make it so." *Every individual name which belongs to a given model set belongs also to every model set to which the first is* H-*alternative*. In QA,

∀x ∀λ ∀μ C K Bxλ Hλμ Bxμ.

We call this the *Principle of Imaginary Expansion of Population* (*PIEP*). With *PIEP*, if you wish for that new car, it exists. Every individual in a possible world exists also in the real world.

The Barcan Formula is a theorem of every formal system that has *PIEP* as an assumption in its Quantificational Analogue, and the converse of the Barcan Formula is also a theorem of every **CS5**-based system that has *PIEP*.

5.3 Sixteen Systems of Quantified Modal Logic. If the four assumptions concerning singular existence were combined with the various assumptions concerning H-alternativeness mentioned in IV.6, the result would be a plethora of systems of quantified modal logic: the fifteen systems of IV.6 multiplied by the number of possible combinations of the assumptions concerning singular existence. For the moment, we will content ourselves with descriptions of only sixteen systems of quantified modal logic, two of which will turn out to be equivalent. These will be quantificational extensions of **CMn**, **CM**, **CS4** and **CS5** generated by adding *PUP*, *PHEP and PIEP* as antecedent assumptions (*PUPI* and *PUP* generate equivalent systems, so we omit *PUPI*).

The procedure for determining whether or not a given formula is a truth of logic with respect to certain assumptions on the model is the same as that given in IV.7. Set out the formula in QA, and construct a conditional having that formula as consequent and a conjunction of appropriate antecedent assumptions as antecedent. In Table 13, all systems which allow *none* of the new assumptions on singular existence are called '**Q1**' systems: **CMnQ** is now **CMnQ1**, etc. Systems which allow the assumption of *PHEP* are called '**Q3**' systems. The result of adding the assumption *PHEP* to the assumptions for the system **CM** is the new system **CMQ3**, and so on. The **CM** assumption is *Rf*, '**∀λ Hλλ**'. To determine whether or not the Barcan Formula is a theorem of **CMQ3**, using the Quantificational Analogue, we set up the conditional having *Rf* and *PHEP* conjoined in the antecedent, and having the Quantificational Analogue of the Barcan Formula as consequent. (The procedures for making the same determination in the object language are described in §5.4.)

Rf PHEP

C K ∀λ Hλλ ∀x∀λ∀μ C K Bxλ Hμλ Bxμ

QA: C M ∃x Fx ∃x M Fx

C ∃λ K Hλ0 ∃x K Bxλ Fxλ ∃x K Bx0 ∃λKHλ0 Fxλ

The formula does not cancel to closed branches, and the Barcan Formula is not a theorem of a formal system whose model has only these assumptions.

5.3.1 The Barcan Formula and its Converse Under the First Interpretation. In Table 13, we indicate the basic system of modal logic in the left-hand column, and the quantificational variant in the top horizontal row. The name of each formal system can be put together by adding '**Q1**', '**Q2**', etc. to the name of the basic modal system. The word "Yes" in the table indicates that the Barcan Formula (BF) is a theorem of the system in question, or that its converse (CBF) is. "No" indicates that the formula is not a theorem.

TABLE 13

	Q1 No new assumption		**Q2** Assume PUP		**Q3** Assume PHEP		**Q4** Assume PIEP	
	BF	CBF	BF	CBF	BF	CBF	BF	CBF
CMn	No	No	Yes	Yes	No	Yes	Yes	No
CM	No	No	Yes	Yes	No	Yes	Yes	No
CS4	No	No	Yes	Yes	No	Yes	Yes	No
CS5	No	No	Yes	Yes	Yes	Yes	Yes	Yes

The systems **CS5Q3** and **CS5Q4** can easily enough be seen to be equivalent systems, since the **CS5** symmetry assumption *Sm* taken together with either of *PHEP* or *PIEP* implies the other.

Exercise

Show that **QA: C M ∃x Fx ∃x M Fx** cancels to closed branches in **CS5Q3**. (The antecedent assumptions *PHEP* and *Sm* are sufficient.)

5.4 Object Language Closure Conditions. The formal systems of modal logic which constitute the object language reflections of our most recent manipulations of the model will have closure conditions that allow the manipulation of quantifier indices as well as modal indices. We now consider a *pair of opposites* in a branch to be a basic subformula and its negation with identical *modal* index columns (after replacement of dummies by constants and modal closure) and *no occurrences of (individual) dummies elsewhere in the index matrix of a formula that are not replaced by occurrences of constants from the matrix of the opposing formula.* Individual constants will be allowed to replace only those individual dummies that appear on the same horizontal line of an index matrix as the constant. In the example in Figure 6, '1' cannot replace 'b', and there is no closure. The manipulations

$$
\begin{array}{l}
\text{C}\ \exists x\ \text{M}\ \text{F}x\ \text{M}\ \exists x\ \text{F}x \\
\quad 1\ 2 \qquad\quad a\ b \\
\not{\text{C}}\ \not{\bar{\exists}}x\ \not{\bar{\text{M}}}\ \bar{\text{F}}x\ \not{\text{M}}\ \not{\exists}x\ \text{F}x \\
\qquad\quad 01 \qquad\quad 0 \\
\qquad\quad 2 \qquad\quad\ \ ab \\
\\
\qquad\quad \bar{\text{F}},\ a,\ b,\ \text{F} \\
\qquad\quad 01 \qquad 0 \qquad [a = 2] \\
\qquad\quad 2 \qquad\ \ ab
\end{array}
$$

FIGURE 6

of the index matrices for the **Q2** and **Q3** systems will permit the closure of the branch in Figure 6. Without further manipulation, however, it remains open. (The formula is of course CBF.)

5.4.1 Quantifier Closure Conditions. The following manipulations of occurrences of individual dummies and constants in index matrices reflect the assumptions *PUP*, *PHEP* and *PIEP* discussed in 5.3 above. They are labeled according to the quantificational variant which is produced by their use.

Q1-*closure. No further manipulation is permitted.*

Q2-*closure. An individual constant occurring in an index may be repeated on every horizontal line above and below its original occurrence.* (cf. the **Q2** assumption *PUP*)

An example of **Q2**-closure is given in Figure 7.

Q3-*closure. An individual constant occurring in an index matrix may be repeated on every horizontal line* below *its original occurrence.* (See Figure 8 for an example.)

Q4-*closure. An individual constant occurring in an index matrix may be repeated on every horizontal line* above *its original occurrence* (see Figure 9).

5.4.2 Modal Closure Conditions. The manipulations of the modal index column are unchanged. Such deletions as are permissible under the **CM**,

```
C M ∃x M Fx ∃x M M Fx

  1  2   3       a  b  c
C̸ M̸̄ ∄̄x M̸̄ F̄x ∄ x M̸ M̸ Fx
             0             0a
             12             b
             3              c
             F̄ ,  a, b, c, F
             0             0a      [No closure]
             12            b
             3             c
..................................................................
             F̄ ,  a, b, c, F       Repeat '2' at all
             02            0a      positions, by
             12            b       Q2-closure.
             32            c
                                   [a = 2, b = 1, c = 3
                                          closed       ]
```

FIGURE 7

```
C ∃x M M M Fx M ∃x M M Fx

   1  2  3  4      a   b  c  d
C̸ ∄̄ x M̸̄ M̸̄ M̸̄ F̄x M̸ ∄x M̸ M̸ Fx
              01             0
              2              ab
              3              c
              4              d
              F̄ , a, b,c, d,  F
              01             0
              2              ab      [No closure]
              3              c
              4              d
..................................................................
              F̄ , a, b, c, d,  F
              01             0       Repeat '1' at all
              21             ab      positions below original,
              31             c       by Q3.
              41             d
                                     [a = 2, b = 1,
                                      c = 3, d = 4,
                                         closed    ]
```

FIGURE 8

```
C M M M ∃x Fx M ∃x M M Fx
  1  2  3  4      a    b c    d
Ȼ M̸̄ M̸̄ M̸̄ ∄̄x F̄x M̸ ∄x M̸̄   M̸Fx
              0               0
              1               ab
              2               c
              34              d
              F̄ , a, b, c, d,  F
              0               0
              1               ab      [No closure]
              2               c
              34              d
-----------------------------------------------------------
              F̄ , a, b  c, d,  F       Repeat '4' at all
              04              0       positions above the
              14              ab      original, by Q4.
              24              c
              34              d       [a = 1, b = 4]
                                      [c = 2, d = 3]
                                      [   closed   ]
```

FIGURE 9

CS4 and **CS5** closure conditions are carried out *after* quantifier conditions are used. The one closure rule which involves a complication is the **CM** closure rule, which permits the deletion of a dummy in a modal index column. As we noted in Chapter III, this has the effect of repeating a constant from the index column of the opposing formula and then setting the dummy equal to that constant and deleting repetitions. In the cancellation of an unquantified formula, like 'CpMp', there is no special reason for choosing the dummy-deletion version of **CM** closure over the constant-repetition version, except that the dummy-deletion version is more similar to the closure rules for **CS4** and **CS5**. The rule in both applications looks like this:

CM closure as dummy deletion	**CM** closure as constant repetition
C p M p a Ȼ p̄ M̸ p 0 0 a̸ [**CM**; closed]	C p M p a Ȼ p̄ M̸ p 0 0 0 a [**CM** a = 0, closed]

When modal indices have individual dummies and constants next to them, however, we must be careful not to blast individuals out of existence in applying our closure rules. Following is a simple quantificational version of 'C p M p':

$$\begin{array}{l} C\,\exists x\,Fx\,M\,\exists x\,Fx \\ \quad 1 \quad a \;\; b \\ \not{C}\,\bar{\not\exists}x\,\bar{F}x\,\not{M}\,\not\exists x\,Fx \\ \qquad 01 \qquad 0 \\ \qquad\qquad\;\; ab \end{array} \qquad\qquad \begin{array}{l} C\,\exists x\,Fx\,M\,\exists x\,Fx \\ \quad 1 \quad a \;\; b \\ \not{C}\,\bar{\not\exists}x\,\bar{F}x\,\not{M}\,\not\exists x\,Fx \\ \qquad 01 \qquad 0 \\ \qquad\qquad\;\; ab \end{array}$$

The problem is to apply **CM** closure without losing individuals. The solution is relatively simple for each of the two versions of **CM** closure. In the dummy-deletion version, simply move any individual variables or constants from the line of the modal dummy being deleted to the next higher line. In the constant-repetition version, simply repeat any individual dummies or constants when you repeat the modal dummy or constant from the original line:

$$\begin{array}{l} \bar{F}\;,a,b,F \\ 01 \qquad 0b \\ \qquad\quad\;\; \not{a}\not{b} \\ [b = 1;\ \text{closed}] \end{array} \qquad\qquad \begin{array}{l} \bar{F}\;,a,b,F \\ 01 \qquad 0 \\ 01 \qquad ab \\ \left[\begin{array}{c} a = 0,\ b = 1 \\ \text{closed} \end{array}\right] \end{array}$$

5.4.3 Summary of Procedures for Cancelling Quantified Modal Formulas. The cancellation procedure is still relatively straightforward, but with the various indices and closure conditions, there is more to do when cancelling a formula to determine whether or not it is a theorem of a given system, or, as will more often be the case, to determine in which systems of modal logic it is a theorem. So long as the cancellation and closure procedures are applied in the correct order, no problems will arise. Here, then, is the order in which the procedures are to be applied.

(1) Apply the cancellation rules to the original formula, moving from left to right through the formula until all cancellable elements have been attended to.

(2) Construct the branch chart below the cancelled formula, guided by the split markers. If no split markers occur in the cancellation, this step is not really necessary. The cancelled formula can serve as its own branch chart.

(3) Apply the *quantifier* closure conditions appropriate to the formal system being used (5.4.1).

(4) Apply the *modal* closure conditions appropriate to the formal system being used (Chapter III and 5.4.2).
(5) Replace dummies by constants as permitted within the Uniformity and Order Requirements.

Exercise

Perform cancellations to determine whether or not '**C∀x L C Fx Gx L∀x C Fx Gx**' is a theorem of **CMnQ1**, **CMnQ2**, **CMnQ3** and **CMnQ4**. One cancellation of the original formula will do; then apply the four different sets of closure conditions.

6 FURTHER INTERPRETATION OF BASIC FORMULAS

The First Interpretation of basic formulas is not very satisfying to the intuition. The question whether or not an individual name belongs to a given model set is left entirely to the quantifier binding an individual variable. This question is, as we have been considering it, similar to the question whether or not an individual exists in a given possible world. When a modal operator intervenes between a quantifier and a basic formula, the result is rather indefinite without further interpretation. Consider '∀x L Fx' in the Quantificational Analogue:

∀x C Bx0 ∀λ C Hλ0 Fxλ.

"All individuals in the real world are F's in every alternative to the real world." Should '∀x L Fx' require, like *PHEP*, that every existing individual be in every possible world? Or should '∀x L Fx' simply require that every existing individual be an F in every *possible* state of affairs in which it exists? Or perhaps that it never be a non-F, which may or may not be the same thing as being an F? The open questions left by quantified formulas are not the worst of it. When we consider formulas containing individual constants in Chapter VIII, we will have to interpret, in quantified modal logic, initial well-formed formulas like 'M F2': "It is possible that (the individual) 2 is an F." Must constants in such formulas always be assigned to existing individuals (i.e., to individual names that belong to the fixed model set), or can we say "It is possible that Sherlock Holmes was a detective" without committing ourselves to the existence of Holmes in the real world?

In this section we will consider four different ways of interpreting basic formulas on the model system. The four interpretations will differ with respect to the kind of requirement placed on individual names in order for an atomic formula or the negation of an atomic formula to be interpreted as true with respect to a given model set. Each of the four interpretations has a sense of rightness about it, and we will present each in the most favorable intuitive light. Three of the four interpretations also have certain formal shortcomings, however, which we will spell out by showing how a given interpretation affects the provability of certain key formulas.

It is worth repeating here that there is not, and never will be, a single correct system of formal logic. The four different interpretations of basic formulas we are about to consider can be imposed on non-modal formulas as well as on modal formulas, and the results will be similar. There simply *are* more ways than one of interpreting the basic formulas of a predicate calculus, and the choice of the best interpretation must rest upon the way in which the formal system is being used.

The instrument for considering the various interpretations of atomic formulas will be the Quantificational Analogue. In order to focus the issues of interpretation where they belong—on basic formulas—we will expand the translation rules for the Quantificational Analogue in such a way that any negation bars in an object language formula are driven as deep into the formula as they will go in the process of translation.

6.1 Expanded Mapping Rules. The following rules for translating object language formulas into the Quantificational Analogue are precisely equivalent to those given at IV.7.2 and 4.1 above, with the exception of the rules for translating basic formulas. There are no rules for basic formulas in this section; several alternative sets of rules for basic formulas will be considered in the following sections, but the rules here will be used with *all* the alternative rules for basic formulas. As before, we begin the translation of an object language formula by enclosing it in parentheses and writing '0' to the right of the parentheses. The rules are then rules for the removal of parentheses.

6.1.1 '$(\mathrm{K}\alpha\beta)\boldsymbol{\mu}$' is translated to '$\mathbf{K}(\alpha)\boldsymbol{\mu}(\beta)\boldsymbol{\mu}$'.
6.1.2 '$(\mathrm{A}\alpha\beta)\boldsymbol{\mu}$' is translated to '$\mathbf{A}(\alpha)\boldsymbol{\mu}(\beta)\boldsymbol{\mu}$'.
6.1.3 '$(\mathrm{C}\alpha\beta)\boldsymbol{\mu}$' is translated to '$\mathbf{A}(\bar{\alpha})\boldsymbol{\mu}(\beta)\boldsymbol{\mu}$.
6.1.4 '$(\bar{\mathrm{K}}\alpha\beta)\boldsymbol{\mu}$' is translated to '$\mathbf{A}(\bar{\alpha})\boldsymbol{\mu}(\bar{\beta})\boldsymbol{\mu}$'.
6.1.5 '$(\bar{\mathrm{A}}\alpha\beta)\boldsymbol{\mu}$' is translated to '$\mathbf{K}(\bar{\alpha})\boldsymbol{\mu}(\bar{\beta})\boldsymbol{\mu}$'.
6.1.6 '$(\bar{\mathrm{C}}\alpha\beta)\boldsymbol{\mu}$' is translated to '$\mathbf{K}(\alpha)\boldsymbol{\mu}(\bar{\beta})\boldsymbol{\mu}$'.
6.1.7 '$(\exists x\,\Phi x)\boldsymbol{\mu}$' is translated to '$\boldsymbol{\exists}\mathbf{x}\,\mathbf{K}\,\mathbf{Bx}\mu\,(\Phi x)\boldsymbol{\mu}$'.
6.1.8 '$(\forall x\,\Phi x)\boldsymbol{\mu}$' is translated to '$\boldsymbol{\forall}\mathbf{x}\,\mathbf{C}\,\mathbf{Bx}\mu\,(\Phi x)\boldsymbol{\mu}$'.
6.1.9 '$(\bar{\exists} x\,\Phi x)\boldsymbol{\mu}$' is translated to '$\boldsymbol{\forall}\mathbf{x}\,\mathbf{C}\,\mathbf{Bx}\mu\,(\bar{\Phi} x)\boldsymbol{\mu}$'.
6.1.10 '$(\bar{\forall} x\,\Phi x)\boldsymbol{\mu}$' is translated to '$\boldsymbol{\exists}\mathbf{x}\,\mathbf{K}\,\mathbf{Bx}\mu\,(\bar{\Phi} x)\boldsymbol{\mu}$'.
6.1.11 '$(\mathrm{M}\alpha)\boldsymbol{\mu}$' is translated to '$\boldsymbol{\exists\lambda}\,\mathbf{K}\,\mathbf{H}\boldsymbol{\lambda}\mu\,(\alpha)\boldsymbol{\lambda}$'.
6.1.12 '$(\mathrm{L}\alpha)\boldsymbol{\mu}$' is translated to '$\boldsymbol{\forall\lambda}\,\mathbf{C}\,\mathbf{H}\boldsymbol{\lambda}\mu\,(\alpha)\boldsymbol{\lambda}$'.
6.1.13 '$(\bar{\mathrm{M}}\alpha)\boldsymbol{\mu}$' is translated to '$\boldsymbol{\forall\lambda}\,\mathbf{C}\,\mathbf{H}\boldsymbol{\lambda}\mu\,(\bar{\alpha})\boldsymbol{\lambda}$'.
6.1.14 '$(\bar{\mathrm{L}}\alpha)\boldsymbol{\mu}$' is translated to '$\boldsymbol{\exists\lambda}\,\mathbf{K}\,\mathbf{H}\boldsymbol{\lambda}\mu\,(\bar{\alpha})\boldsymbol{\lambda}$'.

That these translation rules involve no change from those given earlier may be seen easily enough by translating a formula under both sets of rules and cancelling the results. The rules given here have the effect of confining negation to atomic formulas in QA.

The material conditional 'C' has not disappeared from the Quantificational Analogue. It continues to appear in the translations of quantifiers and modal operators, and in the various antecedent assumptions. Translation rules 6.1.3 and 6.1.6 simply require that *object language* occurrences of 'C' be translated into the analogue as equivalent expressions containing alternation and negation, in order to push all object language negations inward to atomic formulas. In §4.1 above, we translated **T-61** into QA using the "unexpanded" translation rules. In Figure 10 is a step by step translation of the same formula under the expanded rules.

C ∀xLFx∃x M Fx	
(C ∀x L Fx ∃x M Fx)**0**	
A ($\bar{\forall}$x L Fx)**0** (∃x M Fx) **0**	6.1.3
A ∃x K Bx0 ($\bar{\text{L}}$ Fx)**0** (∃x M Fx)**0**	6.1.10
A ∃x K Bx0 ∃λ K Hλ0 ($\bar{\text{F}}$x)**λ** (∃x M Fx)**0**	6.1.14
A ∃x K Bx0 ∃λ K Hλ0 ($\bar{\text{F}}$x)**λ ∃x K Bx0** (M Fx)**0**	6.1.7
A ∃x K Bx0 ∃λ K Hλ0 ($\bar{\text{F}}$x)**λ ∃x K Bx0 ∃λ K Hλ0** (Fx)**λ**	6.1.11

The Quantificational Analogue translation of the same formula arrived at in 4.1 was as follows (with parentheses restored around basic formulas):

C ∀x C Bx0 ∀λ C Hλ0 (Fx)**λ ∃x K Bx0 ∃λ K Hλ0** (Fx)**λ**.

The two formulas are equivalent up to this point. The expanded version, with negation confined to the atomic formula 'Fx', is the version to be used in considering further interpretations of basic formulas.

The Barcan Formula and its converse will be of particular interest in the interpretations to follow. Because of the confinement of negation to basic formulas under the new translation rules for QA, the formulas will appear in an unfamiliar form. It might be helpful to do the confinement and the elimination of 'C' in the object language, to get clear that the new translation rules are nothing more than ways of confining negation, and will not interfere with equivalences unless the interpretation of basic formulas makes them do so.

The Barcan Formula, as we have been considering it, takes the form 'C M ∃x Fx ∃x M Fx'. This is equivalent in the object language to 'A L ∀x $\bar{\text{F}}$x Ex M Fx', wherein the negation is confined. By the translation rules above, both of these formulas come over into the analogue as follows:

A ∀λ C Hλ0 ∀x C Bxλ ($\bar{\text{F}}$x)**λ ∃x K Bx0 ∃λ K Hλ0** (Fx)**λ**.

The converse of the Barcan Formula is 'C ∃x M Fx M ∃x Fx', which is equivalent to the confined 'A ∀x L F̄x M ∃x Fx'. Both of these yield the following in QA:

A ∀x C Bx0 ∀λ C Hλ0 (F̄x)**λ ∃λ K Hλ0 ∃x K Bxλ** (Fx)**λ**.

6.2 Four Interpretations of Basic Formulas. Each of the interpretations to be offered is reasonable *from a point of view* concerning assertion and denial. Whether or not there is a single overall point of view that sorts out the interpretations and brings them together will not concern us for now. The immediate question is how we are to interpret '(Fx)λ' and '(F̄x)λ' in such a way as to have a satisfactory quantified modal (and non-modal) logic. We count the interpretation of basic formulas at 4.1.3. as the First Interpretation, although it may be regarded as no interpretation at all, since it leaves the question of belonging up in the air.

First, we present the four remaining interpretations without comment. In the following sections we will discuss the plausibility of each interpretation and its effect on the formal systems. We use 'Ψx' to represent a formula which contains no logical constants—an atomic formula if 'x' is replaced by an individual constant; an atomic subformula if 'x' is an individual variable.

6.2.1 The Second Interpretation of Basic Formulas.
'(Ψx)**μ**' is translated to '**K Bxμ Ψxμ**'.
'(Ψ̄x)**μ**' is translated to '**K Bxμ Ψ̄xμ**'.

6.2.2 The Third Interpretation of Basic Formulas.
'(Ψx)**μ**' is translated to '**K Bxμ Ψxμ**'.
'(Ψ̄x)**μ**' is translated to '**K̄ Bxμ Ψxμ**'.

6.2.3 The Fourth Interpretation of Basic Formulas.
'(Ψx)**μ**' is translated to '**C Bxμ Ψxμ**'.
'(Ψ̄x)**μ**' is translated to '**C̄ Bxμ Ψxμ**'.

6.2.4 The Fifth Interpretation of Basic Formulas.
'(Ψx)**μ**' is translated to '**C Bxμ Ψxμ**'.
'(Ψ̄x)**μ**' is translated to '**C Bxμ Ψ̄xμ**'.

Exercise

Translate **T-58**, **T-62** and **T-70** into QA, using the expanded mapping rules and the Second through Fifth Interpretations of bsfs.

6.3 The Plausibility of the Interpretations. The easiest way to get a sense of the four new interpretations to to symbolize some statements and translate the symbolizations into the Quantificational Analogue. "Someone got a parking ticket and will possibly have to pay a fine," symbolized with 'Gx' representing "x got a parking ticket" and 'Fx' representing "x will have to pay a fine," will give us an interesting configuration. First, in the object language:

∃x K Gx M Fx.

Now, using the rules in 6.1, we interpret everything but the basic formulas:

∃x K Bx0 K (Gx)**0 ∃λ K Hλ0** (Fx)**λ.**

"There is an individual, x, that belongs to the real world *and* got a parking ticket (in the real world) *and* there is a possible world wherein x will have to pay a fine." Now, under the new interpretations of basic formulas:

Second and Third: **∃x K Bx0 KK Bx0 Gx0 ∃λ K Hλ0 K Bxλ Fxλ**
Fourth and Fifth: **∃x K Bx0 KC Bx0 Gx0 ∃λ K Hλ0 C Bxλ Fxλ**

Under the Second and Third Interpretations, an individual must belong to a world in which he has a property. To say that there is a possible world in which the hapless ticketee might have to pay a fine is to say that there is a possible world in which he exists *and* will have to pay the fine. The second occurrence of '**Bx0**' is merely a stutter, and needn't concern us.

Under the Fourth and Fifth Interpretations, an individual must belong to a world in order to have a property. To say that there is a possible world in which our friend might have to pay his fine is, on these interpretations, to say that there is such a world, and *if* he exists in that world, he will have to pay his fine (or, to cite familiar equivalences, that he either doesn't exist there or else will have to pay, or that it isn't the case that he both exists there and won't have to pay).

The Second and Third Interpretations differ with respect to the way in which they treat negation. "Someone got a parking ticket and possibly *won't* have to pay a fine," symbolized '∃x K Gx M $\bar{\text{F}}$x', appears thus under the two interpretations:

Second: **∃x K Bx0 K K Bx0 Gx0 ∃λ K Hλ0 K Bxλ $\bar{\textbf{F}}$xλ**
Third: **∃x K Bx0 K K Bx0 Gx0 ∃λ K Hλ0 $\bar{\textbf{K}}$ Bxλ Fxλ**

On the Second Interpretation, to say that an individual doesn't have a property in a given world is to say that he exists in that world *and* doesn't have that property. The Third Interpretation simply negates the whole expression that interprets "affirmation:" if 'Fx' is taken to mean that x both exists and is an F, then '$\bar{F}x$' is taken to mean precisely the opposite: that it is not the case that x both exists and is an F. The negation of an atomic formula under the Fourth and Second Interpretations is equivalent, and under the Third and Fifth.

The Fourth and Fifth Interpretations differ from each other in the same way that the Second and Third differ from each other: under the Fourth Interpretation, negation is precisely the opposite of affirmation; under the Fifth, the negation is pushed to the predicate, with the existence condition kept as it was for affirmation. The formula above appears as follows under the Fourth and Fifth Interpretations:

Fourth: $\exists x\ K\ Bx0\ K\ C\ Bx0\ Gx0\ \exists\lambda\ K\ H\lambda 0\ \bar{C}\ Bx\lambda\ Fx\lambda$

Fifth: $\exists x\ K\ Bx0\ K\ C\ Bx0\ Gx0\ \exists\lambda\ K\ H\lambda 0\ C\ Bx\lambda\ \bar{F}x\lambda$

It may be more illuminating to translate a simple theorem into QA under each of the interpretations, now that we have worked our way through a complicated formula. The following should be—but will not be—provable under each interpretation: '$\forall x\ C\ \bar{L}\ \bar{F}x\ M\ Fx$'. This is one side of the definitional relation between 'M' and 'L'.

Translated into QA with negations confined, the formula is as follows:

$$\forall x\ C\ Bx0\ A\ \forall\lambda\ C\ H\lambda 0\ (\bar{F}x)\lambda\ \exists\lambda\ K\ H\lambda 0\ (Fx)\lambda.$$

With basic formulas interpreted under *all five* interpretations this time, the formula takes on this variety of configurations.

First: $\forall x\ C\ Bx0\ A\ \forall\lambda\ C\ H\lambda 0\ \bar{F}x\lambda\ \exists\lambda\ K\ H\lambda 0\ Fx\lambda$

Second: $\forall x\ C\ Bx0\ A\ \forall\lambda\ C\ H\lambda 0\ K\ Bx\lambda\ \bar{F}x\lambda\ \exists\lambda\ K\ H\lambda 0\ K\ Bx\lambda\ Fx\lambda$

Third $\forall x\ C\ Bx0\ A\ \forall\lambda\ C\ H\lambda 0\ \bar{K}\ Bx\lambda\ Fx\lambda\ \exists\lambda\ K\ H\lambda 0\ K\ Bx\lambda\ Fx\lambda$

Fourth: $\forall x\ C\ Bx0\ A\ \forall\lambda\ C\ H\lambda 0\ \bar{C}\ Bx\lambda\ Fx\lambda\ \exists\lambda\ K\ H\lambda 0\ C\ Bx\lambda\ Fx\lambda$

Fifth: $\forall x\ C\ Bx0\ A\ \forall\lambda\ C\ H\lambda 0\ C\ Bx\lambda\ \bar{F}x\lambda\ \exists\lambda\ K\ H\lambda 0\ C\ Bx\lambda\ Fx\lambda$

Under the Second Interpretation, the formula does not cancel to closed branches without further assumptions. Under all other interpretations, it does. Here, then, is the first basis for a selection of a standard interpretation of basic formulas. A system in which the equivalence is broken between '$\bar{L}\bar{F}x$' and 'M Fx' might be worth pursuing further, but we must count it as a departure from the mainstream and set it aside. As we will shortly note in a more formal way, the formula above *does* cancel to closed branches under the Second Interpretation in all **Q2** and **Q3** systems, and in **CS5Q4**. Under the other interpretations, it cancels to closed branches without any antecedent assumptions.

There is a further point to be made while the formulas above are still before us. The confinement of negation in the Quantificational Analogue (including the implicit negation in object language 'C') makes it difficult to spot theorems without actually going through the cancellation procedure. After working through the preceding chapters and the first five sections of the present chapter, one has a feeling for what will cancel to closed branches and what will not. The feeling should not be trusted with respect to QA with confined negation.

The question whether an object language formula, α, implies another, β, is different from the question whether or not **QA:$\boldsymbol{\alpha}$** implies **QA:$\boldsymbol{\beta}$**, although up to now the answers to these two questions have always been the same (assuming that "implies" has a meaning in QA as well as in the object language). An object language formula, α, implies another, β, in a formal system, S, if $C\alpha\beta$ cancels to closed branches in S. This is the case if and only if **QA: C $\boldsymbol{\alpha\beta}$** cancels to closed branches in **QA: S.** With negations confined and basic formulas treated under any interpretation but the First, we may no longer assume that **QA: C $\boldsymbol{\alpha\beta}$** and **C (QA:$\boldsymbol{\alpha}$) (QA:$\boldsymbol{\beta}$)** are the same formula. **QA: C $\boldsymbol{\alpha\beta}$** may cancel to closed branches when the conditional having **QA:$\boldsymbol{\alpha}$** as antecedent and **QA:$\boldsymbol{\beta}$** as consequent does not, and vice versa.[2]

Exercise

Cancel the five interpretations of '$\forall x C \bar{L} \bar{F}x M Fx$' shown above, without taking any antecedent assumptions in QA.

6.4 The Choice of an Interpretation. We now turn to sorting out the interpretations of basic formulas with an eye toward choosing a Standard Interpretation. Such a choice does not entail discarding the other interpretations totally. Some interesting and possibly useful variant formal systems can be generated by taking a different interpretation of basic formulas than the one that will be the Standard Interpretation here, and we will make at least passing suggestions here and in Chapter VIII as to how such variants might appear. But, having set aside all systems other than **CMn**, **CM**, **CS4** and **CS5** a few sections back in order to reduce the

[2] The point can best be dramatized—and some hasty and wrong decisions avoided—by considering two cases that fool the feeling one has for the cancellation systems by now, both of which involve individual constants. If you had to make a fast bet on whether or not the object language formula '$CFlFl$' cancelled to closed branches in QA under the Second Interpretation, you would probably bet that it would. You would lose. The translation rules at 6.1 require that the formula be translated to '**A** ($\bar{F}$l)**0** (Fl)**0**' before the basic formulas are translated. When the translation is completed, under the Second Interpretation, the resulting formula does not cancel to closed branches. Second case: '$CFl\exists xFx$' under the Fifth Interpretation. It *does* turn out to be provable.

number of variant quantified modal logics to sixteen when the four **Q** variants were added, we are now faced with five times that number of formal systems (minus, in each case, an allowance for systems which turn out to be equivalent). We cannot pursue all of the possible formal systems here, nor can we give object language closure rules for all of them. We can and will, however, take brief note of the properties imparted to the object language formal systems by each interpretation.

The scythe to be used in cutting down three of the four new interpretations in this: the Standard Interpretation must preserve, in its **Q1** variants, all of the theorems **T-57** through **T-82** of the **Q1** systems under the First Interpretation. We take these theorems to be basic. The same must hold for **Q2** variants, and so on, whatever new theorems may be added by the Standard Interpretation. Moreover—and this will amount to the same thing—the Standard Interpretation must preserve the definitional object language equivalences among truth-functional operators, among quantified formulas, and among modal formulas that we have taken for granted in the cancellation rules.[3]

It should be noted at the outset that with the assumption *PUP*, '$\mathbf{\forall x \forall \lambda Bx\lambda}$', all five interpretations are equivalent. The requirement that an individual belong to a model set in order to have (or fail to have) a property, or that an individual have a property *if* it belongs to a model set, becomes hollow if every individual belongs to every model set. If *PUP* is an appropriate assumption for an application of the formal systems—that is, if it is appropriate to assume that every possible state of affairs has precisely the same population—then there is no choice to be made among the five interpretations. The **Q2** systems, which have both the Barcan Formula and its converse as theorems, will appear in all five interpretations exactly as they appeared in the First.

Assuming that a **Q1**, **Q3** or **Q4** variant is preferable to a **Q2** variant—assuming, that is, that we want to allow ourselves to treat individuals that are possible but not actual, or that exist without existing in all possible worlds—we turn now to the Second through Fifth Interpretations in these variants.

6.4.1 Rejection of the Second Interpretation. The Second Interpretation makes too strong a requirement on basic formulas; strong enough that some of the desirable equivalences are lost. As noted above, the formula '$\forall x C \bar{L} \bar{F} x M F x$' is not a theorem of every **Q** variant system under the Second Interpretation. The formula fails in all **Q1** variants and in all the **Q4** variants except **CS5Q4**. Worse yet, '$C \exists x M F x \exists x M F x$', which

[3] But see Saul Kripke [V.11] for a discussion of the way in which such equivalences might be broken.

should be provable in any formal system, fails in the same systems where the equivalence fails, as does '$C\forall xLFx\exists xMFx$', which is **T-61**. In the **Q2** variants (with *PUP* assumed in QA), the **Q3** variants (with *PHEP* assumed), and in **CS5Q4** (with *PIEP* and *Sm* yielding a system equivalent to **CS5Q3**), the Second Interpretation produces results similar to the First. As a general interpretation of basic formulas in quantified modal logic, however, it must be rejected.

6.4.2 Rejection of the Third Interpretation. The Third Interpretation, which interprets atomic formulas like the Second, but interprets the negations of atomic formulas differently, produces an interesting anomaly. The converse of the Barcan Formula is provable in all **Q** variants in one form, but not in another:

> '$CM\exists xFx\exists xMFX$' is a theorem of all sixteen systems;
> '$CM\exists x\bar{F}x\exists xM\bar{F}x$' is not;

and the contrapositive equivalents of the two formulas above produce the same results:

> '$C\forall xL\bar{F}xL\forall x\bar{F}x$' is a theorem of all sixteen systems;
> '$C\forall xLFxL\forall xFx$' is not.

In general, if a formula containing an occurrence of α is provable, then so is the formula which is the same as the first except that all occurrences of α are replaced by occurrences of $\bar{\alpha}$, and vice versa. This is not the case in all formal systems on the Third Interpretation. The systems where this principle fails for the converse of the Barcan Formula (among other formulas, of course) are, again, the **Q1** variants and the **Q4** variants except **CS5Q4**.

'$CCM\exists xFx\exists xMFxCM\exists x\bar{F}x\exists xM\bar{F}x$' is a theorem of every **Q1** system on the First Interpretation. It is not a theorem of any **Q1** system on the Third Interpretation. We take this to be sufficient grounds for setting aside the Third Interpretation.

The Barcan Formula, in all of its versions, appears as it does under the First Interpretation.[4]

[4] The Third and Fourth Interpretations, which may be called asymmetrical with respect to negation, may have an appeal to those philosophers who hold that some properties are inherently primitive. If one sees philosophical ground to make out such a case, he holds for example that there is something basic about "— is blue" as distinct from "— is not blue" that cannot be captured by introducing a new atomic predicate "— is eulb" in place of the negated predicate "— is not blue" and letting "— is not eulb" take the place of "— is blue." This is a difficult case to make out, and it requires a linguistic theory of predication that can distinguish between primitive and non-primitive predicates before going into the formal system. For discussions of this sort of problem, see, among others, A. J. Ayer [V.2], Romane Clark [V.5] and Sidney Axinn [V.1].

6.4.3 Rejection of the Fourth Interpretation. The Fourth Interpretation, like the Third, is asymmetrical with respect to negation. The odd result with respect to the converse of the Barcan Formula occurs, but in the opposite cases. '$C\exists x M \bar{F}x M E x \bar{F}x$' is provable in all sixteen systems, but '$C\exists x M F x M \exists x F x$' is not, and so on. Anomalies with respect to unquantified formulas will also appear, as might be expected.

Again, in rejecting the Fourth Interpretation, we are not suggesting that it is not worthy of further consideration for special uses of the formal machinery, only that it produces "non-standard" systems of formal logic.

6.4.4 The Fifth, and Standard, Interpretation. The Fifth Interpretation, finally, preserves all the theorems of the First Interpretation and brings with it several bonuses, not all of which we can make use of at the moment.[5] The most important bonus is the converse of the Barcan Formula, which is provable, in all of its various forms, in all sixteen systems. The Barcan Formula is provable only where it was provable on the First Interpretation. The important result of this is that there is now a system based on **CS5** that has the converse of the Barcan Formula provable without having the Barcan Formula itself. The system is **CS5Q1**, for which none of the assumptions *PUP*, *PHEP* or *PIEP* are taken in QA.

The Fifth Interpretation is now the *Standard Interpretation* of basic formulas. Finally, then, we can get back to the object language and introduce an extra cancellation step to capture the sense of the Standard Interpretation.

6.5 Cancellation under the Standard Interpretation. In the object language cancellation of quantified formulas, we add the following step, justified by the Standard Interpretation of basic formulas, to the *cancellation* procedure. In the summary of procedures for cancelling quantified modal formulas at 5.4.3, this new step should be taken under part (1) of the procedure.

6.5.1 All individual dummies and constants in the index matrix of a bsf are repeated under their original occurrences in the index matrix, in the bottom line of the index matrix, to the right of the last entry in the modal index column. (See Figure 11 for examples.) All other cancellation and closure procedures remain the same as before.

[5] The added bonuses have to do with the way in which identity and individual reference work out on this interpretation. See Chapter VIII.

```
   C∃x M Fx M ∃x Fx

     1 2      a  b
   Ȼ ∄̄x M̸̄ F̄x M̸ ∄x Fx
           01       0
           2        ab

           F̄, a, b, F      ('1' is repeated, by 6.5.1)
           01      0
           21      ab      [b = 1; closed]

C ∀x L C Fx Gx L C ∀x F x ∀x Gx

   a b          1    c     2
Ȼ ∀̸̄x L̸̄ Ȼ Fx Ḡx L̸ Ȼ ∀̸̄x F̄x ∀̸x Gx
         | 0a 0a        0     0
         | b  b         1c    12
         |___/
         |
         | F ,      c,  F̄,    G
         | 0a           0     0      ('a' repeated, by 6.5.1)
         | ba           1c    12     [ b = 1, a = c = 2      ]
  a, b,  |_____________________      [closes bottom position ]
         |    Ḡ ,   c,  F̄,    G
         |    0a        0     0      [Upper occurrence of 'a']
         |    ba        1c    12     [ remains open, by 5.4  ]
```

FIGURE 11

6.6 The Barcan Formula and its Converse Under the Standard Interpretation. Table 14, like Table 13, shows the sixteen basic systems of modal logic, and answers the question for each whether or not the Barcan Formula (BF) and its converse (CBF) are provable.

TABLE 14

	Q1		Q2 Assume PUP		Q3 Assume PHEP		Q4 Assume PIEP	
	BF	CBF	BF	CBF	BF	CBF	BF	CBF
CMn	No	Yes	Yes	Yes	No	Yes	Yes	Yes
CM	No	Yes	Yes	Yes	No	Yes	Yes	Yes
CS4	No	Yes	Yes	Yes	No	Yes	Yes	Yes
CS5	No	Yes	Yes	Yes	Yes	Yes	Yes	Yes

7 THE BARCAN INTERPRETATION OF '∃'

It is possible to interpret the existential quantifier '∃' in such a way that it carries no existential commitment, but rather carries the sense that "for some value of x" the formula following the quantifier holds. This, it seems, is the sense of the existential quantifier that Ruth Barcan Marcus had in mind when she developed the systems which include the Barcan Formula (cf. [V.15]). Interpreting the quantifier in this way would seem to obligate us to introduce "x exists" as a new constant in the vocabulary, which involves no special difficulty. A full exploration of such systems is outside the scope of the present consideration, but we can suggest directions that such an exploration might take.

The existential quantifier can be modified in several ways. The one that comes closest to Mrs. Marcus' sense is the following, shown as a translation rule for QA.

'(∃x Φx)**λ**' is translated to '**∃x K ∃μ Bxμ** (Φx)**λ**'.

"For some value of x, x is Φ," related to a model set λ, becomes "There is an x that belongs to some model set μ such that x is Φ in λ." This can be tightened up in several ways, by inserting before '**Bxμ**' any of '**K Hμλ**', '**K Hλμ**', or '**K A Hμλ Hλμ**', so that for x to be Φ in λ requires that x belong to a model set that is related to λ by the alternativeness relation. Under any of these interpretations of '∃', the Barcan Formula becomes a theorem of additional formal systems.

8 EXISTENCE AS A PREDICATE

If any of the variant interpretations of '∃' suggested in the preceding section is adopted, or if the Second, Third or Fourth Interpretations of basic formulas are adopted, it would seem worthwhile to have on hand an object language constant for existence. This is easily enough done with the present machinery, although a more reasonable way can be suggested when identity is introduced in Chapter VIII. The new constant is not quite a predicate constant under the present interpretation, and cannot be cancelled in the object language as if it were. We introduce it here to make a point regarding the **Q** variant systems that can be made without devising yet another object language closure condition for the present.

We introduce the constant 'Q' into the object language, as a "quasi-predicate" constant of degree one. 'Qx' is read "x exists," and is interpreted in the Quantificational Analogue in the following way:

'(Qx)**μ**' is translated to '**Bxμ**'.

This new constant can be incorporated into any of the formal systems of quantified modal logic discussed in this chapter, and even into quantified first-order logic without modality as described in Chapter II. Under the standard interpretation of the existential quantifier, '∃x Qx' is somewhat redundant, but there is nothing wrong with it formally.

The exploration of formal systems involving variant interpretations of the quantifiers and a constant for existence is outside the scope of the present chapter.[6] For our present purpose, we will use the constant 'Q' only to compare the **Q** variant systems. Its redundancy makes it easy to evaluate the various assumptions on singular existence that differentiate the variants.

Although special applications of the formal systems might, as we have repeatedly noted, call for or permit unusual situations with respect to singular existence, we must regard the following four formulas as unacceptable in general. That is, any system of quantified modal logic which has the following formulas provable must be regarded as unacceptable as a *generic* system of alethic modal logic, however useful it might be under other interpretations.

Anti-Theorem 1. ∀x C Qx L Qx
Whatever exists, exists necessarily.

This formula is provable for all **Q2** and **Q3** systems of quantified modal logic discussed in this chapter.

Anti-Theorem 2. ∀x C M Qx Qx
Whatever possibly exists, actually exists.

This formula is provable for all **Q2** and **Q4** systems discussed.

Anti-Theorem 3. ∀x C M Qx L Qx
Whatever possibly exists, necessarily exists.

This formula is provable for all **Q2** systems discussed.

Medieval notions of existence aside, none of the Anti-Theorems belongs in a generic system of quantified alethic modal logic. The formulas above, with existence interpreted on '**B**——' produce Quantificational Analogue formulas which cancel to closed branches under the antecedent assumptions of the systems indicated. What remains for general use as systems of alethic modality in the generic sense is the set of **Q1** systems.

9 THE MULTIPLICITY OF FORMAL SYSTEMS

In the Introduction we suggested that there are a number of different contexts that can be regarded as modal: alethic, deontic, epistemic,

[6] In Chapter VIII, where existence presuppositions are considered, a way of specifying existence in the object language will be developed as part of the Standard Interpretation.

temporal, perhaps others. We suggested also that all of these notions are related in that they behave somewhat analogously. In Chapters III, IV and V we have constructed or at least described well over one hundred distinct systems of modal logic, all of which are related formally. It remains to match up formal system with modal notion, or to consider further minor variations of the formal systems to capture particular modal notions. To this we devote the following chapter.

With respect to alethic modal logic, which is regarded as the main-stream area of work in modal logic, it seems clear that the **Q1** systems are preferable to others considered, and that one of **CMQ1**, **CS4Q1** or **CS5Q1** under the *Standard Interpretation* is the best choice for a generic system.

Exercise

Using the additional cancellation step for the Standard Interpretation given in §6.5.1, p. 160, cancel 'C L ∀x C Fx Gx ∀x L C Fx Gx' and show the replacements needed for closure.

BIBLIOGRAPHY—CHAPTER V

[V.1] Axinn, Sidney, "Ayer on Negation," *Journal of Philosophy* **61** (1964), pp. 74–75.

[V.2] Ayer, A. J. "Negation," *Journal of Philosophy* **49** (1952), pp. 797–815.

[V.3] Carnap, Rudolph. "Modalities and Quantification," *Journal of Symbolic Logic* **11** (1946), pp. 33–64.

[V.4] Church, Alonzo, "Notes on Existence and Necessity," *Journal of Symbolic Logic* **8** (1943), pp. 45–47.

[V.5] Clark, R. L. "More on Negation," *Philosophical Studies* **4** (1953), pp. 81–87.

[V.6] Hintikka, Jaakko, "Form and Content in Quantification Theory." *Acta Philosophica Fennica* **8** (1955), pp. 7–55.

[V.7] ———. "Existential Presuppositions and Existential Commitments," *Journal of Philosophy* **56** (1959), pp. 125–137.

[V.8] ———. "Modality and Quantification," *Theoria* (*Lund*) **27** (1961), pp. 119–128.

[V.9] ———. "Studies in the Logic of Existence and Necessity," *The Monist* **50** (1966), pp. 55–76.

[V.10] ———. "Individuals, Possible Worlds and Epistemic Logic," *Nous* **1** (1967), pp. 33–62.

[V.11] Kripke, Saul. "Semantical Considerations in Modal Logic," *Acta Philosophica Fennica* **16**, (1963), pp. 83–94.

[V.12] Leblanc, Hugues, and T. Hailperin. "Nondesignating Singular Terms,' *Philosophical Review* **68** (1959), pp. 239–243.

[V.13] Marcus, Ruth Barcan. "A Functional Calculus of First Order Based on Strict Implication," *Journal of Symbolic Logic* **11** (1946), pp. 1–16.

[V.14] ———. "The Identity of Individuals in a Strict Functional Calculus of Second Order," *Journal of Symbolic Logic* **12** (1947), pp. 12–15.

[V.15] ———. "Interpreting Quantification," *Inquiry* **5** (1962), pp. 252–259.

[V.16] Prior, Arthur Norman. "Modality and Quantification in *S5*," *Journal of Symbolic Logic* **21** (1956), pp. 60–62.

[V.17] ———. *Formal Logic.* (Clarendon Press, Oxford, 1955)

[V.18] ———. *Time and Modality.* (Clarendon Press, Oxford, 1957)

[V.19] von Wright, G. H. *An Essay in Modal Logic.* (North Holland, Amsterdam, 1951)

Chapter VI

SPECIES AND GENERA OF MONADIC MODALITY

1 INTRODUCTION

In the Introduction we took note of von Wright's suggestion that the study of modal logic is the study of a *genus* which has several *species*. We have made informal appeal not only to several species of alethic modality —logical possibility and necessity, theoretical possibility and necessity, and perhaps other species of possibility and necessity—but also to other genera of modality: deontic, epistemic and temporal. In a later chapter, similar distinctions will be suggested with respect to the dyadic notions of entailment and compatibility. We turn now to ways of differentiating the species and genera of monadic modality formally, using the semantic devices developed in Chapter IV, and to a consideration of how various species and varieties of modality can be interrelated.

The aim of this chapter is not to arrive at definitive and final systems of alethic, temporal, deontic and epistemic modality, but rather to show how, on certain understandings of these modal notions, formal systems of each sort may be developed. We will argue for a standard system of alethic modality, but there are other ways of understanding alethic modality than the one we will adopt here, and given these other understandings the standard system argued for here may not be appropriate. Temporal and deontic logics will be sketched in simple forms, and a system of deontic logic will be developed as a case study in building a formal system to meet certain conditions. An epistemic system will be developed for one sense of "know" and "believe," but clearly there are other senses that are subject to formalization.

The formal systems described and developed in the present chapter, then, should be regarded as case studies in the application of the formal resources to specific sorts of contexts. In some cases, they will reflect the philosophical prejudices of the author; in other cases not. In no case should these formal systems be construed as the sole correct way of articulating alethic, temporal, deontic or epistemic concepts, for there is no such thing as the sole correct way of articulating any concept.

2 SPECIES OF ALETHIC MODALITY

In this section we will lay out semantic rules to articulate alethic modalities of different strengths, distinguishing them and indicating in what ways they are related. As examples of different strengths, we will consider three species of alethic modality, with a sketch of what will be called "logical possibility," "conceptual possibility" and "theoretical possibility." The distinctions will then be brought to bear on the choice of a "standard" system of alethic modality from among **CM**, **CS4** and **CS5**.

2.1 Logical Possibility and Necessity. Whether or not a formula is true is determined by interpreting it on the fixed model set. Whether or not a formula is possible is determined by interpreting it on a model system which includes other model sets alternative to the fixed model set. The various species of alethic modality will be distinguished by attending to the sorts of conditions which a model set must fulfill in order to be an alternative to another set.

Logical possibility is the minimal sense of possibility, requiring only that no logical rules be violated. Model sets which constitute alternatives to the fixed model set in this sense are required to obey only the conditions governing logical constants that are obeyed by the fixed model set. Such alternatives we call *logical alternatives*. A model set which is a logical alternative to the fixed model set may not contain any atomic sentence *and* its negation, may not contain a conjunction without also containing each of its conjuncts, may not contain an alternation without containing at least one member of that alternation, and so on. '$H_{log}\, \mu\, \lambda$' is an abbreviation for "(Model set) μ is a logical alternative to (model set) λ."

2.1.1 Conditions on Logical Alternatives. Any model set μ which is a logical alternative to another model set λ must meet the following conditions:

(a) All conditions on truth-functional operators given in IV.2.1.

(b) The Standard Interpretation of quantifiers and atomic formulas given in V.3.1 and V.6.4.4.

2.1.2 Reflexivity of Logical Alternativeness. Whatever the outcome of our considerations of **CS4** and **CS5** within and among the various species of alethic modality, it is clear that any acceptable system of alethic modal logic must contain the system **CM** for whatever species of possibility or necessity is present. The **CM** distinguishing formula, 'C p M p', is clearly a law of modal logic in the alethic mode. We therefore adopt the condition *Rf* for logical alternativeness: *Every model set in the model system is a logical alternative to itself.* In QA, '$\forall\lambda\, H_{log}\lambda\lambda$'.

2.1.3 Logical Possibility. It can easily be demonstrated that such compounds as '$K p \bar{p}$', '$K \exists x K Fx Gx \forall x \bar{F}x$', '$K K \exists x Fx \forall x C Fx Gx \forall x \bar{G}x$' are false when interpreted on any model set in the model system described. These formulas violate logical rules; their negations are tautologies; they are, in a clear sense, logically impossible.

We now define logical possibility with respect to the model system in the following way:

It is logically possible that $\alpha =_{df} [\alpha]$ meets the conditions for membership in a logical alternative to the fixed model set.

2.1.4 Conditions Governing Modal Operators. We now introduce into the object language operators for logical possibility (M_{log}) and logical necessity (L_{log}) with the following conditions.

> $[M_{log}\ \alpha] \in \mu$ if and only if there is in Ω at least one model set λ such that $H_{log}\lambda\mu$ and $[\alpha] \in \lambda$.
> $[L_{log}\,\alpha] \in \mu$ if and only if for every model set λ in Ω such that $H_{log}\lambda\mu$, $[\alpha] \in \lambda$.

These conditions are precise reproductions of the conditions governing the model system for the formal system **CMQ1**. All theorems of **CMQ1** hold for logical modality when 'M' is understood as 'M_{log}' and 'L' is understood as 'L_{log}'.

2.2 Conceptual Possibility and Necessity. Species of alethic modality other than the logical can be articulated in a similar way by imposing other groups of conditions on allowable alternatives to the fixed model set. Suppose, to take a classic example, that two of the predicate constants in our vocabulary represent the predicates "—is a bachelor" and "—is married." We might then impose upon any alternative to the fixed model set the requirement that if for some individual name, "Sam," the model set contains the sentence "Sam is a bachelor," it may not also contain the sentence "Sam is married." Such a stipulation is a *conceptual* stipulation as distinct from a logical stipulation, since it has to do with an articulation of the concept "bachelor." A model set which contains the conjunction "Sam is a bachelor and Sam is married" would not be a conceptually permissible model set under such a stipulation, and the object language formula associated with this sentence would be *conceptually impossible.* A model set which contains the conjunction " Sam is a bachelor and Sam is a redhead," in the absence of stipulations which forbid it, would be conceptually possible—permissible in a conceptual alternative to the fixed model set.

Conceptual alternatives to the fixed model set are those for which stipulations have been made with respect to the marks of the model associated with certain predicate constants of the object language. We noted earlier that beyond a certain point of complexity, there is no way of deciding once and for all what counts as a truth of logic—the decision must be made with respect to a given formal system and its model, and there is no single overall formal system. Even less final and comprehensive is any list of conceptual truths. The aim of the present distinctions is to develop formal machinery for expressing and interpreting claims of necessity and possibility that are conceptual. In any applications of the formal systems, what is and what is not conceptually possible will be decided for a particular context. It cannot be decided once and for all. Common-sense examples like the "bachelor"—"married" example will serve to illustrate the sort of stipulations that can be made, however. Presumably such stipulations would forbid, for predicates that are to be articulated, combinations like the following.

Alice punched Bill in the mouth at 3 P.M.
Alice and Bill were twenty miles apart at 3 P.M.

Carlos is a koala bear.
Carlos is not a marsupial.

Schulz committed murder.
Schulz did not commit a crime.

These pairs of statements are incompatible only because of conventions governing the terms "punch," "koala bear," "murder" as they relate to certain other terms. There is a looseness about such conventions, although in some respects they may be taken to be determinate. The use of a term like "punch" can be determined for certain carefully delineated groups, say adolescents from middle-income families in suburban Philadelphia. Or more to the point, a legally correct use of terms like "arson," "mayhem," "rape," "contract," "murder" and perhaps even "punch," as they relate to other terms, may be determined within the jurisdiction of a given legal code.

With respect to the "koala bear"–"marsupial" example, there is at least one set of conventions that may be determined: those which come into play when a taxonomist identifies Carlos as a koala bear. Such conventions are not wholly uniform within any discipline, of course, and they are not strictly speaking definitional. While the connection between being a koala bear and being a marsupial may be uniformly agreed upon by modern taxonomists, the cluster of characters that would enter into the identification or "diagnosis" of Carlos as a koala bear are not.

Such conventions are not part of the *logical* structure of natural or artificial language in the sense that conventions governing the use and interrelations of logical particles are taken to be. It is possible and sometimes useful, however, to make certain minimal stipulations regarding the use of a given set of predicate terms in certain formal uses of language. Such stipulations are rarely definitional in the sense of providing equivalent and intersubstitutable expressions. While conceptual stipulations cannot be given for a whole natural language, or even for a whole scientific language, they can be given for some limited segment of a language. We call such a segment of a language the *articulated segment* of that language.

A formal explication of the sense in which the pairs of statements above are impossible pairs hinges upon supposing that the terms involved belong to a conceptually articulated segment of a language. Not all the terms in a language need to be articulated, and articulation need not involve equivalence or intersubstitutability. The entire set of predicate constants in a formal language may be taken to be primitive, even though some of them are related by semantic conditions.

2.2.1 Conditions on Conceptual Alternatives. In a later section we will articulate the vocabulary of "miniature universes" as used in earlier chapters for exercises. For the purpose of considering the interrelations of the species of alethic modality, however, it is not necessary to select a set of predicate constants, or to make stipulations on such a set of predicate constants, or even to decide on a particular form that such stipulations must take. The predicate constants to be articulated will of course depend upon the context under analysis. Stipulations on predicate constants might take any of the following forms, or other forms, for a model set μ which is a conceptual alternative to another model set.

If $[\Phi x] \in \mu$, then $[\Psi x] \in \mu$.
If $[\Phi x] \in \mu$, then $[\Psi x] \notin \mu$.
$[\forall x\, C\, \Phi x\, \Psi x] \in \mu$.

For any model set which is a conceptual alternative to the fixed model set with respect to a given articulated segment of the object language, there will be a series of such conditions *in addition to* the conditions on logical constants cited above.

2.2.2 Conceptual Possibility. Any complex formula which violates conceptual rules is judged to be conceptually impossible. If one of the stipulations on predicate constants is

$$\text{If } [(\text{koala bear})\ x] \in \mu \text{ then } [(\text{marsupial})\ x] \in \mu,$$

then "Carlos is a koala bear but not a marsupial" is conceptually impossible. Conceptual possibility can then be defined with respect to the model system once such conditions have been imposed on the alternativeness relation for a given set of predicate constants:

It is conceptually possible that $\alpha =_{df}$ $[\alpha]$ meets the conditions for membership in a conceptual alternative to the fixed model set.

The various alternativeness relations suggested here are of course interrelated, but they are not equivalent. Any conceptual alternative to the fixed model set must also be a logical alternative, but not all logical alternatives need be conceptual alternatives. Whatever is conceptually possible is also logically possible, but not everything that is logically possible is conceptually possible. "Alice punched Bill in the mouth while they were twenty miles apart" violates conceptual, not logical, rules.

2.2.3 Conditions Governing Modal Operators. The following conditions reflect the relationship between conceptual and logical modality and interpretation of object language operators for conceptual possibility (M_{con}) and conceptual necessity (L_{con}).

If $H_{con}\lambda\mu$ then $H_{log}\lambda\mu$.
$[M_{con}\alpha] \in \mu$ if and only if there is in Ω at least one model set λ such that $H_{con}\lambda\mu$ and $[\alpha] \in \lambda$.
$[L_{con}\alpha] \in \mu$ if and only if for every model set λ in Ω such that $H_{con}\lambda\mu$, $[\alpha] \in \lambda$.

To preserve the links between modality and truth in the conceptual species of modality, we assume that *every model set in the model system is a conceptual alternative to itself.*

2.2.4 Conceptual Modality in a Miniature Universe. In previous chapters, we have used for exercises "miniature universes" with limited populations of limited kinds of objects—dots, stars and wavy lines. These can now serve to illustrate a way in which stipulations can be imposed upon a model system. The vocabulary of the miniature universes is as follows:

Let 'Di' represent "i is a dot";
Let 'Si' represent "i is a star";

Let 'Wi' represent "i is a wavy line";
Let 'Oij' represent "i is over j";
Let 'Bijk' represent "i is between j and k";
Let 'Tij' represent "i is next to j";
Let 'Iij' represent "i is the same object as j".

There is nothing in the *logical* conditions imposed upon the model system that prevents a single object from being both a dot and a wavy line. '$\exists x\, K\, K\, Dx\, Wx\, Sx$' is impossible on intuitive grounds—nothing could be a dot *and* a wavy line *and* a star—but there is nothing in the conditions imposed on the model system up to this point to make such a formula impossible on formal grounds. We can now supply such conditions, however, for the seven predicate constants used to describe miniature universes.

We turn first to the "classificatory" predicates, 'D', 'S' and 'W'. We need to rule out the possibility of any object being more than one of a dot, a star or a wavy line. We may, but need not, also require that every object be *precisely* one of these. Following are several ways in which the stipulations might be made, for every model set μ which is a conceptual alternative to another model set with respect to this vocabulary.

(1) For every individual name [i] such that $B[i]\mu$, precisely one of [Di], [Si] or [Wi] $\in \mu$.
(2) If [Di] $\in \mu$, then neither [Si] $\in \mu$ nor [Wi] $\in \mu$.
If [Si] $\in \mu$, then neither [Di] $\in \mu$ nor [Wi] $\in \mu$.
If [Wi] $\in \mu$, then neither [Di] $\in \mu$ nor [Si] $\in \mu$.
(optional) $[\forall x\, A\, A\, Dx\, Sx\, Wx] \in \mu$.
(3) $[\forall x\, A\, A\, K\, K\, Dx\, \overline{Sx}\, \overline{Wx}\, K\, K\, \overline{Dx}\, Sx\, \overline{Wx}\, K\, K\, \overline{Dx}\, \overline{Sx}\, Wx] \in \mu$.
(4) $[\forall x\, A\, A\, A\, K\, K\, Dx\, \overline{Sx}\, \overline{Wx}\, K\, K\, \overline{Dx}\, Sx\, \overline{Wx}\, K\, K\, \overline{Dx}\, \overline{Sx}\, Wx\, K\, K\, \overline{Dx}\, \overline{Sx}\, \overline{Wx}] \in \mu$.

In the following discussion, we will assume condition (1).

In addition to the classificatory predicates, there are three relational predicates in the vocabulary of miniature universes. These can be articulated also by formal conditions, some of which give properties of the relations represented by the predicate constants, others of which make conceptual connections among the three relations. It should be noted, however, that the sorts of conditions we will lay down will not *uniquely* determine "meanings" for the predicate constants in our vocabulary. The condition at (1) above gives no sense of the meanings of 'D', 'S' and 'W'; it simply specifies that no object can be more than one of these, and would apply equally well if the three classificatory predicates represented horses, chickens and cows as dots, stars and wavy lines. What is important for the present sketch is that condition (1) *does* apply to the three predicates as we are interpreting them, whatever other interpretations it might also apply to.

The relation "over" has been used in the sense of "somewhere over" rather than "immediately over." That is, in the following world, the star is over both the wavy line and the dot:

The relation is transitive:

(5) If $[Oij] \in \mu$ and $[Ojk] \in \mu$, then $[Oik] \in \mu$.

The relation is *a*symmetrical; two things cannot be over each other:

(6) If $[Oij] \in \mu$, then $[Oji] \notin \mu$.

The relation is *ir*reflexive; nothing is over itself.

(7) For every individual name [i] that belongs to μ, $[Oii] \notin \mu$.

To capture the sense of "immediately over" rather than "somewhere over," the following condition might be used:

(8) $[\forall x \forall y\, C\, Oxy\, \bar{\exists} z\, K\, Ozy\, Oxz] \in \mu$.
"If x is over y, then there is no z such that z is over y and x is over z."

We will not include (8) in our conditions on the relation.

There are many relations that meet conditions (5), (6) and (7) besides "over." Virtually any relation that orders the items it relates will have the properties of transitivity, asymmetry and irreflexivity: "larger than," "more expensive than," "beneath," "to the left of," and other similar relations.

The relation "same object" is an "identity relation:" it has properties in common with "same size as," "same price as," and other relations that include a notion of sameness. The most important properties of "same object" are reflexivity, transitivity and symmetry.

(9) For every individual name [i] that belongs to μ, $[Iii] \in \mu$.
(10) If $[Iij] \in \mu$ and $[Ijk] \in \mu$, then $[Iik] \in \mu$.
(11) If $[Iij] \in \mu$, then $[Iji] \in \mu$.

Before we set down any conditions relating the various relational predicates, there is already an object language formula involving these predicates that we can claim as a theorem. Given the conceptual conditions on 'O' and 'I' at (5), (6), (7), (9), (10) and (11), it is conceptually impossible for x to be over y if x is the same object as y:

$$\bar{M}_{con} \exists x \exists y \, K \, Oxy \, Ixy.$$

The three-place relation "between" can, for our purposes, be wholly defined in terms of "over," although this will not do as a general definition of "between":

(12) $[Bijk] \in \mu$ if and only if either $[K \, Oij \, Oki] \in \mu$ or $[K \, Oik \, Oji] \in \mu$.

"i is between j and k if and only if either i is over j and k over i, or else i is over k and j over i." (12) will not do as a general definition of "between" because it assumes only one ordering relation, "over." There is nothing in our stipulations so far to rule out the positioning of objects in a world horizontally:

so that the star is between the two dots in an obvious way without any of the three objects being over any other. In general, we can make the following connection regarding "between" and "over," however.

(13) If either $[K \, Oij \, Oki] \in \mu$ or $[K \, Oik \, Oij] \in \mu$, then $[Bijk] \in \mu$.

This condition does not constitute a full explication of "between," and it allows for objects to be placed in arrangements other than a single vertical line. Another condition in addition to (13) is needed for "between," stipulating that there always be *three* distinct objects involved in the relation:

(14) If $[Bijk] \in \mu$, then $[Iij] \notin \mu$, $[Iik] \notin \mu$ and $[Ijk] \notin \mu$.

The relation "next to" can be wholly defined in terms of "between:"

(15) $[Tij] \in \mu$ if and only if $[\forall x \, \bar{B}xij] \in \mu$.

Further conditions relating the predicate constants can be given, although enough are already present to derive a number of conceptual truths with respect to the miniature universes. The following, for example, might be added to (13) and (14) in explicating "between."

(16) If $[Bijk] \in \mu$, then $[K \, Oij \, Oik] \notin \mu$ and $[K \, Oji \, Oki] \notin \mu$.

"If i is between j and k, then it cannot be the case that i is over both of j and k, nor that both of j and k are over i."

In obvious ways, the vocabulary and the stipulations can be further elaborated. Object language theorems can be detected, and means can be devised for detecting such theorems by cancellation of object language formulas if special closure conditions for articulated predicate constants are added.[1]

With respect to an articulated segment of the object language, we may even delineate certain "conceptual implications" [cf. VI. 25; pp. 220–221], allowing '$L_{con}C\alpha\beta$' to represent "'α' conceptually implies 'β'." "x is a dot" conceptually implies that x is not a star, for example: $\forall x\, L_{con}C\, Dx\bar{S}x$. Conceptual implication can then be distinguished from logical implication, which can be represented as '$L_{log}C\alpha\beta$'.

2.3 Theoretical Possibility and Necessity. In some cases, claims are made for possibility or necessity that hinge on neither the sheer logical structure of a statement nor meaning relations among predicates within a statement, but on matters of physical or scientific theory. To recall an earlier example, it is neither logically impossible nor conceptually impossible for me to walk unaided on the ceiling. It is impossible, nonetheless. This third, and weaker, sense of impossibility we will call *theoretical.* It has also been called *physical* and *causal.*

As was the case with conceptual modality, theoretical modality is articulated by making stipulations on predicate constants. The distinction between conceptual and theoretical modality, if there is reason to observe it, has more to do with the reasons for making such stipulations than with the stipulations themselves. The stipulations for theoretical modality would take the same forms as those for conceptual modality, except that they would involve reference to *theoretical alternatives* to the fixed model set rather than conceptual alternatives. Given a series of such stipulations—and we will shortly suggest some for the miniature universe—theoretical possibility can be defined in the same manner as logical possibility and conceptual possibility:

It is theoretically possible that $\alpha =_{df} [\alpha]$ meets the conditions for membership in a theoretical alternative to the fixed model set.

2.3.1 Conditions Governing Modal Operators. The conditions on theoretical

[1] For example, condition (9) would justify allowing a bsf containing an occurrence of 'I' followed by two occurrences of the same constant to close a branch: 'I22' would close a branch all by itself. Condition (6) would justify allowing '$\bar{O}12$, $\bar{O}21$' to close a branch. For a full discussion of 'I', see Chapter VIII, §2.

modal operators and on the relationship between theoretical alternatives and conceptual alternatives parallel precisely the conditions in 2.2.3.

> If $H_{the}\lambda\mu$, then $H_{con}\lambda\mu$.
> $[M_{the}\alpha] \in \mu$ if and only if there is in Ω at least one model set λ such that $H_{the}\lambda\mu$ and $[\alpha] \in \lambda$.
> $[L_{the}\alpha] \in \mu$ if and only if for every model set λ in Ω such that $H_{the}\lambda\mu$, $[\alpha] \in \lambda$.

Again, we assume reflexivity for the alternativeness relation: *every model set in the model system is a theoretical alternative to itself.*

2.3.2 Three Ranked Species of Alethic Modality. What is developing here is not a series of formally distinct species of modality, but at best a series of modalities which are ranked formally according to strength. The strongest sense in which a statement may be said to be impossible (and its negation necessary) is the logical sense. The weaker senses depend upon stipulations that are made for reasons having to do with matters other than the sheer logical structure of statements. Assuming stipulations on predicate constants for an articulated segment of the object language such that conceptual and theoretical modality are delineated, it seems natural to arrange the three species of modality under discussion as we have done semantically. Whatever is theoretically possible is also conceptually possible (but not conversely), and whatever is conceptually possible is logically possible (but not conversely).

$$M_{the}\alpha \rightarrow M_{con}\alpha \rightarrow M_{log}\alpha.$$

The expected and intuitively correct contrapositive result also holds, given the conditions set up relating the three alternativeness relations. Whatever is logically *im*possible is impossible in every sense, and whatever is conceptually impossible is theoretically impossible.

$$\bar{M}_{log}\alpha \rightarrow \bar{M}_{con}\alpha \rightarrow \bar{M}_{the}\alpha$$

2.3.3 Theoretical Modality in a Miniature Universe. In many cases, there is not a clear and straightforward way of deciding what is conceptually possible as distinct from what is theoretically possible. Is the statement "All crows are black" necessary at all? If so, is it conceptually necessary or only theoretically necessary? The latter question is a question about *meaning*. Is "x is black" part of what we mean when we say "x is a crow"? On the view of most contemporary philosophers, there is no single answer to such a question. What is meant by a term must be determined to a large extent with respect to the context in which it occurs.[2] It is often

[2] For a further discussion of this subject—and analyticity is the subject in large measure—see W. V. O. Quine [VI.41].

only when we are analyzing a given theory that we are able, *for* that theory, to make the decision as to which connections are conceptual and which are theoretical. The distinction with respect to our miniature universe will be somewhat arbitrary and artificial, therefore, but not wholly so.

The first assumption concerning the miniature universe that might properly be called theoretical has to do with the arrangement of objects within each world in the universe. If we are allowing the dots, stars and wavy lines to be arranged only in a single vertical line within each world, surely we cannot claim this is done on either logical or conceptual grounds. There is nothing in either logical structure or meanings of predicates that demands just this arrangement. We may therefore take as our first theoretical condition that must be met by every model set μ that is a *theoretical* alternative to another model set with respect to this segment of the language, the following:

(1) For any two individual names [i] and [j] which belong to μ, either [Oij] $\in \mu$ or [Oji] $\in \mu$.

Given any two objects, one of them must be over the other. The condition simply requires that the basic ordering relation of the vocabulary apply to every pair of objects in every world. To put it another way, the condition requires that there be no objects in a world that cannot be ordered by this relationship.

Theoretical conditions may also demand that certain sorts of items be accompanied by others in every case, not because of definitional relations but for other reasons. Lightning is accompanied by thunder, paresis by syphilis, clouds by silver linings. We might require for worlds in the miniature universe that every wavy line have a star over it:

(2) if [Wi] $\in \mu$, then [$\exists$x K Sx Oxi] $\in \mu$.

In the following diagram, consider the left-most world as the real world:

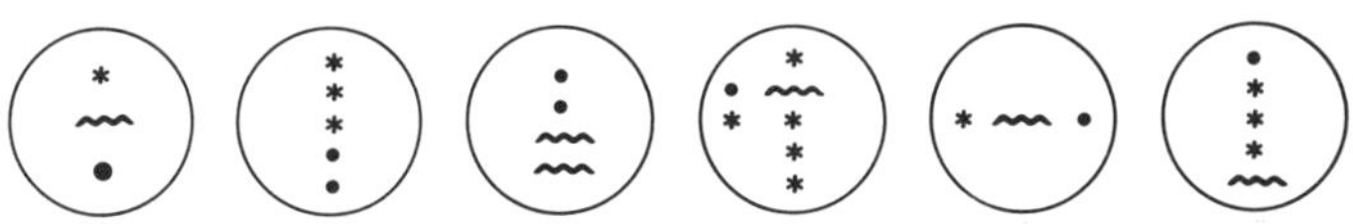

0 1 2 3 4 5

All six worlds are logically and conceptually possible; that is, the sets of statements describing them in the vocabulary of miniature universes would violate no logical or conceptual conditions.[3] The real world is of

[3] There is some philosophical significance in the fact that we cannot draw a picture of what is logically or conceptually impossible, while we can draw a picture of what is theoretically impossible.

course also theoretically possible—the set of statements describing it is a theoretical alternative to itself (2.3.1). For the rest, only worlds **1** and **5** are theoretically possible with respect to conditions (1) and (2) above. World **2** violates condition (2). The set of statements describing it would contain [Wi] for *two* individual names, [i], but would not contain [∃x K Sx Oxi] for either individual name. Worlds **3** and **4** violate condition (1), and **4** violates (2) as well.

2.4 Iterated Modalities Within Each Species. Within each of the three species of alethic modality sketched here, the formal systems **CM**, **CS4** and **CS5** may be developed. We have already assumed that every model set is a logical, conceptual and theoretical alternative to itself, obtaining by these assumptions models for each species of the system **CM**. The distinguishing formula of the generic system **CM** is 'C p M p' or, equivalently, 'C L p p'. Under the assumption of reflexivity of all the alternativeness relations, these formulas are truths of logic with respect to the model system for each of the three species of alethic modality: whatever is true is theoretically possible, conceptually possible and logically possible; whatever is theoretically necessary, conceptually necessary or logically necessary is true.

The distinguishing formula of the generic system **CS4** is 'C M M p M p'. The model system described becomes a model for **CS4** within each species of modality if we add to the assumptions already made the further assumptions that each of the three alternativeness relations is transitive:

If $H_{log}\lambda\mu$ and $H_{log}\mu\nu$, then $H_{log}\lambda\nu$.
If $H_{con}\lambda\mu$ and $H_{con}\mu\nu$, then $H_{con}\lambda\nu$.
If $H_{the}\lambda\mu$ and $H_{the}\mu\lambda$, then $H_{the}\lambda\nu$.

The distinguishing formula of the generic system **CS5**, 'C M L p L p', becomes a truth of logic with respect to the model system for each species of alethic modality if we make the further assumption that each of the three alternativeness relations is symmetrical.

Part of the difficulty with the **CS4** and **CS5** formulas becomes clearer if we restrict the discussion for the moment to cases where all the modal operators in a given formula are of the same species. The verbal arguments noted in Chapter III, you will remember, all seemed to be cases where species of modality were mixed. These, cases like '$C\,M_{con}\,M_{log}\,p\,M_{con}\,p$', we will momentarily set aside, and consider only unmixed cases like '$C\,M_{con}\,L_{con}\,p\,L_{con}\,p$'.

In the semantic definitions of logical, conceptual and theoretical modality given above, the only criterion for one model set's being alternative to another in a given sense is that the two sets be constructed according to the same group of conditions. All model sets which are logical alternatives to other sets are constructed according to a given group of conditions governing logical constants. All model sets which are conceptual alternatives to other sets are constructed according to the same group of stipulations governing the interrelations of predicate constants in the articulated segment of the object language, and so are all model sets which are, in the sense outlined, theoretical alternatives to other sets.

If what we require for a model set to be alternative to another in a given sense is that they be constructed according to the same group of stipulations—the group appropriate to a given species of modality—then certain things become clear concerning the properties of the alternativeness relation. "Constructed according to the same conditions," like "same object" and "same size" is an *identity* relation. It is reflexive—every set is constructed according to the same conditions as itself. It is also transitive and symmetrical. *It has precisely the properties of the* **CS5** *alternativeness relation.*

Interpreting the antecedent and consequent of the **CS5** distinguishing formula on the model system makes the point clearer. Where we have formerly used "—is an alternative to —" we now use "— is constructed according to the same conditions as—." We assume, of course, that we are working within a single species of modality and a single set of stipulations on the alternativeness relation.

(i) 'M L p' is true just in case [M L p] is a member of the fixed model set.

(ii) But this is true just in case [L p] $\in \mu$, where μ is some model set constructed according to the same conditions as the fixed model set; and

(iii) [L p] $\in \mu$ for some μ if and only if [p] $\in \lambda$ for every model set λ which is constructed according to the same conditions as μ.

(iv) *But if μ is constructed according to the same conditions as the fixed model set, then every λ which is constructed according to the same conditions as μ is also constructed according to the same conditions as the fixed model set.*

(v) It follows, then, that if [M L p] is a member of the fixed model set, then [p] is a member of every set constructed according to the same conditions as the fixed set, which is to say that [L p] is a member of the fixed set.

(vi) Therefore, if 'M L p' is true, so is 'L p'.

The crucial step, of course, is (iv). But if we spell out the alternativeness relation in terms of logical, conceptual and theoretical stipulations that a possible world must have in common with the real world, there seems to be no other choice but to accept the **CS5** properties for that relation. The alternativeness relation as described in Chapter IV is shrouded in mystery: it requires that logical conditions be met, but there is no indication what other conditions must be met. In making these more clear, we end up compelled to accept **CS5** as the generic system of alethic modality —the one which best represents truths of modal logic *within each of the several species.*

2.5 Iterated Modalities of Mixed Species. A separate line of argument must be taken with respect to the **CS4** and **CS5** equivalences where species of modality are mixed within a formula. We will consider here only two species of modality, the logical and the conceptual. The relative strength of the species of modality will be the primary factor in considering formulas of mixed species, and the following arguments can easily enough be extended to accommodate a third species of modality.

For two species of modality there are eight different versions of the **CS4** equivalence between 'MMp' and 'Mp', and eight different versions of the **CS5** equivalence between 'MLp' and 'Lp'. In Tables 15 and 16, the two conditionals which make up each equivalence are considered separately, under (a) and (b). After each formula is the number of the section in which it is argued for or against. Where the argument is in favor of accepting the formula as a truth of modal logic, the section number is preceded by "+." In most cases we will be able to justify accepting or rejecting a formula by making reference to object language formulas already accepted or rejected.

TABLE 15. THE **CS4** EQUIVALENCE WITH TWO SPECIES OF MODALITY

	(a)		(b)	
(1)	$C M_{log} M_{log} p M_{log} p$	+2.5.1	$C M_{log} p M_{log} M_{log} p$	+2.5.2
(2)	$C M_{con} M_{log} p M_{log} p$	+2.5.3	$C M_{log} p M_{con} M_{log} p$	+2.5.2
(3)	$C M_{log} M_{con} p M_{log} p$	+2.5.3	$C M_{log} p M_{log} M_{con} p$	2.5.7
(4)	$C M_{con} M_{con} p M_{log} p$	+2.5.3	$C M_{log} p M_{con} M_{con} p$	2.5.5
(5)	$C M_{log} M_{log} p M_{con} p$	2.5.5	$C M_{con} p M_{log} M_{log} p$	+2.5.3
(6)	$C M_{con} M_{log} p M_{con} p$	2.5.5	$C M_{con} p M_{con} M_{log} p$	+2.5.3
(7)	$C M_{log} M_{con} p M_{con} p$	2.5.7	$C M_{con} p M_{log} M_{con} p$	+2.5.2
(8)	$C M_{con} M_{con} p M_{con} p$	+2.5.1	$C M_{con} p M_{con} M_{con} p$	+2.5.2

TABLE 16. THE **CS5** EQUIVALENCE WITH TWO SPECIES OF MODALITY

	(a)		(b)	
(9)	$C\,M_{log}\,L_{log}\,p\,L_{log}\,p$	+2.5.1	$C\,L_{log}\,p\,M_{log}\,L_{log}\,p$	+2.5.2
(10)	$C\,M_{con}\,L_{log}\,p\,L_{log}\,p$	+2.5.3	$C\,L_{log}\,p\,M_{con}\,L_{log}\,p$	+2.5.2
(11)	$C\,M_{log}\,L_{con}\,p\,L_{log}\,p$	2.5.6	$C\,L_{log}\,p\,M_{log}\,L_{con}\,p$	+2.5.4
(12)	$C\,M_{con}\,L_{con}\,p\,L_{log}\,p$	2.5.6	$C\,L_{log}\,p\,M_{con}\,L_{con}\,p$	+2.5.4
(13)	$C\,M_{log}\,L_{log}\,p\,L_{con}\,p$	+2.5.4	$C\,L_{con}\,p\,M_{log}\,L_{log}\,p$	2.5.6
(14)	$C\,M_{con}\,L_{log}\,p\,L_{con}\,p$	+2.5.4	$C\,L_{con}\,p\,M_{con}\,L_{log}\,p$	2.5.6
(15)	$C\,M_{log}\,L_{con}\,p\,L_{con}\,p$	2.5.7	$C\,L_{con}\,p\,M_{log}\,L_{con}\,p$	+2.5.2
(16)	$C\,M_{con}\,L_{con}\,p\,L_{con}\,p$	+2.5.1	$C\,L_{con}\,p\,M_{con}\,L_{con}\,p$	+2.5.2

2.5.1 The formula at (1)(a) and (8)(a) do not mix species of modality and are instances of the **CS4** formula which we take to hold *within* each species of modality. The formulas at (9)(a) and (16)(a) are similar instances of the **CS5** formula.

2.5.2 From the truth of a statement it follows that the statement is possible in every sense (cf. §2.4). '$C\alpha M\alpha$' is taken to hold for each species of modality. If 'α' happens to contain modal operators, it still holds. Instances of this principle in Tables 15 and 16 are (1)(b), (2)(b), (7)(b), (8)(b), (9)(b), (10)(b), (15)(b) and (16)(b).

2.5.3 Whatever is conceptually possible must also be logically possible. Taken together with the considerations of the two preceding paragraphs, this principle leads us to accept (2)(a), (3)(a), (4)(a), (10)(a), (5)(b) and (6)(b). Consider (3)(a), for example. The antecedent is '$M_{log}\,M_{con}\,p$'. From '$M_{con}\,p$' it follows that '$M_{log}\,p$'; and from the antecedent of (3)(a) it then follows that '$M_{log}\,M_{log}\,p$'. Given that we have accepted (1)(a), it follows from '$M_{log}\,M_{log}\,p$' that '$M_{log}\,p$'.

2.5.4 Given the interdefinability of possibility and necessity within each species of modality, whatever is logically necessary is necessary in every sense. This, taken together with the preceding considerations, leads us to accept (13)(a), (14)(a), (11)(b) and (12)(b).

That the claims of this and the preceding sections are justified semantically can be demonstrated by translating the formulas into QA and cancelling, using as antecedent assumptions *Rf*, *Tr* and *Sm* for both H_{log} and H_{con}, as well as the principle which relates the two alternativeness relations, '$\forall\lambda \forall\mu \, C \, H_{con}\lambda\mu \, H_{log}\lambda\mu$'. The appeal to object language principles seems clear enough for the formulas considered so far, however.

2.5.5 Most of the formulas not marked with "+" on the tables must be rejected, given that certain others are already accepted as truths of logic with respect to the model system. Rejecting a formula as a truth of logic does not mean that it can *never* be true; it simply means that this formula is not a truth of logic—it is not always true. For example, we have rejected "Whatever is logically possible is conceptually possible," and our rejection of some of the formulas on the tables will be affected by this. But certainly *some* formulas which are logically possible are also conceptually possible. We simply cannot accept it as a truth of logic, for to do so would be precisely to collapse the distinction between conceptual and logical modality.

Again, the formulas we are about to reject can be translated into QA and cancelled to see that they do not produce closed branches. Instead, we will make the appeal through the object language. If '$C \, M_{log} \, p \, M_{con} \, p$' is not a truth of logic, then we cannot have a pair of formulas, both of which *are* truths of logic, which have the configurations '$C \, M_{log} \, p \, \alpha$' and '$C \, \alpha \, M_{con} \, p$'. For this reason, we must reject (5)(a) on Table 15, given that we have accepted (1)(b). Moreover, (6)(a) must be rejected given (2)(b), and (4)(b) must be rejected given (8)(a).

2.5.6 We have also rejected as a general principle "Whatever is conceptually necessary is logically necessary," '$C \, L_{con} \, p \, L_{log} \, p$'. As in the preceding argument, we must reject certain formulas on the tables because others have been accepted, whatever other semantic reasons we might also have for rejecting them. (11)(a) must be rejected given (15)(b); (12)(a) must be rejected given (16)(b); (13)(b) must be rejected given (9)(a); and (14)(b) must be rejected given (10)(a).

2.5.7 The formulas not accounted for are (7)(a), (15)(a) and (3)(b). None of these are truths of logic with respect to the model system developed in the preceding sections. On the other hand, no one of these formulas poses a threat to the distinctions we wish to maintain. If (7)(a) and (3)(b) were *both* truths of logic with respect to the model system, we would be in difficulty. The situation decribed in §2.5.5 would be present, and the distinction between logical and conceptual modality would collapse:

$$C \, M_{log} \, p \, M_{log} \, M_{con} \, p \qquad C \, M_{log} \, M_{con} \, p \, M_{con} \, p$$

(3)(b) (7)(a)

There is no reason to accept any of the three formulas as truths of logic—i.e., no reason to alter the model system in such a way as to make them truths of logic. We have assumed that in the articulation of conceptual modality, any number of different sets of stipulations on predicate constants may be imposed on the model system. Both (7)(a) and (15)(a) would militate against there being more than one acceptable set of conceptual conditions if they were made truths of logic, and this may be taken as reason enough to reject them. (3)(b), on the other hand, seems to *require* that there be more than one set of conceptual stipulations on the model—that there be more than one sense of conceptual modality. This becomes clearer if we translate (3)(b) into QA:

$$\mathbf{C}\exists\lambda\,\mathbf{K}\,\mathbf{H}_{\text{log}}\lambda 0\,\mathbf{P}\lambda\,\exists\lambda\,\mathbf{K}\,\mathbf{H}_{\text{log}}\lambda 0\,\exists\mu\,\mathbf{K}\,\mathbf{H}_{\text{con}}\mu\lambda\,\mathbf{P}\mu.$$

Roughly, "If there is a logically permissible model set which contains [p], then there is a logically permissible model set with respect to which there is a *conceptually* permissible model set which contains [p]." Put another way, "If 'p' violates no logical rules, then 'it is conceptually possible that p' violates no logical rules."

A verbal example should help here. Suppose 'p' is "Schulz is a married bachelor," and that conceptual stipulations on "married" and "bachelor" have been made in such a way as to make 'p' conceptually impossible. "Schulz is a married bachelor" still violates no *logical* conditions, and is logically possible. The antecedent of (3)(b) is thus fulfilled. For the consequent of (3)(b) to be fulfilled also, there would have to be a logical alternative to the fixed model set which had in turn a conceptual alternative for which the stipulations forbidding "Schulz is a married bachelor" did not hold. Perhaps in this second set of stipulations the terms "married" and "bachelor" would not occur at all.

If (3)(b) simply requires us to allow that there be more than one logically consistent set of stipulations on predicate constants, there is no harm in allowing it to become a truth of logic with respect to the model system. We can do so, in the most straightforward way, by taking the QA translation of (3)(b) as an assumption on the alternativeness relation. Whether or not we do so is not crucial to the distinction among species of modality, nor is it crucial to the arguments for **CS5** as the standard system of monadic modality within each of the alethic species. None of the distinctions important to the present discussion are damaged by admitting it.

2.5.8 Summary. With respect to formulas involving varying strengths of modality, we have the following results: First, the **CS4** and **CS5** equivalences hold *within* each species of modality—cases (1), (8), (9) and (16) on the tables. In addition, the conditions on the model system that we have accepted make truths of logic of the equivalences on lines (2) and (10): '$M_{con} M_{log} p$' is equivalent to '$M_{log} p$', and '$M_{con} L_{log} p$' is equivalent to '$L_{log} p$'. Finally, if (3)(b) is accepted as a truth of logic, the equivalence on line (3) holds, and '$M_{log} M_{con} p$' is equivalent to just '$M_{log} p$'. No other biconditionals hold, although all of the conditionals marked with '+' hold. There is, then, no general collapsing of iterated modalities when more than one species of modality is present.

The verbal counterexamples to the **CS4** and **CS5** equivalences that were mentioned in III.7 can now be examined in the light of our results concerning iterated modalities of mixed species. The first argument, which purported to be against the **CS4** distinguishing formula, held that a statement could be (theoretically) impossible while at the same time it was (logically) possible for it to be (theoretically) possible. Theoretical possibility, compared only with logical possibility, would produce the same results that we found for conceptual possibility in Tables 15 and 16. The formula at stake is (7)(a), '$C M_{log} M_{con} p M_{con} p$'; or, replacing "con" with "the," '$C M_{log} M_{the} p M_{the} p$'. The formula was rejected on semantic grounds. The argument has no bearing on the question whether or not the **CS4** formula is a truth of logic *within* a species of modality; it simply gives us further corroboration of our semantic rejection of (7)(a).

In a similar way, (15)(a) is precisely the formula which is at stake in Erik Stenius' arguments concerning truths of mathematics. Understood in our present notation, Stenius' argument is that '$M_{log} L_{con} p$' can be the case without its also following that '$L_{con} p$' is the case. Again, the argument has no bearing upon whether or not the **CS5** formula is acceptable *within* a given species of modality, and since we have rejected (15)(a) as a truth of logic, the counterexample is accommodated.

In summary, our present arguments lead us to accept the **CS4** and **CS5** equivalences so long as there is no equivocation on the sense of possibility and necessity present within a given formula. If, however, two or more senses of possibility and necessity are mixed within a given formula, one must determine the relative strengths of the senses of modality before determining whether or not the **CS4** and **CS5** equivalences are acceptable.

2.6 Object Language Cancellation of Two Strengths of Modality. An axiomatic deductive system of monadic modal logic in which modalities of two species are integrated has been developed by A. W. Burks [VI.15]. Following an excellent discussion of the motivation for such a system, Burks devises a logical calculus containing a monadic modal operator which may be interpreted "It is logically or theoretically necessary that—."[4] The "logical or theoretical" operator is then integrated with an operator that represents "It is logically necessary that—." The theoretical implication of 'q' by 'p' is expressed by substituting the theoretical operator for the logical operator in a strict implication formula: 'L_{the} C p q'. The relation between strict implication and theoretical implication is such that if 'L_{log} C p q' is true, then so is 'L_{the} C p q'.

Nicholas Rescher, in his review of the Burks article [VI.42], suggests introducing a third modality, defined by the first two: "It is theoretically, but not logically, necessary that—." We might term Rescher's modality *exclusive* theoretical necessity, since it excludes the case where a formula is logically necessary; and Burks' modality *inclusive* theoretical modality, since it does not exclude such a case. The sense of theoretical necessity described semantically in the preceding sections is, in this sense, inclusive.

It should be pointed out that neither of the Burks or Rescher systems (nor their cancellation versions, following) is peculiarly physical or causal. What we have here are strong and weak modalities, which may be interpreted as, respectively, logical and theoretical modalities. The choice is not predetermined by the formal system itself. If special closure conditions are added for certain predicate constants, as suggested earlier, the formal system will come closer to reflecting a species of modality that is clearly other than, as well as weaker than, the logical.

To simplify the presentation of an object language system of modal logic which contains two species of modality, we will drop the subscripts 'log', 'con' and 'the', the deal simply with two modalities, one of which is stronger than the other. The stronger of the two modalities we will represent with the familiar modal operators 'L' and 'M', and the weaker with these same operators primed, 'L′' and 'M′'. If the stronger operator is taken to represent logical modality, the weaker may be either conceptual or theoretical. If the stronger operator is conceptual, then the weaker may be taken to be theoretical.

2.6.1 Cancellation Rules. The cancellation rules for the four modal operators are simply reproductions of the cancellation rules for 'L' and 'M' given in Chapter III. The only complicating feature is that a primed modal operator introduces a primed dummy or constant, and an unprimed modal operator introduces an unprimed dummy or constant (see Figure 1).

2.6.2 Closure. Replacement of dummies by constants is the same as for the system **CM** (III.4.1), except that *dummies marked by primes may*

[4] We substitute "theoretical" where Burks uses "causal."

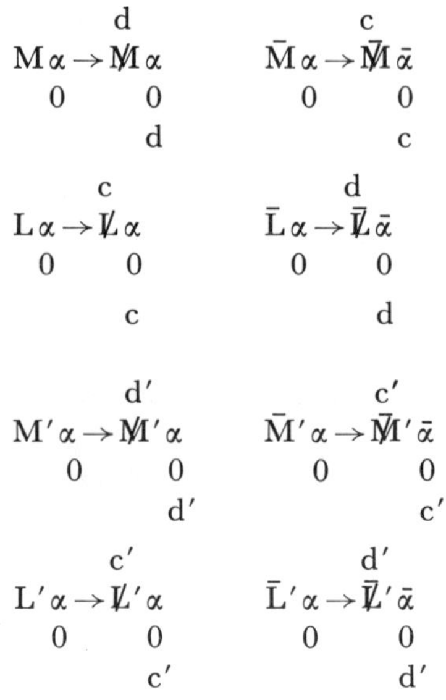

FIGURE 1

be replaced only by constants marked by primes. Unprimed dummies may be replaced by primed or unprimed constants. The special constant '0' is the one exception: it may replace any dummy.

This closure condition reflects the relationship between the stronger and weaker modalities treated semantically in the earlier sections. Suppose the unprimed operators are logical modality and the primed operators conceptual. Primed dummies and constants would then be assigned *conceptual* alternatives to the set assigned the constants immediately above them, and unprimed dummies and constants would be assigned *logical* alternatives to the sets assigned the constants immediately above them. The restriction on replacement of primed dummies, but not on replacement of unprimed dummies, reflects our semantic assumption that all conceptual alternatives to a given model set must also be logical alternatives, but not all logical alternatives must be conceptual alternatives. Examples of cancellation involving the two strengths of modality will be given in the following section.

2.6.3 Theorems. It is obvious that the counterparts of all theorems of **CM** are theorems of the integrated system *within each species of modality.* If 'L' is replaced everywhere by 'L′' and 'M' is replaced everywhere by 'M′' in the theorems of **CM**, the results will be theorems of the new system. The special theorems of this system are those in which the two species of modality are mixed. We will interpret the primed operators here as "logical or conceptual," and the unprimed as "logical."

IT-1[5] CLpL′p

What is logically necessary is logically or conceptually necessary.

$$\begin{array}{llll}
\mathrm{C\,L\,p\,L'p} & & & \\
\quad\mathrm{a}\quad \mathrm{1'} & & & \\
\not{\mathrm{C}}\,\bar{\not{\mathrm{L}}}\,\bar{\mathrm{p}}\,\not{\mathrm{L}}'\mathrm{p} & & & \\
\quad 0 \quad\; 0 & & & \left[\begin{array}{l}\mathrm{a} = 1' \\ \text{closed}\end{array}\right] \\
\quad \mathrm{a} \quad\; 1' & & &
\end{array}$$

The converse of this formula is not a theorem. 'CL′pLp' does not cancel to closed branches:

$$\begin{array}{llll}
\mathrm{C\,L'p\,L\,p} & & & \\
\quad\mathrm{a'}\quad 1 & & & \\
\not{\mathrm{C}}\,\bar{\not{\mathrm{L}}}'\bar{\mathrm{p}}\,\not{\mathrm{L}}\,\mathrm{p} & & & \\
\quad 0 \quad\; 0 & & & \left[\begin{array}{c}\mathrm{a'}\text{ cannot be replaced by 1 (2.6.2)} \\ \text{No closure}\end{array}\right] \\
\quad \mathrm{a'} \quad\; 1 & & &
\end{array}$$

IT-2 CM′pMp

What is logically or conceptually possible is logically possible.

IT-3 CKL′pLCpqL′q

If a statement is logically or conceptually necessary and logically implies a second, then the second is logically or conceptually necessary.

In general, this theorem reflects a principle from Aristotle's modal logic: A conclusion can have no stronger modality than the weakest premise needed to prove it. If 'L′q' is replaced by 'Lq' in **IT-3**, for example, the cancellation does not close.

2.6.4 **CS4** *and* **CS5**. The **CS4** and **CS5** closure conditions from Chapter III apply in the integrated systems, with only the need to restrict them when a given modal index column contains both primed and unprimed entries. The **CS4** closure condition permits the deletion of the upper of any two constants in an index column. So long as both constants are primed, or both unprimed, the condition is as before. When only the upper of two constants is primed, it may be deleted; when only the lower is primed, the upper constant may be deleted, *but the lower constant then loses its prime*. The three cancellations given in Figure 2 illustrate the rules.

[5] '**IT**' for "Integrated Theorem," so as not to confuse theorems of this system with theorems of the "generic" systems of modal logic.

IT-4 C M′ M′ p M p (cf. formula (4)(a) on Table 15)

1′ 2′ a
Ȼ M̸̄′ M̸̄′ p̄ M̸ p
0 0
1̸′ a [a = 2′ closed]
2′

IT-5 C M M′ p M p (cf. formula (3)(a) on Table 15)

1 2′ a
Ȼ M̸̄ M̸̄′ p̄ M̸̄ p
0 0 [a = 2 closed]
1̸ a
2′

IT-6 C M′ M p M p (cf. formula (2)(a) on Table 15)

1′ 2 a
Ȼ M̸̄′ M̸̄ p̄ M̸ p
0 0 [a = 2 closed]
1̸′ a
2

FIGURE 2

If the consequent of **IT-5** were 'M′ p', the cancellation would not close. The branch chart would contain 'p̄' and 'p', and the replacement of a
0 0
2 a′
primed dummy by an unprimed constant is forbidden (cf. formula (7)(a)).

The **CS5** closure rule, adapted to integrated modalities, follows essentially the same pattern as the **CS4** closure rule: Deletions may be made freely if all entries are primed or all unprimed, but the deletion of an unprimed entry robs all lower entries of their primes.

2.7 A Standard System of Alethic Modality. Given the arguments concerning quantified alethic modal logic in Chapter V and the arguments concerning semantic interpretations of alethic modal logic in the present chapter, the *standard* system of quantified alethic modal logic *within each species of modality* seems to be **CS5Q1**. If it is desirable to integrate more than one species of alethic modality in a single formal system, a **Q1** extension of the systems just described would appear to be appropriate.

By a "standard" system we can only mean a formal system which best captures the sense of "possibility" and "necessity" that is most commonly used in talking about concrete individuals in ordinary situations. Where the set of individuals being discussed is other than an ordinary set of concrete individuals, another quantificational variant might be appropriate. Where a very specialized sense of alethic modality is involved, or where "possibility" and "necessity" are interpreted in a way different from the way in which we have interpreted them here, one of the modal variants **CMn**, **CM** or **CS4** might be appropriate.

Exercises

Using primed modal operators for logical or conceptual modality and unprimed operators for logical modality, show that the following formulas cancel to closed branches in the integrated version of **CS5** described in §2.6.4.

IT-7 CM′LpLp
IT-8 CLpM′Lp
IT-9 CLpM′L′p
IT-10 CLpL′L′p
IT-11 CL′pML′p

3 TEMPORAL MODALITY

In the Introduction and throughout the discussion in the ensuing chapters, we have made verbal appeal to the simple temporal notions "always" and "sometimes," maintaining a parallel between "It is always the case that p" and "It is necessary that p," and between "It is sometimes the case that p" and "It is possible that p." We can now at least take note of more sophisticated treatments of temporal notions and the way in which they parallel the alethic modal notions.

We have considered the model set device as a way of generating descriptions of possible worlds other than the real world. It can as well be taken as a way of constructing descriptions of *this* world at points in time other than the present instant. Depending upon how and why one is using the model system temporally, any number of different ways might be chosen for hacking up the temporal continuum into instants. The choice will be arbitrary, of course. But given such a set of increments of time, it is possible to consider alternative model sets—descriptions of the world at other instants—as *ordered* temporally. The properties which it is appropriate to ascribe to the alternativeness relation will depend upon how we conceive the alternative "temporal points" (model sets) to be ordered.

A. N. Prior has developed a number of systems of temporal logic, among them one in which the operator 'M' is interpreted "It is true at some (past, present or future) time that—" and the operator 'L' is interpreted "It is true at all (past, present and future) times that—."[6] This, of course, is precisely our naive interpretation of 'L' as "always" and 'M' as "sometimes," and it demands no ordering at all of alternative temporal points. Not surprisingly, Prior's consideration of temporal logic in this simple sense leads him to the same set of modal principles arrived at here for alethic modality, since alternative model sets are not ordered. The appropriate modal system is **CS5**. This is easy to see if, following Prior, we verbalize the distinguishing formulas of **CMn**, **CM**, **CS4** and **CS5** in the following way:

CLpMp	If it is true at all times that p, then it is true at some time that p.
CpMp	If it is true (now) that p, then it is true at some time that p.
CMMpMp	If it is true at some time that it is true at some time that p, then it is true at some time that p.
CMLpLp	If it is true at some time that it is true at all times that p, then it is true at all times that p.

All the principles, including the **CS5** principle, are intuitively correct.

Prior argues against the Barcan Formula, and properly so, given the primary interpretation of quantifiers and atomic formulas. Consider the following verbalizations, first of the Barcan Formula, then of its converse:

CM∃xFx∃xMFx	If at some (past, present or future) time someone will fly to Venus, then there is (now) someone such that at some (past, present or future) time he will fly to Venus.[7]
C∃xMFxM∃xFx	If there is (now) someone who will at some (past, present or future) time fly to Venus, then at some (past, present or future) time someone will fly to Venus.

The Barcan Formula is inappropriate; not every individual who exists at some past, present or future time exists in the present. The converse, however, is appropriate. If someone existing now will do something at some future time, it follows that he will exist at that future time.

[6] [VI.40, pp. 18–28]. See also [VI.38].

[7] Prior has someone flying to the moon. His book was written in 1957 when, as a matter of fact, Lovell, Anders and Borman all *did* exist, making his counterexample to the Barcan Formula (as we now know) ineffective. It is hoped that Venus will remain appropriate at least until this book goes to press. If not, for "Venus" read "Aldebaran."

In fairness to Ruth Barcan Marcus, we should note that the Barcan Formula is unacceptable here (and elsewhere) only on the standard interpretation of the existential quantifier.

3.1 Ordered Temporal Modality. A second relatively simple temporal interpretation treated by Prior is the so-called *Diodoran* modal logic. On this interpretation, alternative model sets (points in time) all represent the future. The past is fixed: whatever has been the case *is* the case. The operator 'L' is interpreted "It is and will always be the case that—" and the operator 'M' is interpreted "It is or will be the case that—." Temporal reference is assumed to be built into statements, so that, for example, it is and will always be the case that Julius Caesar was assassinated on the Ides of March, 44 B.C. The fixed model set is taken to include the accumulation of present and past truths. Alternative sets represent future possibilities [cf. VI.40, pp. 8–17].

On this interpretation of the modal operators, the appropriate set of assumptions is that for **CS4**. Again, we verbalize the distinguishing formulas:

CLpMp	If it is and will always be the case that p, then it is or will be the case that p.
CpMp	If it is the case that p, then it is or will be the case that p.
CMMpMp	If it is or will be the case that it is or will be the case that p, then it is or will be the case that p.
CMLpLp	If it is or will be the case that it is and will always be the case that p, then it is and will always be the case that p.

The last formula, the **CS5** formula, is not acceptable on this interpretation. This is easy enough to see if we reword the verbal rendering in such a way as to emphasize that the temporal (modal) operator "sets the time" for the formula it ranges over: "If it will be the case at some time that *from then on* it will always be the case that p, then it is and will always be the case that p."

A quantified version of this second temporal interpretation of modality seems to demand the Principle of Historical Expansion of Population (*PHEP*) discussed at V.5.2 above. The appropriate quantified modal logic is then **CS4Q4**, under either the Standard Interpretation or the First Interpretation.

In the metalanguage and the Quantificational Analogue, the properties ascribed to '**H**' and '**B**' under this interpretation are clear. '**H**$\lambda\mu$' is understood "(model set) λ describes a point in time the same as or later than (model set) μ," and the relation is clearly transitive and reflexive but *not* symmetrical. '**Bxλ**' is understood "(the individual) x belongs to (exists at) the point in time described by (model set) λ, *or some earlier time*." *PHEP* is clearly the appropriate assumption here.

Prior's systems of temporal logic, as well as those of Nicholas Rescher and others,[8] are not as simple as we have represented them here. Once we begin discussing points in time, the temptation to quantify in the object language is almost irresistible. Systems of temporal logic often involve forms like '$\forall$t Rt (p)'—"For all times, t, 'p' is realized at t" (Rescher); and '$\exists$t Ut p'—"There is a time, t, such that p will happen at t" (Prior). These, interpreted as quantification over model sets (where a model set is a list of the things which are true at a given instant of time), are easily enough accommodated to the machinery we have developed.

Systems of ordered temporal logic other than the Diodoran have been developed, of course, and with sufficient jockeying of the alternativeness relation there is no reason why a formal structure cannot be developed with time branching in both directions from the present.

4 DEONTIC MODALITY

Deontic modality, as we have treated it in the preceding chapters, reflects in a very elementary way the notions "It is obligatory that—" and "It is permissible—." We can now schematize these notions as we have been using them, and suggest somewhat more sophisticated ways of schematizing obligation, permission and related notions. The value of deontic logic in particular is not, of course, that deontic principles will give us ways of demonstrating what is and what is not obligatory. It is rather that deontic logic provides a way of systematizing reasoning about obligation, so that the consequences of certain sorts of arguments can be spelled out. Consider, for example, what may be called the "permissive parent" view of obligation and permission: "Whatever my child does is all right with me." It takes no great facility with the formal machinery to see that this principle is just the **CM** principle with 'M' interpreted "It is permissible that—." If the **CM** principle is accepted, and the definitional relation between strong and weak operators is preserved, so that "It is obligatory that p" is equivalent to "It is not permissible that not-p," a rather bizarre ethical structure results: the system **CM** interpreted as a deontic system. Whatever happens is permissible, and whatever is obligatory happens.

But it is clear, as we have noted earlier, that for any serious attempt to capture the deontic notions of obligation and permission, the link between

[8] See the bibliography at the end of this chapter.

modality and truth must be removed, and a system similar to **CMn** must be adopted. Actually, as we will shortly show, a somewhat more subtle system than **CMn** can be developed by using the Quantificational Analogue.

The pioneering work in contemporary deontic logic was done by G. H. von Wright.[9] Von Wright's deontic operators function very much like the operators 'M' and 'L' in one of the non-logical species of alethic modality, except that, on von Wright's account, they form statements not when prefixed to simpler statements, but when prefixed to "names of acts." For the present sketch, however, we will treat deontic operators as statement operators.

Whether or not **CMn** is a plausible formal system for deontic logic can be tested easily enough by re-reading theorems **T-1** through **T-28** with the following interpretations of modal operators (after von Wright):

M p	p is permissible
$\bar{\text{M}}$ p	p is forbidden (not permissible)
L p	p is obligatory
K M p M $\bar{\text{p}}$	p is morally indifferent
$\bar{\text{M}}$ K p q	p and q are morally incompatible
L C p q	p morally implies q (p commits us to doing q)

Many of the theorems of **CMn** are plausible enough on this interpretation. The following, among others, seem intuitively correct.

T-1	C L p M p	What is obligatory is permissible.
T-2	A M p M $\bar{\text{p}}$	One or the other of a pair of opposites is permissible.
T-7	C M K p q K M p M q	If two statements are morally compatible, then each of them is permissible.
T-9	C L C p q C L p L q	If p commits us to doing q, then if p is obligatory, so is q.
T-10	C L C p q C M p M q	If p commits us to doing q, then if p is permissible, so is q.

As noted in Chapter III, however, paradoxes of strict implication make moral implication very implausible indeed:

T-19	C $\bar{\text{M}}$ p L C p q	Doing what is forbidden commits us to doing everything.
T-20	C L q L C p q	What is obligatory is morally implied by everything.

[9] See [VI.55]. See also [VI.54, pp. 36–41] and [VI.56].

T-19 is particularly vicious on the deontic interpretation. Suppose we take promise-keeping as a clear-cut example of something that is obligatory. Breaking a promise is then forbidden. According to **T-19**, "If it is forbidden to break a promise then breaking a promise commits us to robbing from the poor (or anything else we feel like doing)." Commit one forbidden act and you are committed to doing everything, whether it is permissible or not. The dyadic modal notions of entailment and compatibility, insofar as they are paradox-free, will provide more plausible formal accounts of moral commitment and moral compatibility.[10] These will be treated in the following chapter. Setting aside dyadic relations, then, we can consider what sort of formal system is appropriate for the simpler monadic notions of obligation and permission as applied to single "acts" or states of affairs.

4.1 An Exercise in System Building: Collapsed Iterations Without *Rf.* If we accept **CMn** as the basic formal system of deontic logic, we are faced with a problem. Iterated modalities cannot be reduced in **CMn**, and the system will contain irreducible formulas like 'L M L L M p', which are sheer nonsense on the deontic interpretation. On the other hand, **CS4** and **CS5**, which allow the collapsing of iterated modalities, are extensions of **CM**. Both will have what are in the deontic mode the "permissive parent" theorems, which we have already rejected.

The problem can be laid out a little more clearly if we list the formulas we want to be theorems of deontic logic and the formulas we *do not* want to be theorems, along with the properties of the H-relation already considered that make them truths of logic with respect to the model system (see Table 17).

Here, then, is the problem: In order to get formulas that we want provable, we need the assumption *Rf*; but *Rf* will also bring into the system theorems that we do not want. A compromise is possible, of course. In Table 11 at IV.6, the systems numbered 5 through 9 all possess *Ex* and lack *Rf*. They would avoid the unwanted theorems and allow at least the first of the wanted theorems. Systems 5 and 6 would allow two more of the wanted theorems, since they possess *Tr*. But there is no combination of the properties on the table that will give us the fully collapsed iterations without *Rf*.

Here is a situation where the Quantificational Analogue comes to our rescue. First, we must get clear what the H-alternativeness relation is to represent in the model for a deontic system. Then we can see which

[10] My colleague C. E. Dyke calls **T–19** the Principle of Diabolical Puritanism. For further discussion, see [VI.7].

TABLE 17

Formulas Wanted	Properties	Formulas Not Wanted	Properties
CLpMp	Ex or Rf	CLpp	Rf
CLpLLp	Tr	CpMp	Rf
CMMpMp	Tr		
CLLpLp	Rf		
CMpMMp	Rf		
CMLpLp	Sm and Qe		
CMpLMp	Sm and Qe		
CLpMLp	Rf		
CLMpMp	Rf		

properties it is plausible to attribute to the H-relation and try them in QA to see if they give the results we want.

To say that a model set or "world description" is an "ethical alternative" to another model set is to say that the alternative set describes a state of affairs that is desirable from an ethical point of view. *Which* ethical point of view determines this is a question we can leave open here. Presumably such a question would be determined by stipulations similar to those one might lay down for the theoretical species of alethic modality —e.g., we might require that if "Schulz makes a promise" is a member of a given model set which is an ethical alternative, then "Schulz keeps the promise" is also a member.

Several things now become apparent with respect to ethical alternatives.

- With respect to a given set of atomic sentences, there can be a number of different ethically desirable states of affairs.
- The "real world" model set is morally neutral: it is not directly affected by what ought to be the case. The theoretical stipulations place a limitation on what can occur in ethically governed worlds, but not this one.
- Any model set which *is* constructed according to theoretical stipulations is an alternative to any other set that is constructed according to the same stipulations, and therefore to itself.

The alternativeness relation must then have the following properties: reflexivity, transitivity and symmetry as it applies to all sets which are ethical alternatives to other sets, but *ir*reflexivity as it applies to the real world set, since that set is constructed so as to describe what is, not what ought to be. We will take the following assumptions in QA:

∀λ∃μ Hμλ (*Ex*) Every model set has at least one ethical alternative.

$\overline{\text{H}}$00 The fixed model set is not an ethical alternative to itself. More cynically, we might assume '∀λ $\overline{\text{H}}$0λ'— the fixed set is not an ethical alternative to *any* set. The present assumption will do the job, however, and the more cynical one can be derived from it and the following assumptions in any case.

∀λ C∃μ Hλμ Hλλ Any model set which is an ethical alternative to another model set is an ethical alternative to itself.

∀λ ∀μ ∀ν C K Hλμ Hμν Hλν (*Tr*) If a model set is an ethical alternative to a second model set, and that one to a third, then the first is an ethical alternative to the third.

∀λ ∀μ C K Hμμ Hλμ Hμλ If a model set is an ethical alternative to itself, and a second model set is an ethical alternative to that set, then that set is an ethical alternative to the second set. Despite the complexity of the verbal reading, this is the simplest version of limited symmetry that will give the results we want: symmetry among all sets that are ethical alternatives to other sets. Given to the preceding conditions, this is equivalent to

∀λ ∀μ C K ∃ν Hλν ∃ν Hμν C Hλμ Hμλ

The version we have adopted saves us some quantifier cancellation in QA and has the same effect. The restriction on symmetry is needed to supplement the restricted version of reflexivity adopted above. Without this restriction, the fixed set would become an alternative to any set which was alternative to it, and hence an alternative to itself.

∀λ ∀μ ∀ν C K Hλμ Hνμ Hλν (*Qe*).

Using these assumptions in QA, we can easily enough show that all of the "wanted" theorems are truths of logic with respect to a model system where the H-relation has these properties, and none of the "unwanted" theorems are.

The object language effects are relatively simple to spell out. The **CM** closure condition, in either version, can be applied *except when the constant involved is* '0'. The **CS5** closure condition carries the same restriction, in effect: Ignore all entries but the bottom entry, but no dummy may be replaced by '0'. In Figure 3 are applications of both rules, showing on the left-hand side an unwanted theorem which does not close, and on the right-hand side a wanted theorem involving the same closure condition, which does close. The first example involves the **CM** closure condition, and the second example the **CS5** closure condition.

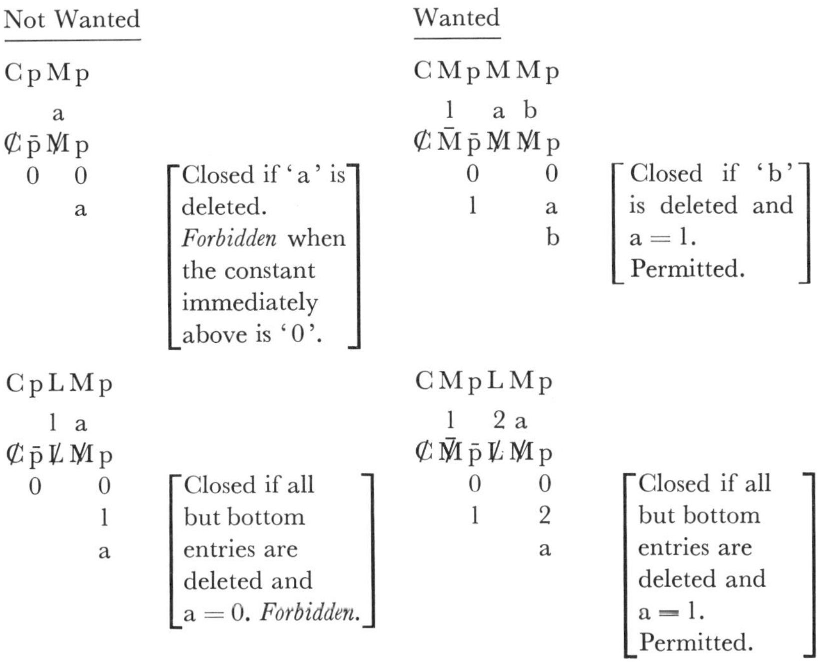

FIGURE 3

The formal system we have developed is somewhat limited, to be sure. It gives no satisfactory account of the dyadic notion of moral commitment. But no formal system containing only monadic operators does. The system does, however, fulfill the conditions we set out for it. It avoids the distinguishing formulas of **CM** while at the same time providing for full collapsing of iterated modalities, as in **CS5**.

This excursion into building a system of modal logic to meet a particular need should indicate that while it is not the case that a formal system can be constructed having *whatever* combination of properties we can dream up, there is obviously a great flexibility in the machinery available, and formal systems *can* be developed which meet rather specialized conditions. The present system is a case in point.

It remains to relate our new formal system to systems of alethic modal logic. Clearly, the deontic operators can be treated in much the same way as theoretical alethic modalities. Integrating these operators with logical or conceptual alethic operators, however, poses a new problem. The integrated system of 2.6 above treated the weaker mode in an inclusive sense, so that the operator 'M' would have to be read something like "It is morally permissible *and* logically possible that—." Under this interpretation, the operator 'L' would range over some strange formulas: all tautologies of truth-functional logic become logically necessary or obligatory, for example. The solution to the new problem is relatively simple and obvious: Develop an *exclusive* operator. If we restore subscripts to modal operators in order to distinguish the different modes, the method of developing a satisfactory deontic mode can be described briefly: Let 'M_{log}' and 'L_{log}' represent the logical operators, as before; let 'M_{eth}' and 'L_{eth}' represent *inclusive* modal operators in the deontic mode. *Exclusive* deontic operators can then be defined as follows:

$$L_{deo}\,\alpha =_{df} K\, L_{eth}\,\alpha\, \bar{L}_{log}\,\alpha.$$

What is obligatory is both obligatory-or-necessary and not necessary.

$$M_{deo}\,\alpha =_{df} C\, M_{log}\,\alpha\, M_{eth}\,\alpha$$

What is permissible is that which, if it is possible, is both possible and permissible.

4.2 Other Deontic Concepts. We have mentioned at several points the *dyadic* notion of obligation, "commits one to doing" or "morally entails." This we must set aside until a formal foundation has been laid for dyadic systems in the following chapter. There are other concepts related to ethical considerations, however, that have been subjected to formal scrutiny. Imperatives, and conditional imperatives like "If you want to do what is morally right, then do so-and-so," have received considerable attention in philosophical literature.

"Practical Reason," or reasoning which leads to action, involves statements of the form "s(omeone) judges that p," "s decides to bring it about that p," "s causes it to be that p," and related statement forms. A formalization of practical reasoning, coupled with a formalization of deontic logic, might permit an examination of the connection, if any, between "s judges that it is obligatory that p" and "s decides to bring it about that p" (cf. [VI. 13]).

A considerable amount of formal work has been done in "theory of value," which may also be seen as related to ethical considerations. The consideration of which sorts of states of affairs are preferable to others bears an intuitive as well as a formal similarity to the model structure just discussed for the simple deontic notions.

Finally, obligation and permission may be taken in a legal, rather than a moral, sense, with perhaps a somewhat different set of formal principles than we have developed here for moral obligation.

Exercises

1. Show that the following are theorems of the deontic logic described above:
   (a) If it is obligatory that p and it is permissible that q, then 'p' and 'q' are morally consistent.
   (b) If 'p' and 'q' are not morally consistent, then if it is obligatory that p, it is obligatory that not-q.

2. Show in the Quantificational Analogue, using the antecedent assumptions for the system of deontic logic described above, that the following are truths of logic with respect to the model system for deontic logic.

CMLpLp
CMpMMp

5 EPISTEMIC MODALITY

Epistemic logic is concerned with knowledge, belief and related notions. The rather simple sense of epistemic modality that we made use of in Chapter III by no means gives a clear sense of the amount of formal analysis that has been done recently in the area. In this section we will sketch a formal account of the *simple* epistemic modalities "It is known that—" and "It is believed that—," along with related constructions. In §5.4 we will describe ways in which more sophisticated analyses of these important constructions have been developed.

There are obviously many different senses of the term "know." There is, first of all, the sense of the term in which it is perfectly reasonable to say "I *knew* that it was Bette Davis who played Scarlett O'Hara, but when I saw the movie it turned out to be Vivien Leigh." This sense of "know" has reference only to the epistemic *attitude* of a person toward a statement—a feeling of certainty—and has no direct connection with the question whether or not the statement is actually true. It occurs most often (but not always) in first-person uses of "know." There is a second, stronger sense of "know" that may be described as "being aware of something which is true." In this sense, the speaker above couldn't have *known* that it was Bette Davis who played the part. He only believed or "thought" that it was Bette Davis. Yet a third sense of "know" can be squeezed out of this situation, when the speaker says "Of course, I knew all along that it was Vivien Leigh, because I remember that she won the Academy Award for the part. I must have just forgotten." This third sense involves knowing *without* always being aware of what one knows. Finally, there is a sense of knowing which might be described as "responsible knowing:" suppose that our movie buff knew that "Gone With the Wind" was made by Metro-Goldwyn-Mayer, and also knew that at that time Bette Davis was under exclusive contract to Warner Brothers. Whether or not it ever occurred to him, he *also* knew, in this fourth sense, that Bette Davis did not appear in "Gone With the Wind": he knew two statements which, taken together, imply that she did not. To take a more schematic case, if someone knows that p, and also knows that C p q, then he knows that q, in this sense of "know," whether or not it ever occurred to him that q. He is, in a sense, *responsible for* knowing the consequences of what he knows.

The most straightforward sense of "know"—the easiest one to formalize—is the fourth sense, responsible knowledge. It is the one we will attack here. There are, of course, yet other senses of the term that we have not mentioned. In the psychoanalytic sense, for example, one may be said to know or believe something firmly enough that one consistently acts in accordance with it, and yet be unwilling to assent to it.

In the presentation of the first few theorems of **CMn** in Chapter III it was noted that "It is known that—" and "It is believed that—" are not related definitionally in the same way that "It is necessary that—" and "It is possible that—" are related. Knowledge and belief are two epistemic modalities of different strengths, analogous in their relationship to, say, conceptual and theoretical necessity in the alethic mode. We will turn first to "know" as an epistemic operator.

We will use 'L_k' for an epistemic operator which represents "It is known that—" and 'M_k' for an epistemic operator which is definitionally related to 'L_k' in the usual way:

$$M_k \alpha =_{df} \bar{L}_k \bar{\alpha}.$$

Clearly, '$M_k \alpha$' cannot be interpreted "It is believed that α." "It is not known that not-α" is not belief in α. I do not know that there are no diamonds on Mars. This is certainly not to say that I believe that there *are* diamonds on Mars. In a more complex system than the present one, Jaakko Hintikka interprets the construction we represent 'M_k' as "It is possible for all (someone) knows that—." To say that I don't know that there are no diamonds on Mars is to say that, for all I know, there could be—there is nothing in what I *do* know that rules out diamonds on Mars. We will use the expression "It is credible that—" for 'M_k', with the understanding that for something to be credible is different from its being believed.[11]

In laying down semantic conditions for the interpretation of 'L_k' and 'M_k' we should keep in mind that we are selecting one particular sense of "know" from among many. We are not capturing *the* meaning of "know" for there is no such thing. The best we can claim is an articulation of *one* clear sense of the term. But once we have this articulation, we can

[11] Von Wright uses "verified" in place of "known" and "not falsified" in place of "credible" in a system of epistemic logic much like the one presented here. See [VI.54, pp. 29–35].

use it as a point of reference from which to discuss other senses of the term.[12]

Now for the features of "know" in the simple sense that we will formalize here:

- We will take '$L_k p$' in the strong sense of know: what is *known* is true; if it is not true, it is not in this sense really known. '$CL_k pp$' will be a theorem of this epistemic logic.
- As an immediate consequence of the first feature, it follows that what is true is also credible. Suppose 'q' is true. Then it is not known that not-q. But this, given the definition above, is just to say '$M_k q$'. '$CpM_k p$' will be a theorem of this epistemic logic.
- If two statements that are known imply a third, then the third statement is also known. This is the sense of "responsible knowledge" described above. To take a paradigm case, if it is known that p and it is also known that Cpq, then it is known that q. '$CKL_k pL_k CpqL_k q$' will be a theorem, as will similar formulas.

So far, it is clear that what is emerging is a very elementary epistemic interpretation of a system of modal logic that resembles **CM**. The principles '$CL_k pp$' and '$CpM_k p$' take us this far. The last-mentioned feature makes this sense of "know" even more like alethic necessity. Now we must face a choice. If we simply adopt one of the alethic systems of modal logic as a satisfactory interpretation of responsible knowing, then we must allow that every logical truth is known. That is, if α is a provable statement, then so is $L_k \alpha$, unless we choose to break the analogy at this point. (See III.3.2.1.) Given that responsible knowing is a rather peculiar sense of "know" to start with, we will not break the analogy here. We have already adopted the principle that one is responsible for the logical consequences of what one knows, and a defense of the principle that all *truths of logic* are known might be attempted. Rather than defend the principle, we will simply admit that we are adopting it for the sake of simplicity. The ploy used in §4 to prevent logical truths from being obligatory will *not* work here, however. There we simply specified that if α is logically necessary, α is not morally obligatory. We cannot, of course, say that if α is logically necessary then it is not *known*. The trick is to avoid

[12] Jaakko Hintikka's book *Knowledge and Belief* [VI.32] is perhaps a classic case of formal "conceptual analysis." Hintikka's interpretation of the epistemic operators will be discussed in a later section, but his work has affected the present section also. The book has come under severe critical fire for failing to do the impossible—for failing to give a formal articulation of "know" which accommodates *all* of the various senses of that term. What it does accomplish is a thorough articulation of one sense of "know," "believe" and related notions. We are now able to say in discussion "I am using 'know' in Hintikka's sense," or "I am using 'know' in a sense that differs from Hintikka's in the following way: —." Providing a point of reference for discussion of the epistemic concepts which is that clear is a significant accomplishment indeed.

the requirement that *every* logically necessary statement be known, if it seems desirable to do so.

A certain amount of selection has to enter into any formal analysis of knowledge and the contents of knowledge. In addition to assuming complete rationality and responsibility, we must also assume that we can fix a moment in time and discuss what is known at precisely that moment, whatever else may come to be known an instant later. The sense of 'L_K' that we are developing is one in which *forgetting* is not allowed. Once a thing becomes known, it remains known.

Before we turn to the model system, it might be helpful to develop further the idea of what is known in an instant of time. Think of a questionnaire of enormous length, consisting of statements followed by boxes to be checked. The boxes are labeled "known," "credible but not known" and "not credible." If you fill out this questionnaire completely, you give (with respect to that set of statements) a list of what is known on that occasion. But we are not finished yet. The following week you get the same questionnaire to fill out again, to give an account of what is known on *that* occasion. The "no forgetting" rule will be enforced in the following way: Once a statement has been marked "known," the box will be already checked when you receive the questionnaire on subsequent occasions. You may come to know more on subsequent occasions than you did on the first—you may check additional boxes marked "known"—but you may not forget or contradict something that you have marked "known" on a prior occasion.

The features of the model system are fairly straightforward, given this notion of *occasion*. A model set is a set of statements which are true on a given occasion, and may include statements of the form '$L_K \alpha$'. The fixed model set is assumed to include statements of the form '$L_K \alpha$' as well as statements which do not fall under epistemic operators. The statements in a model set which do have the form '$L_K \alpha$' can be thought of as delineating a list of things that are known on a given occasion. Alternative model sets will contain everything on the original list of what is known plus other things that might *come to be known* on a subsequent occasion. Such alternatives we call (after Hintikka) *epistemic* alternatives. We will use the form '$H_K \lambda \mu$' for "(Model set) λ is an epistemic alternative to (model set) μ."

All of the requirements concerning truth-functional operators and quantifiers that were imposed on model sets for alethic modal logic are imposed on model sets which are epistemic alternatives to other model sets. A set of statements about what is known must obey all the basic logical rules. Given our assumption that what is known is also true, the alternativeness relation will be reflexive. Given the further assumption that what is known *remains* known on all subsequent occasions, the alternativeness relation will also be transitive. It is not the case, of course, that what might be known on some subsequent occasion is known now: the alternativeness relation is *not* symmetrical. The system so far is just **CS4** interpreted epistemically.

The **CS4** assumptions on the Quantificational Analogue, along with the translation rules for modal operators, produce reasonable results. '$L_k \alpha$' is translated to '$\forall\lambda\, C\, H_k\lambda 0\, (\alpha)\lambda$' (cf. IV.7.2.2). "It is known that α" becomes "On this and every subsequent alternative occasion, α" (*this* occasion, because of reflexivity). Given transitivity, if it is known that α, then on this and every subsequent alternative occasion it is known that α: the translation of '$L_k \alpha$', taken together with the transitivity of the H_k-relation, implies '$\forall\lambda\, C\, H_k\lambda 0\, (L_k \alpha)\lambda$'.

'$M_k \alpha$' is interpreted just like possibility: '$\exists\lambda\, K\, H_k\lambda 0\, (\alpha)\lambda$'. "It is credible that α" becomes "There is a subsequent alternative occasion on which α." If α is credible, then there is a model set alternative to the set of true statements *which contains everything that is known* (as all alternative sets do), *and also contains* α. Since the conditions which require that a model set be logically consistent are imposed on epistemic alternatives, this means that if α is credible, it must be logically consistent with everything that is known.

The semantic development for epistemic logic is just precisely the semantic development for the system **CS4** (including the Quantificational Analogue for **CS4**) with '$L\alpha$' and '$M\alpha$' replaced everywhere by '$L_k \alpha$' and '$M_k \alpha$', and '$H\lambda\mu$' replaced everywhere by '$H_k\lambda\mu$'.

In addition to the formulas already cited, all of the theorems of **CS4** which were developed in Chapter III will hold for responsible knowledge, with the appropriate replacement of operators. Following are a few happy results of such replacement:

T-1	$C\, L_k p\, M_k p$	If it is known that p, then it is credible that p.
T-6	$C\, L_k K p q\, K\, L_k p\, L_k q$	If it is known that both p and q, then it is known that p and it is known that q.
T-10	$C\, L_k C p q\, C\, M_k p\, M_k q$	If it is known that p implies q, then if it is credible that p, it is credible that q.
T-13	$C\, K\, L_k p\, M_k q\, M_k K p q$	If it is known that p and it is credible that q, then it is credible that both p and q.
T-29	$C p\, M_k p$	If p then it is credible that p.

T-31	$C L_k p p$	If it is known that p, then p.
T-43	$C L_k L_k p L_k p$	If it is known that it is known that p, then it is known that p.
T-46	$C L_k p L_k L_k p$	If it is known that p, then it is known that it is known that p.

With respect to quantification, the principle *PHEP* seems to be the most appropriate assumption on individual existence (cf. V.5.2). Individuals which are known to exist continue to be known to exist in all epistemic alternatives, and additional individuals may come to be known to exist. The system **CS4Q4**, on the epistemic interpretation, seems to be the most appropriate system of quantified epistemic logic with respect to responsible knowledge.

The Barcan Formula takes on an interesting coloration in an epistemic system, although we must reject it. We have discussed the Barcan Formula always in the form '$C M \exists x Fx \exists x M Fx$'. An equivalent formula, given the definitions of 'M' and 'L' and of '$\forall$' and '$\exists$', is '$C \forall x L Fx L \forall x Fx$'. On the epistemic interpretation, the Barcan Formula becomes a backhanded "Principle of Induction:" If everything is known to be F, then it is known that everything is F. Suppose that there are twelve dragons in the world, and that each of them is known to be green. This fulfills the antecedent of the Barcan Formula. Let 'Dx' represent "x is a dragon" and 'Gx' represent "x is green." We then have '$\forall x L_k C Dx Gx$'. Does '$L_k \forall x C Dx Gx$' follow? That is, can we infer from "Everything which is a dragon is known to be green" that "It is known that all dragons are green"? We cannot. We would also have to know that those twelve green dragons are all the dragons there are, or the inference doesn't follow.

5.1 Rational Belief. There are at least as many different senses of "believe" as there are of "know." The sense of "believe" that is the proper companion to responsible knowledge we call *rational belief.* Without exploring all the other senses of belief that might come to mind, we can lay down some general principles concerning rational belief and its relationship to responsible knowledge.

- One cannot rationally believe what is logically absurd.

For example, a person cannot, on one and the same occasion, believe both a statement and its negation.

● What is known is also believed; and what is believed by a person on a given occasion must be consistent with everything that is known by that person on that occasion.

● Whether or not a statement is true is irrelevant to whether or not it is believed; false statements may be believed, and true statements may fail to be believed.

We will represent "It is believed that—" as 'L_b'. The corresponding 'M'-operator will be defined in the usual way:

$$M_b \alpha =_{df} \bar{L}_b \bar{\alpha}.$$

The operator 'M_k', related to knowledge, we called "credibility," and to say "It is credible that α" carried the sense that α was something that might come to be known. The operator 'M_b', related to belief, will be interpreted in a similar way. '$M_b \alpha$' carries the sense that α might come to be believed, and will be read "It is plausible that α." The terms "credible" and "plausible" may be almost indistinguishable in some of their natural language uses; here they are quite distinct. What is true is credible; it may not be plausible. The diagram in Figure 4 shows the interrelations of the four epistemic operators. What is known is also true, believed, plausible and credible. What is believed (rationally) is both plausible and credible. What is plausible is credible. What is true is credible.

The formal features of the two belief operators involve no really new strategies. We have already constructed formal systems that include more than one strength of modality, in sorting out logical, conceptual and theoretical alethic modality. Clearly there will be a second alternativeness relation which is not equivalent to the epistemic one. We have already constructed a formal system that lacks the link between the modal operators and truth, in setting up the system of collapsed iterations for deontic logic, as well as the non-collapsing system **CMn** with which we began the formal consideration of modality in Chapter III. What makes

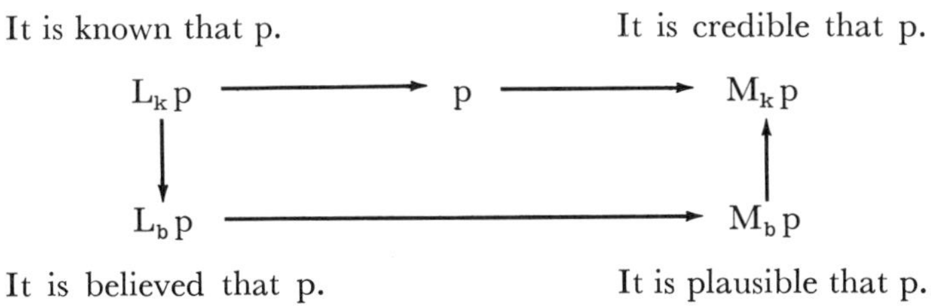

FIGURE 4

the epistemic modalities more complicated than the preceding systems is the fact that the two strengths of modality differ with respect to their properties: one—knowledge—maintains the link with truth, while the other—belief—does not.

The one decision that remains to be made before we set up the semantic machinery for the belief operators is whether or not a version of **CMn** will suffice. If we wish to collapse iterations, so that "It is believed that it is believed that p" is equivalent to just "It is believed that p," then it will be necessary to construct a formal system which allows for such collapsing without including the system **CM.** This was already done in developing the deontic system in §4 above, where a system much like **CS5** was constructed which lacked the link between modality and truth.

The collapsing of similar operators seems desirable: '$L_b L_b p$' should be equivalent to just '$L_b p$' and '$M_b M_b p$' should be equivalent to just '$M_b p$'. The collapsing of dissimilar operators, after the fashion of **CS5**, does not seem appropriate for the belief operators. '$M_b L_b p$'—"It is plausible that it is believed that p"—should not be equivalent to '$L_b p$'—"It is believed that p." The system will resemble the deontic system of §4, then, but will lack the semantic condition which makes the **CS5** formula a truth of logic. The new alternativeness relation will not be symmetrical.

The alternativeness relation which serves as the means for interpreting 'L_b' and 'M_b' will be written 'H_b', following the pattern we have set. '$H_b\lambda\mu$' is read (again, after Hintikka) "(Model set) λ is a *doxastic* alternative to (model set) μ." As was the case with the interpretation of epistemic operators, the doxastic operators are interpreted on model sets which constitute a list of what is believed on a given *occasion*. What may be included in such a list of beliefs is limited by several things: first of all, the consistency conditions that are imposed on all model sets—this is, after all, rational belief, and we have decided not to allow belief in contradictions. Secondly, what is rationally believed is limited by what is known. I cannot rationally believe what I know to be false, nor can I rationally believe something that is inconsistent with what I know to be true. We have already set up a system of *epistemic* alternatives in such a way that every model set which is an epistemic alternative to the fixed model set contains everything that is known, and has logical conditions imposed upon it to insure that nothing which is inconsistent with what is known can be included in an alternative set. We now need simply require that every *doxastic* alternative to the fixed model set also be an epistemic alternative: every model set that includes a list of what is rationally believed must also include a list of what is responsibly known, *and* continue to fulfill the conditions on truth-functional operators and quantifiers that ensure its logical integrity.

5.2 An Integrated System of Knowledge and Belief. The simplest way to lay down the formal systems of knowledge and belief that we have been developing is in the Quantificational Analogue. This will involve giving the translation rules for the four epistemic operators, and setting out the assumptions on the epistemic and doxastic alternativeness relations that we have decided are appropriate.

5.2.1 The "Knowledge" Operators. The operators 'L_k' and 'M_k'—"It is known that —" and "It is credible that—"—are translated into QA using the epistemic alternativeness relation 'H_k'.

'$(L_k \alpha)\mu$' is translated to '$\forall\lambda\, C\, H_k\lambda\mu\, (\alpha)\lambda$'.
'$(M_k \alpha)\mu$' is translated to '$\exists\lambda\, K\, H_k\lambda\mu\, (\alpha)\lambda$'.

5.2.2 The "Belief" Operators. The operators 'L_b' and 'M_b'—"It is believed that —" and "It is plausible that—"—are translated into QA using the doxastic alternativeness relation 'H_b'.

'$(L_b \alpha)\mu$' is translated to '$\forall\lambda\, C\, H_b\lambda\mu\, (\alpha)\lambda$'.
'$(M_b \alpha)\mu$' is translated to '$\exists\lambda\, K\, H_b\lambda\mu\, (\alpha)\lambda$'.

5.2.3 The Remaining Operators. Truth-functional operators and atomic formulas are translated according to the rules given in Chapter IV at 7.2.3, 7.2.4 and 7.2.5.

5.2.4 Conditions on Epistemic Alternatives. The antecedent assumptions permissible in QA for the epistemic alternativeness relation are those of the system **CS4**:

Reflexivity	$\forall\lambda\, H_k\lambda\lambda$
Transitivity	$\forall\lambda\, \forall\mu\, \forall\nu\, C\, K\, H_k\lambda\mu\, H_k\mu\nu\, H_k\lambda\nu$

5.2.5 Conditions on Doxastic Alternatives. The antecedent assumptions permissible in QA for the doxastic alternativeness relation are those developed for ethical alternativeness in §4.1 above, omitting the symmetry condition.

Limited Reflexivity	$\forall\lambda\, C\, \exists\mu\, H_b\lambda\mu\, H_b\lambda\lambda$
Transitivity	$\forall\lambda\, \forall\mu\, \forall\nu\, C\, K\, H_b\lambda\mu\, H_b\mu\nu\, H_b\lambda\nu$

5.2.6 Condition Relating Doxastic and Epistemic Alternatives. All doxastic alternatives to a given model set are also epistemic alternatives to that model set.

$$\forall\lambda\, \forall\mu\, C\, H_b\lambda\mu\, H_k\lambda\mu$$

5.3 Some Truths of Epistemic Logic. In addition to the formulas of **CS4** cited above as holding for "It is known that—," the following formulas are truths of logic with respect to the epistemic-doxastic model structure just described. To demonstrate that they are, one may translate them into the Quantificational Analogue, take the antecedent assumptions listed in 5.2.4, 5.2.5 and 5.2.6, and cancel. They are labeled '**ET**' for "Epistemic Theorem."

ET-1	$C L_k p L_b p$	What is known is believed.
ET-2	$C L_k p p$	What is known is true.
ET-3	$C L_k p M_k p$	What is known is credible.
ET-4	$C L_k p M_b p$	What is known in plausible.
ET-5	$C p M_k p$	What is true is credible.
ET-6	$C L_b p M_b p$	What is believed is plausible.
ET-7	$C M_b p M_k p$	What is plausible is credible.
ET-8	$C L_k L_b p L_b p$	What is known to be believed is believed.
ET-9	$C L_k L_k p L_k p$	What is known to be known is known.
ET-10	$C L_b M_b p M_b p$	What is believed to be plausible is plausible.

In addition to **ET-9**, every case where the same operator is iterated will imply and be implied by a formula where the operator occurs just once.

5.4 More Complicated Epistemic Modality. The system of epistemic logic just presented has to do with "what is known" and "what is believed" on a given occasion. Tacitly, the system has to do with what is known and believed *by a given person* on a given occasion. Beginning with the work of Jaakko Hintikka already mentioned, there has been considerable work done with the systems of epistemic logic where the "knower" and "believer" can be brought into the formalization. The epistemic operators become relational operators which represent "x knows that—" "x believes that—," "It is credible to x that —" and "It is plausible to x that—." Model sets are arranged in terms of an alternativeness relation that is indexed to a given person, and represent what is known or believed by that person on an occasion. Hintikka's system is essentially the system just developed, but it is considerably more flexible and subtle because of the relational character of the operators.

It is possible to construct variants of the system of epistemic logic above, or of a system like Hintikka's, which give a formal pattern that resembles another sense of "know." In the sense just formalized, one can know something, as a consequence of other things he knows, without being aware of it. Knowledge as "awareness" or knowledge as an "epistemic attitude" toward a statement whether or not the statement is true, can also be treated formally. The systems will be somewhat more complicated than the systems of "responsible knowledge" and "rational belief," but they are within the capabilities of the formal machinery available.

Exercise

Using the translation rules and antecedent assumptions given in §5.2, show *in QA* that the following formulas from §5.3 are truths of logic with respect to the model system for epistemic logic: **ET-1, ET-4, ET-5, ET-10**.

6 OTHER NON-ALETHIC MODALITIES

The conception of a model set as a set of statements partially describing a possible world or a possible state of this world suggests an interesting approach to intentional contexts and fictional contexts. At this level of discussion, it is best to think of a possible world—a model set alternative to the fixed set—as one which is fully possible—*theoretically* possible, in the vocabulary of §2 above. This imposes logical, conceptual and theoretical limitations upon what can be included in such a model set. Apart from fantasies and nursery fairy tales, most works of fiction as well as most wishes and hopes can be viewed as descriptions of theoretically possible worlds or as the construction of sets of statements which are theoretically possible, and some of which are true.

To do justice formally to intentional contexts such as wishing and hoping might well involve constructing a formal system that has almost as many distinct modalities as has natural language. For example, one would want a principle something like this to govern hoping:

> If x hopes that p, then x believes that it is theoretically possible that p.

Or, to introduce yet another reasonable modality:

> If x hopes that p, then x believes that it is theoretically possible that now or at some future time, p.

Wishing, of course, does not involve such a requirement. A man falling off a cliff might well wish for the suspension of the law of gravitation, without taking (or having) the time to consider whether or not such a thing is theoretically possible.

We will have more to say of fictional individuals in the final chapter, but it is clear that talk of fictional individuals like Sherlock Holmes can be treated in much the same way that individual names are treated which appear in some H-alternative to the fixed model set without appearing in the fixed set itself. The fictional may be viewed as the theoretically possible in most cases. "Realistic" or "historical" fiction may be viewed as both theoretically possible and consistent with what is known to be true.

7 SUMMARY OF FORMAL SYSTEMS

In this chapter, we have suggested a number of variants of the basic formal systems of modal logic, each designed for a specific purpose. The easiest way to sort out the various formal systems is in terms of the conditions imposed on the alternativeness relation used to interpret the formal systems. This, in turn, is most easily done in the vocabulary of the Quantificational Analogue. The antecedent assumptions we have considered do not exhaust the possibilities. We have actually considered only two properties of the alternativeness relation beyond those discussed in Chapter IV—limited reflexivity and limited symmetry. Table 18 shows the principal systems of modal logic discussed, in terms of the antecedent assumptions permissible for each system in the Quantificational Analogue.

TABLE 18

System	Ex	LRf	Rf	Tr	LSm	Sm	Qe
CMn	+						
CM	+	+	+				
CS4 (Temporal)	+	+	+	+			
CS5 (Alethic)	+	+	+	+	+	+	+
Deontic	+	+		+	+		+
Epistemic	+	+	+	+			
Doxastic	+	+		+			

Existence (*Ex*)	$\forall\lambda\, \exists\mu\, H\mu\lambda$
Limited Reflexivity (*LRf*)	$\forall\lambda\, C\, \exists\mu\, H\lambda\mu\, H\lambda\lambda$
Reflexivity (*Rf*)	$\forall\lambda\, H\lambda\lambda$
Transitivity (*Tr*)	$\forall\lambda\, \forall\mu\, \forall\nu\, C\, K\, H\lambda\mu\, H\mu\nu\, H\lambda\nu$
Limited Symmetry (*LSm*)	$\forall\lambda\, \forall\mu\, C\, K\, H\mu\mu\, H\lambda\mu\, H\mu\lambda$
Symmetry (*Sm*)	$\forall\lambda\, \forall\mu\, C\, H\lambda\mu\, H\mu\lambda$
Quasi-Equivalence (*Qe*)	$\forall\lambda\, \forall\mu\, A\nu\, C\, K\, H\lambda\mu\, H\nu\mu\, H\lambda\nu$

BIBLIOGRAPHY—CHAPTER VI

[VI.1] Anderson, Alan Ross. "The Formal Analysis of Normative Systems" (1956); reprinted in *The Logic of Decision and Action*, Nicholas Rescher, ed. (University of Pittsburgh Press, Pittsburgh, 1967)
[VI.2] ———. "Reduction of Deontic Logic to Alethic Modal Logic, *Mind* **67** (1958), pp. 100–103.
[VI.3] ———. "The Logic of Norms," *Logique et Analyse* **1** (1958), pp. 84–91.
[VI.4] ———. "On the Logic of Commitment," *Philosophical Studies* **10** (1959), pp. 23–27.
[VI.5] ———. Review of E. J. Lemmon 1957, *Zentralblatt für Mathematik* **80** (1959), p. 242.
[VI.6] ———. "Reply to Mr. Rescher," *Philosophical Studies* **13** (1962), pp. 6–8.
[VI.7] ———. "Some Nasty Problems in the Formal Logic of Ethics," *Nous* **1** (1967), pp. 345–360.
[VI.8] Aqvist, L. "A Note on Commitment," *Philosophical Studies* **14** (1963), pp. 22–25.
[VI.9] ———. "A New Approach to the Logical Theory of Interrogatives. Part 1," *Analysis* (Uppsala, 1965), p. 174.
[VI.10] ———. "'Next' and 'Ought'. Alternative Foundations for von Wright's Tense-Logic, with an Application to Deontic Logic," *Logique et Analyse* **9** (1966), pp. 231–251.
[VI.11] ———. "Good Samaritans, Contrary-to-Duty Imperative, and Epistemic Obligations," *Nous* **1** (1967), pp. 361–379.
[VI.12] Berg, Jan. "A Note on Deontic Logic," *Mind* **69** (1960), pp. 566–567.
[VI.13] Binkley, Robert. "A Theory of Practical Reason," *Philosophical Review* **74** (1965), pp. 423–448.
[VI.14] Bohnert, H. G. "The Semiotic Status of Commands," *Philosophy of Science* **12** (1945), pp. 302–315.
[VI.15] Burks, Arthur W. "The Logic of Causal Propositions," *Mind* **60** (1951), pp. 363-382.
[VI.16] Castaneda, Hector-Neri. "The Logic of Obligation," *Philosophical Studies* **10** (1959), pp. 17–23.
[VI.17] ———. "Obligation and Modal Logic," *Logique et Analyse* **3** (1960), pp. 40–48.
[VI.18] ———. "On the Logic of Self-Knowledge," *Nous* **1** (1967), pp. 9–21.
[VI.19] ———. "Acts, the Logic of Obligation, and Deontic Calculi," *Critica* **1** (1967), pp. 77–99.
[VI.20] ———. "Ethics and Logic: Stevensonian Emotivism Revisited," *The Journal of Philosophy* **64** (1967), pp. 671–683.
[VI.21] Chisholm, Roderick M. "The Logic of Knowing," *Journal of Philosophy* **60** (1963), pp. 733–795.
[VI.22] ———. "Contrary-to-Duty Imperatives and Deontic Logic," *Analysis* **24** (1963), pp. 33–36.
[VI.23] ———. "Believing and Intentionality," *Philosophy and Phenomenological Research* **25** (1964), pp. 266–269.
[VI.24] ———. "Identity Through Possible Worlds: Some Questions," *Nous* **1** (1967), pp. 1–8.
[VI.25] Clark, R. L. and Paul Welsh. *Introduction to Logic*. (Van Nostrand Reinhold, New York, 1962)

[VI.26] Fisher, Mark. "A System of Deontic-Alethic Modal Logic," *Mind* **71** (1962), pp. 231–236.

[VI.27] ———. "A Contradiction in Deontic Logic?" *Analysis* **25** (1964), pp. 12–13.

[VI.28] Hanson, William H. "Semantics for Deontic Logic," *Logique et Analyse* **8** (1965), pp. 177–190.

[VI.29] ———. "A Logic of Commands," *Logique et Analyse* **9** (1966), pp. 329–343.

[VI.30] Hare, R. M. *The Language of Morals.* (Clarendon Press, Oxford, 1952)

[VI.31] Hintikka, Jaakko. "Quantifiers in Deontic Logic," *Societas Scientiarum Fennica, Commentationes* **22** (1957)

[VI.32] ———. *Knowledge and Belief.* (Cornell University Press, Ithaca, N.Y., 1962)

[VI.33] ———. "Individuals, Possible Worlds and Epistemic Logic," *Nous* **1** (1967), pp. 33–62.

[VI.34] Lemmon, E. J., and P. H. Nowell Smith. "Escapism: the Logical Basis of Ethics," *Mind* **69** (1960), pp. 289–300.

[VI.35] Marcus, Ruth Barcan. "Essentialism in Modal Logic," *Nous* **1** (1967), pp. 91–96.

[VI.36] Pap, Arthur. *Semantics and Necessary Truth.* (Yale University Press, New Haven, Conn., 1958)

[VI.37] Powers, Lawrence. "Some Deontic Logicians," *Nous* **1** (1967), pp. 381–400.

[VI.38] Prior, A. N. *Past, Present and Future.* (Clarendon Press, Oxford, 1968)

[VI.39] ———. "Escapism: The Logical Basis of Ethics," *Essays in Moral Philosophy*, A. I. Melden, ed. (University of Washington Press, Seattle, 1958), pp. 135–146.

[VI.40] ———. *Time and Modality.* (Clarendon Press, Oxford, 1957)

[VI.41] Quine, W. V. "Two Dogmas of Empiricism," in *From a Logical Point of View* (Harvard University Press, Cambridge, 1953), pp. 20–46.

[VI.42] Rescher, Nicholas. Review of Burks' "The Logic of Causal Propositions," *Journal of Symbolic Logic* **16** (1951), pp. 277–278.

[VI.43] ———. "Conditional Permission in Deontic Logic," *Philosophical Studies* **13** (1962), pp. 273–274.

[VI.44] ———. "An Axiom System for Deontic Logic," *Philosophical Studies* **9** (1958), pp. 24–30.

[VI.45] ———. "Assertion Logic," in his *Topics in Philosophical Logic* (Dordrecht, 1968), pp. 250–286.

[VI.46] Robinson, Richard J. *Definition.* (Clarendon Press, Oxford, 1950)

[VI.47] Sellars, Wilfrid. "Reflections on Contrary-to-Duty Imperatives," *Nous* **1** (1967), pp. 303–344.

[VI.48] Sleigh, Robert C., Jr. "On Quantifying into Epistemic Contexts," *Nous* **1** (1967), pp. 23–31.

[VI.49] Smiley, T. J. "The Logical Basis of Ethics," *Acta Philosophica Fennica* **16** (1963), pp. 237–246.

[VI.50] Snyder, Donald Paul. *Causation, Predication and Modal Logic.* (University Microfilms, Ann Arbor, 1964) (Dissertation: Duke University)

[VI.51] Stenius, Erik. *Wittgenstein's 'Tractatus': A Critical Exposition of its Main Lines of Thought.* (Cornell University Press, Ithaca, N.Y., 1960) (Especially Chapter IX)

[VI.52] ———. "The Principles of a Logic of Normative Systems," *Acta Philosophica Fennica* **16** (1963), pp. 247–260.

[VI.53] van Fraassen, Bas C. "Meaning Relations Among Predicates," *Nous* **1** (1967), pp. 161–179.

[VI.54] von Wright, G. H. *An Essay in Modal Logic.* (North Holland, Amsterdam, 1951)

[VI.55] ———. "Deontic Logic," *Mind* **60** (1951), pp. 1–15; reprinted in *Logical Studies.* (Routledge and Kegan Paul, London, 1957)

[VI.56] ———. *Norm and Action: A Logical Enquiry.* (Humanities Press, New York, 1963)

[VI.57] ———. "A New System of Deontic Logic," *Danish Year Book of Philosophy* **1** (1964), pp. 173–182; "A Correction to a New System of Deontic Logic," *Danish Year Book of Philosophy* **2** (1965), pp. 103–107.

Chapter VII

THE DYADIC MODALITIES

1 ENTAILMENT AND COMPATIBILITY

In the Introduction we described *entailment* as an "if-then" relation between statements that is independent of either the truth-value or the modal status of the statements taken individually. The minimal requirement for the entailment of a statement 'q' by another 'p' is that 'p' be relevant to 'q' in some appropriate sense. The relational character of entailment cannot be emphasized too strongly. It is because the entailment relation demands that the entailing statement be relevant to the entailed that many philosophers reject material and strict implication as adequate formalizations of entailment. The "paradoxes" of material and strict implication which we have already discussed, and which we will shortly examine with some care, allow us in some cases to have provable if-then statements where no connection at all holds between the antecedent and the consequent.

Conceived as a modal relation, entailment might be characterized as *necessitation* to underscore the difference between entailment and strict implication, thus: "'p' necessitates 'q'," as distinct from "It is necessary that 'p' materially implies 'q'." This reflects the *dyadic* modal character of entailment, which may be contrasted with the *monadic* modal notions already discussed.

In a similar way, we can contrast *compatibility* with the monadic formulation of *consistency*. To say that 'p' is compatible with 'q' is to say simply that one does not entail the negation of the other—that there is nothing about one that prevents the other's being true. This may be the case when one or the other or both of 'p' and 'q' are impossible, in which cases "It is possible that both p and q" is false.

Before going on to discuss entailment and compatibility, we must at least make proper obeisance to the *use-mention* distinction. It is often pointed out that when we say "'p' materially implies 'q'" we are mentioning 'p' and 'q', whereas when we say "If p, then q" we are *using* 'p' and 'q'. It is further pointed out by those who are careful that the latter, not the former, is an appropriate reading of the formula 'Cpq', since it indicates that 'p' and 'q' function as statements within a larger statement and does not "name" them.

The locutions "—entails—" and "—is compatible with—" mention statements by name without using them. To be perfectly correct, we should express these dyadic relations in such a way that the constituent statements are used rather than mentioned. This might be done by withholding the if-then reading from material implication formulas, as has been suggested (cf. [VII.7, p. 327]), and saving it for the entailment relation. It would be less misleading, however, to read entailment formulas in the subjunctive mood. It is, after all, such subjunctive structures as the counterfactual conditional and lawlike conditionals that have motivated much of the work in modal relations. If we mention two statements when we say that "Today is Friday" entails "Tomorrow is Saturday," we may use them when we say "If today should be Friday, then tomorrow would be Saturday," or, to put the subjunctive force on the conditional itself, "If it should be the case that today is Friday, then it would be the case that tomorrow is Saturday." This is the structure that grammarians call the *theoretical conditional* (cf. [VII.12, p. 425]), and it reflects the stronger force of entailment.

For the sake of brevity, however, we will continue to sacrifice absolute correctness and use the "—entails—" reading. In the presence of purists we can always replace "'p' entails 'q'" with the more respectable but cumbersome "If it should be the case that p, then it would be the case that q."

1.1 Entailment and Relevance. Various definitions of logical entailment, and various criteria for entailment, have been suggested in order to assure in some formal way the relevance demanded of the entailment relation. Some of these involve restrictions on the provability of the material conditional. Von Wright, for example, offers the following definition of entailment (cf. [VII.39]):

> 'p' entails 'q' if and only if 'Cpq' is demonstrable independently of demonstrating either the falsehood of 'p' or the truth of 'q'.

The intention of von Wright's attempt to define entailment is shared by Alan Ross Anderson and Nuel D. Belnap, Jr., in a number of articles presenting formal systems of entailment.[1] Logical entailment is understood as *deducibility* by Anderson and Belnap. A statement 'p' logically entails another, 'q', if and only if 'q' can be deduced from 'p' in a non-trivial way. The restrictions which Anderson and Belnap impose on deductive proof constitute a guarantee of relevance between the entailing and entailed statement: 'p' must be *used* in the proof of 'q', and if 'p' entails 'q' under the Anderson-Belnap criteria, it can be demonstrated that 'p' and 'q' have content in common.[2]

A criterion for entailment that meshes most naturally with the formal work done so far has been suggested by R. W. Binkley. Like von Wright's, Binkley's criterion for the entailment of 'q' by 'p' demands that 'Cpq' be provable under restricted conditions. The proof-detecting technique employed is cancellation, and the criterion is as follows:

> A wff α entails another, β, if and only if '$C\alpha\beta$' cancels to branches, each of which contains some wff γ and also contains $\bar{\gamma}$—i.e., cancels to closed branches—in such a way that precisely one of γ, $\bar{\gamma}$ comes from the antecedent of '$C\alpha\beta$' and the other comes from the consequent.

In applying this criterion for entailment, which we will call *Binkley's Test*, we can keep track of the origin of the bsfs in a cancellation if we simply superscribe 'a' or 'c' to the cancellation results, depending upon whether a given bit of formula comes from the antecedent or the consequent of the original formula.

The formula 'ApKqr' entails 'Apq' according to Binkley's Test. To show this, we form the material conditional and cancel, keeping track of the antecedent and consequent of the conditional by means of superscripts (see Figure 1).

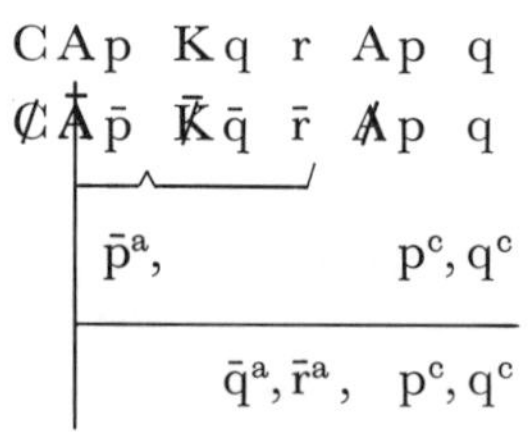

FIGURE 1

[1] One can acquire a good sense of the sort of enterprise that Anderson and Belnap are undertaking in [VII.5]. A list of their articles is given in the Bibliography at the end of this chapter.

[2] In Anderson and Belnap's vocabulary, this requirement takes the form of a stipulation that the entailing and entailed statements have a variable in common. In the present systems, it would take the form of a requirement that they have a statement constant or predicate constant in common.

Both branches of the cancellation close in such a way as to meet the restriction. In the upper branch, '$\bar{p}$' comes from the antecedent of the original conditional and 'p' comes from the consequent; the lower branch has '$\bar{q}$' from the antecedent and 'q' from the consequent.

One of the ways in which the Anderson-Belnap systems of entailment differ from strict implication lies in the formula '$CKp\bar{p}q$'. A contradiction materially implies *and* strictly implies every 'q'. It does not, on the Anderson-Belnap account, *entail* every 'q'. Binkley's Test produces the same results. A contradiction of the form '$K\alpha\bar{\alpha}$' entails each of its members (as does any conjunction):

$$
\begin{array}{llll}
C\,K & p & \bar{p} & p \\
\not{C}\,\not{\bar{K}} & \bar{p} & \not{\bar{\bar{p}}} & p \\
 & \bar{p}^{a}, & p^{a}, & p^{c}
\end{array}
$$

The pair '$\bar{p}^a$' and 'p^c' closes the branch in such a way as to meet the restriction. But a contradiction does not entail every formula:

$$
\begin{array}{llll}
C\,K & p & \bar{p} & q \\
\not{C}\,\not{\bar{K}} & \bar{p} & \not{\bar{\bar{p}}} & q \\
 & p^{a}, & p^{a}, & q^{c}
\end{array}
$$

Although the branch closes on the elements '$\bar{p}$' and 'p', both are from the antecedent of the original formula and the criterion for entailment is not met.

Binkley's Test is not easily adapted to the other paradoxes of material and strict implication, since it functions in its present form only on the main conditional—the left-most sign, if it is 'C'—of a formula. To determine whether or not '$\bar{M}p$' entails that 'p' entails every 'q' would involve a nested test, probably using double superscripts. We can be satisfied with having obtained a sense of *formal relevance* from Binkley's Test, and not complicate our criterion further, since we will shortly develop the formal systems of entailment themselves. Instead, we will legislate concerning the other paradoxes of material and strict implication. *There is nothing about the truth-value or (monadic) modal status of two statements that is sufficient to establish that one of them entails the other.* This is the case because there is nothing about the truth-value or monadic modal status of two statements that is sufficient to guarantee that they are relevant to each other.

1.2 Entailment and Compatibility. We have discussed "compatibility" informally as a relation that obtains when there is nothing about one statement to prevent another's being true. Assuming that a formal account of entailment is forthcoming in the following sections, we can define compatibility in the following way: *A statement* ' p ' *is compatible with another*, ' q ', *if and only if* ' p ' *does not entail* ' $\bar{q}$ '.

We will represent entailment by means of the dyadic modal operator ' Y ', and compatibility by means of the dyadic modal operator ' Z '. Following the usual pattern for introducing elements to the vocabulary, we present them thus:

Y	Entailment	Y p q	' p ' entails ' q '.
Z	Compatibility	Z p q	' p ' is compatible with ' q '.

1.2.1 Definition.

$$Zpq =_{df} = \bar{Y}p\bar{q}$$

1.2.2 Entailment and Compatibility on the DVT. In Chapter III we used the Determinate Value Table (DVT) to compare monadic modalities with truth-functions. The following table shows material and strict implication compared with entailment as described here, and also shows conjunction and consistency compared with compatibility as described here.

The table may be taken as laying down a requirement for entailment and compatibility: on no line of the DVT should we be able to determine that an entailment formula is true or that a compatibility formula is false.[3]

The lines of the DVT where ' T ' can be derived for material and strict implication are precisely the cases which are regarded as "paradoxical." In the following section we will approach the development of the ' Y ' operator by means of singling out those features of the model system which make the "paradoxes" truths of logic, and eliminating them from the model for logical entailment.

2 THE PARADOXES AND THE MODEL SYSTEM

It is helpful to the development of entailment and compatibility operators to consider, in terms of the progression of alethic models, how the various paradoxes are truths of logic in the successive formal systems. We need not concern ourselves with quantified logics for the moment, but rather with truth-functional operators and modal operators only.

[3] The reader who is familiar with G. H. von Wright's dyadic "A New System of Modal Logic" [VII.38] will recognize the extent to which the present account of compatibility is indebted to his account of "relative possibility." Von Wright's system is not amenable to cancellation procedures or to interpretation in terms of model sets. His "relative necessity" operator would produce "paradoxical" occurrences of ' T ' on lines 1, 2 and 3 of the DVT, and "paradoxical" occurrences of ' F ' for relative possibility on lines 13, 14 and 15.

TABLE 19 A DETERMINATE VALUE TABLE OF DEGREE ONE, WITH VALUES SHOWN FOR IMPLICATION AND CONSISTENCY RELATIONS

	Mp	p	Lp	Mq	q	Lq	Cpq	LCpq	Ypq	Kpq	MKpq	Zpq
1	T	T	T	T	T	T	T	T	—	T	T	T
2	T	T	F	T	T	T	T	T	—	T	T	T
3	T	F	F	T	T	T	T	T	—	F	T	T
4	F	F	F	T	T	T	T	T	—	F	F	—
5	T	T	T	T	T	F	T	F	F	T	T	T
6	T	T	F	T	T	F	T	—	—	T	T	T
7	T	F	F	T	T	F	T	—	—	F	—	—
8	F	F	F	T	T	F	T	T	—	F	F	—
9	T	T	T	T	F	F	F	F	F	F	T	T
10	T	T	F	T	F	F	F	F	F	F	—	—
11	T	F	F	T	F	F	T	—	—	F	—	—
12	F	F	F	T	F	F	T	T	—	F	F	—
13	T	T	T	F	F	F	F	F	F	F	F	—
14	T	T	F	F	F	F	F	F	F	F	F	—
15	T	F	F	F	F	F	T	F	F	F	F	—
16	F	F	F	F	F	F	T	T	—	F	F	—

2.1 The Antecedent and Consequent Paradoxes of Material Implication. Truth-functional logic is interpreted on a single model set, which we may regard as the fixed model set, 0. All of the paradoxes of material implication mentioned at the beginning of Chapter III are truths of logic with respect to a model set constructed according to the conditions on truth-functional operators given at IV.2.1. In particular, we will consider the antecedent and consequent paradoxes of material implication. These are, respectively, '$C\bar{p}Cpq$' and '$CqCpq$'. For convenience, and to set things up for a point to be made later, we show the paradoxes in the following way: first, that if we assume [$\bar{p}$] to be a member of the fixed model set, then [Cpq] must also be a member; second, that if we assume [q] to be a member of the fixed model set, then [Cpq] must also be a member.

If we assume that 'p' is false—i.e., that $[\bar{p}] \in 0$—then the condition by which the operator 'A' is interpreted assures us that $[A\bar{p}q] \in 0$ (IV.2.1.3). But if $[A\bar{p}q] \in 0$, then $[Cpq] \in 0$ (IV.2.1.4). The assumption that 'p' is false, then, is sufficient to guarantee the truth of 'Cpq'.

In a precisely parallel way, if we assume that 'q' is true—that $[q] \in 0$—then the condition governing 'A' assures that $[A\bar{p}q] \in 0$, and this in turn assures that $[Cpq] \in 0$. The assumption that 'q' is true is also sufficient to guarantee the truth of 'Cpq'.

2.2 The Antecedent and Consequent Paradoxes of Strict Implication. Monadic modal logic is interpreted on a model system—a set of model sets connected by the H-alternativeness relation. In order for the strict implication formula '$LCpq$' to be true, [Cpq] must be a member of every model set which is an H-alternative to the fixed model set. While the assumption that 'p' is false or the assumption that 'q' is true provides sufficient ground for placing [Cpq] in the fixed model set, such an assumption does not place [Cpq] in every alternative to the fixed model set, and does not, therefore, guarantee the truth of '$LCpq$'. The paradoxes of material implication do not hold for strict implication.

The paradoxes of strict implication, shown as **T-19** and **T-20** in Chapter III, can be shown with respect to the model system in the same way that the paradoxes of material implication were shown above.

For the antecedent paradox of strict implication, we assume that 'p' is impossible—i.e., that $[L\bar{p}] \in 0$. But then $[\bar{p}] \in \lambda$, for every model set λ which is an H-alternative to 0 (IV.3.1.2); and then $[A\bar{p}q] \in \lambda$ and finally $[Cpq] \in \lambda$ for every model set λ which is an H-alternative to the fixed model set (IV.2.1.3 and IV.2.1.4); and if this is the case, then $[LCpq] \in 0$—'$LCpq$' is true.

For the consequent paradox of strict implication, we assume that 'q' is necessary—i.e., that [L q] ∈ 0. This puts [q] in every model set which is H-alternative to 0, and this in turn puts [A p̄ q] in every such model set. Following the same pattern as the antecedent paradox, the assumption that 'q' is necessary guarantees that 'L C p q' is true.

2.3 Expanding the Model System to Avoid Paradoxes. The interpretation of the strict implication formula 'L C p q' on a model system puts strict implication out of the reach of any assumption about the truth or falsity of 'p' or 'q', since truth and falsity are interpreted only on the fixed model set and not on the whole system. In effect, it allows us to "suppose" a state of affairs in which the antecedent of the strict implication formula is true, even though the antecedent happens to be false. So long as such a state of affairs is a possible state of affairs—i.e., so long as we can suppose the antecedent of a strict implication formula to be true with respect to some H-alternative to the fixed model set—paradoxes can be avoided. If the antecedent of a strict implication formula is *impossible*, however, there is no model set which is an H-alternative to the fixed model set and which also allows us to suppose the antecedent to be true.

For the first group of entailment systems, we will use a very simple semantic ploy: we will interpret entailment in such a way as to put it out of reach of any assumption about the possibility of impossibility of its antecedent and consequent. If we introduce a second alternativeness relation, S-*alternativeness*, in addition to the H-alternativeness relation, and play our cards very carefully, we can "suppose" the impossible. We will do this in the following way. First of all, we will require that every model set which is an S-alternative to another model set obey all the conditions on truth-functional operators (and quantifiers, when we bring them back) that are obeyed by H-alternatives. Fulfilling this structural requirement is a necessary condition, but not a sufficient condition, for a model set to be an H-alternative to another set. Second, we will require that every model set which is an H-alternative to a given model set also be an S-alternative, *but not vice versa*—we will allow that (but not require that) there be some model sets which are S-alternatives to a given set without also being H-alternatives. Third, we will continue to interpret the monadic modalities on the H-alternativeness relation. Finally, we will interpret entailment on the S-alternativeness relation, not on the H-alternativeness relation: [Y p q] ∈ μ if and only if [A p̄ q] is a member of every S-alternative to μ.

Now let us see what the expansion of the model system does for us. Specifically, we will consider the antecedent paradox of strict implication. In 2.2 above, we showed this paradox by assuming that 'p' was impossible and finding that, on this assumption, 'L C p q' turned out to be true. But the assumption that 'p' is impossible does not make 'Y p q' true. Following the steps above, the impossibility of 'p' guarantees that [A $\bar{p}$ q] is a member of every H-alternative to the fixed model set. In order for 'Y p q' to be true, however, [A $\bar{p}$ q] must be a member of every S-*alternative* to the fixed model set. Given the way in which S- and H-alternatives are related, the impossibility of 'p' does not guarantee that [A $\bar{p}$ q] is a member of every S-alternative to the fixed model set; only that it is a member of every S-alternative which is also an H-alternative.

The new alternativeness relation allows us to "suppose" a state of affairs in which an impossible statement is true. It does not require us to do so. "Supposing" that there are twenty-six prime numbers under 100 (which is impossible) is analogous to constructing a model set which contains the statement "There are twenty-six prime numbers under 100," and considering what else must be included in the model set as a nontrivial consequence of the inclusion of the impossible statement. If we try to make such a supposition with respect to an H-model system, what we get are trivial consequences of the statement's impossibility. "There are twenty-six prime numbers under 100," which is impossible, entails "There are twenty-five prime numbers under 100," which is also impossible. It also entails "There are twenty prime numbers under 100," which is necessary. It does not entail everything.

The first systems of entailment to be presented formally will be of such a nature that the antecedent and consequent paradoxes of strict implication will not hold for the entailment operator. 'Y p q' will not be materially implied, strictly implied, or entailed by '$\bar{M}$ p' or 'L q'. Virtually all of the "desirable" theorems for strict implication developed in Chapter III will hold for entailment, however: if 'L C' is replaced by 'Y', these formulas will be truths of logic.

The paradox which failed Binkley's Test in §1.1 is *not* eliminated from the system just described. 'Y K p $\bar{p}$ q' is a truth of logic with respect to the model system. This is the case simply because the conditions on truth-functional operators are retained for model sets which are S-alternatives to other model sets. For any model set μ which meets the conditions on truth-functional operators given in IV.2.1, it can be demonstrated easily enough that [A $\bar{K}$ p $\bar{p}$ q] $\in \mu$.

2.4 Expanding the Model System Further. The "paradox" which remains in the first systems of entailment can be eliminated. 'Y K p $\bar{p}$ q' is sometimes called the *Paradox of Consistency.* To avoid it, we must place entailment out of reach of the assumption that a contradiction is unthinkable. For precisely this reason, some philosophers argue that this entailment formula is not paradoxical at all. But it can be eliminated from the formal system without forcing us to contradictions, although certain consequences of its elimination may be deemed undesirable. To eliminate the Paradox of Consistency in the second systems of entailment, we will alter the conditions on model sets which are S-alternatives to other model sets in such a way that a model set which is an S-alternative without also being an H-alternative to another set may contain at least one contradictory pair of statements *without thereby containing all statements*. If certain model sets are allowed to contain a contradictory pair, then it is no longer the case that $[\bar{K}p\bar{p}] \in \mu$ for every model set under discussion, and thus it is no longer the case that $[A\bar{K}p\bar{p}q] \in \mu$. 'YKp$\bar{p}$q' is no longer a truth of logic for every 'p' and 'q'.

With all our stategies out in the open, then, we turn to the formal development of systems of dyadic modality.

3 THE FIRST SYSTEMS OF DYADIC MODAL LOGIC, THE SYSTEMS **CY**

We will present the first entailment systems in a quantified version, first with object language cancellation procedures and a group of theorems, and then interpreted on the quantificational Analogue.

3.1 The Cancellation System CY1.

The system **CY1** is an extension of the system **CM**, described in §3 and §4 of Chapter III.

3.1.1 Vocabulary. The vocabulary of **CY1** consists of the logical constants 'A', 'K', 'C', '‾', '∃', 'V', 'M', 'L' and the two dyadic modal operators 'Y' and 'Z' which were introduced in §2.1. above.

In addition to logical constants, the vocabulary of **CY1** includes the following items described at V.1.1: individual variables, constants, dummy constants, and predicate constants.

Syntactical variables will be used as before: 'α', 'β' and 'γ' to represent wffs, and 'Φ' and 'Ψ' to represent well-formed contexts for individual variables or constants.

The sign '*' will be used in certain of the cancellation rules.

3.1.2 Formation. As before, we will distinguish between initial well-formed formulas (iwffs) and well-formed formulas (wffs). An initial well-formed formula is a formula which meets conditions (i) through (vi) at V.1.3.1, or

(vii) The formula which results from prefixing one of 'Y' or 'Z' to a pair of iwffs.

A well-formed formula is an iwff or a formula resulting from the cancellation or partial cancellation of an iwff. A wff includes any index matrices written below predicate constants within it as the result of applications of the cancellation rules.

3.1.3 Reading Dyadic Formulas. The entailment operator 'Y' may be given either the "if-then" reading, or the subjunctive "if-then" reading mentioned in §1 above, or, more conveniently, the "entails" reading. A "use" verbalization of 'Zpq' does not seem to exist, so it must be read "'p' is compatible with 'q'". The typographical inconvenience of nested quotation marks does not carry over to a verbal inconvenience in reading formulas.

The formula 'YKMpYpqZpq' can be read in several ways. The easiest, perhaps, is "If 'p' is possible and 'p' entails 'q', then 'p' is compatible with 'q'." Although this mixes use (the "if—then—" for the first 'Y') and mention, it seems the most economical and clear sort of verbal reading. This is the pattern we will follow.

3.1.4 Cancellation Rules. In addition to the cancellation rules given in Chapters II, III and V for A, K, C, M, L, ¯, Ǝ and ∀ and their negations, the cancellation rules for dyadic modal logic will contain cancellation rules for the dyadic operators. Note that the rules for dyadic operators are somewhat more complicated than the rules for monadic operators, in that two of them involve the split marker. In the rules given in Figure 2, 'm' is used, as before, to represent the bottom entry of a modal index column when a given rule is to be applied. 'c' and 'd' represent the numerically

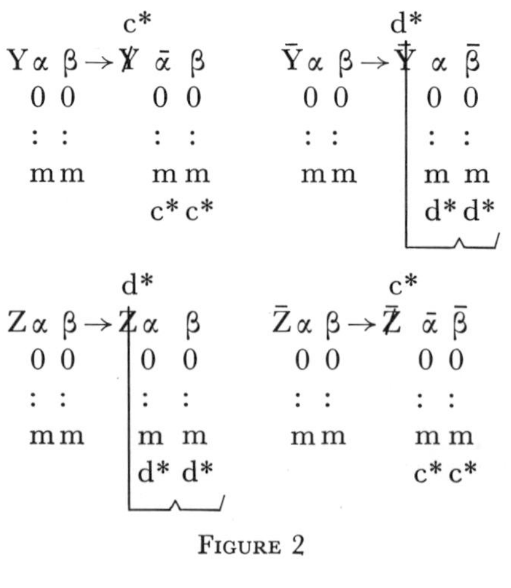

FIGURE 2

first constant and the alphabetically first dummy which has no prior occurrence in the formula. Dummies and constants resulting from the cancellation of dyadic modal operators are marked with a star, '*'. Examples of cancellation using these rules are given in Figure 3. Dummies

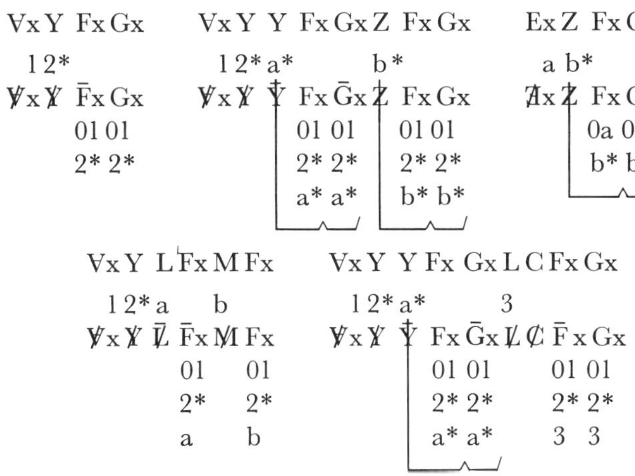

FIGURE 3

and constants resulting from the cancellation of quantifiers and monadic modal operators are placed in the index matrix in precisely the same way as before. All the formulas shown in the examples in Figure 3, and all the formulas in the following few sections, are shown with quantifiers ranging over all modal operators; usually a universal quantifier "out front," with no quantifiers inside modal operators. The integration of quantifiers with dyadic modal operators will be mentioned in a later section.

3.1.5 **CY1** *Closure.* The dyadic modal system **CY1** is similar to the monadic modal system **CM**. The complicating element is, of course, the starred dummies and constants in modal indices. As was the case in the integrated alethic modal systems of Chapter VI, there will be a restriction on which sort of modal dummy can be replaced by which sort of modal constant.

Until we consider the integration of quantifiers with dyadic operators, the dummies and constants resulting from quantifier cancellation can be left alone. The cancellation and closure procedures should be followed in the order given at V.5.4.3 above. For the present, we will omit the repetition of individual dummies and constants on the bottom line of the index matrix, simply to remove some of the complication of index manipulation.

So long as all quantifiers are "out front," this will not have any effect. The *modal closure rules* for **CY1** are as follows.

CY1 I. *Any constant in a modal index column may be repeated immediately below the position in which it first occurs. All (individual) dummies and constants to the right of the modal index column are also repeated.*

CY1 II. *Starred dummies in a modal index column may be replaced by any modal constant in the branch chart.*

CY1 III. *Unstarred dummies in a modal index column may be replaced by unstarred constants* only.

CY1 IV. *A new constant may be introduced to replace two dummies in otherwise opposing formulas. If one of the dummies is unstarred, the new constant may be unstarred.*

CY1 V. *All replacements of dummies by constants are subject to the Uniformity Requirement and the Order Requirement.* (The Order Requirement may be ignored.)

The cancellation and closure procedures are really those of **CM**, with the sole restriction that *unstarred dummies need unstarred constants*. Examples of the **CY1** closure rules will be given in theorem proofs in the following section.

3.1.6 Theorems of **CY1**. The dyadic system **CY1** is not so radical a departure from the monadic systems that we need to explore many of its theorems. Those which are of particular interest will be discussed, and a

DT-1

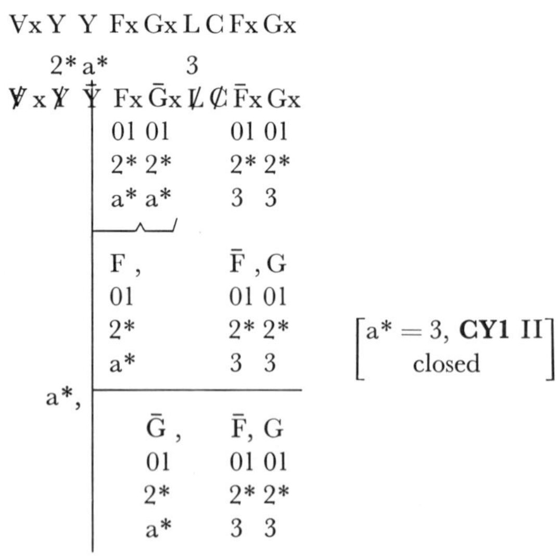

FIGURE 4

list of theorems related to the strict implication and consistency theorems of the monadic modal systems will be given. Theorems of the dyadic system will be labeled '**DT**', for "Dyadic Theorem."

If one statement entails another, then it also strictly implies it. (See Figure 4 for the cancellation proof of **DT-1**.)

DT-2 ∀x Y Y Fx Gx C Fx Gx

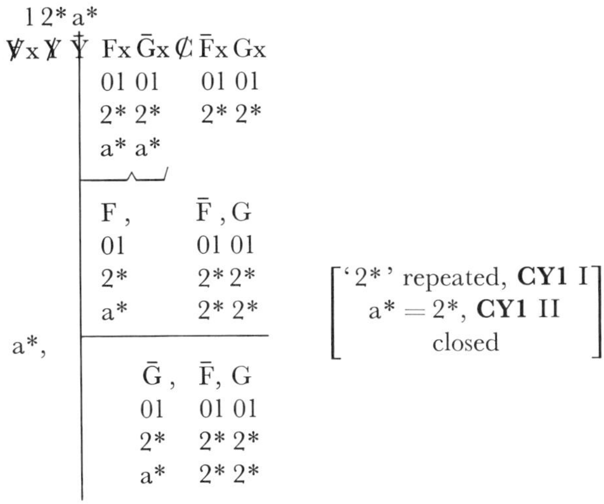

Figure 5

Entailment entails both strict implication (**DT-1**) *and material implication* (**DT-2**). (See Figure 5 for cancellation on proof of **DT-2**.)

The corresponding formulas for compatibility are also theorems of **CY1**.

DT-3 ∀x Y M K Fx Gx Z Fx Gx

DT-4 ∀x Y K Fx Gx Z Fx Gx

If two statements are consistent, then they are compatible; and if two statements are true, they are compatible.

The paradoxes of strict implication are not theorems of **CY1** if the strict implication complex 'L C' is replaced by 'Y'.

∀x Y M̄ Fx Y Fx Gx

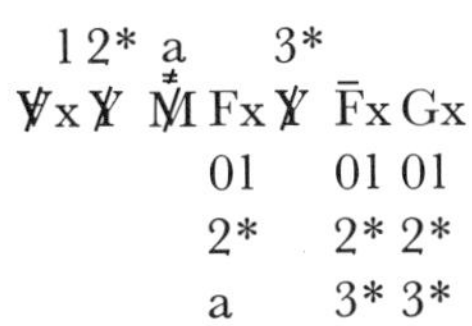

The replacement of 'a' by '3*' is not permitted by the closure rules.

Entailment is stronger than strict implication. The converse of **DT-1** (see Figure 6) is not a theorem of this system:

```
    VxY LCFxGxY FxGx
     12*a        3*
    V̸xY̸ L̸C̸FxḠxY̸ F̄xGx
           |01 01    01 01
           |2* 2*    2* 2*
           |a  a     3* 3*
           |
           |F ,      F̄ ,G
           |01       01 01
           |2*       2* 2*
           |a        3* 3*      [a = 3*, Not permitted]
     a,    |-------------       [     See CY1 III     ]
           |  Ḡ ,    F̄ , G
           |  01     01 01
           |  2*     2* 2*
           |  a      3* 3*
```

FIGURE 6

The *paradox of consistency*, however, is a theorem for the 'Y' operator.

DT-5 VxY KFxF̄xGx

```
    12*
  V̸xY̸ K̸F̄xF̸̄xGx
          01 01 01
          2* 2* 2*      [closed]
```

Entailment, as captured formally in **CY1**, is both reflexive and transitive.

DT-6 VxYFxFx (cf. **T-18**, Chapter III)

DT-7 VxYKYFxGxYGxJxYFxJx (cf. **T-17**)

Compatibility, however, is *not* reflexive. Not every statement is compatible with itself, and 'Zpp' is not in general a theorem. Contradictions, for example, are not self-compatible. 'ZKpp̄Kpp̄' should not be provable, and it is not. This feature of compatibility complicates somewhat the relationship between entailment and compatibility. 'YYpqZpq' is not a theorem of this system. Remember that 'p' is compatible with 'q' if and only if 'p' does not entail 'q̄'. Suppose the antecedent of the entailment formula is a contradiction, 'Kpp̄'. The contradiction certainly entails 'p', but it would be *compatible* with ‘ p ’ only if it did not also entail ‘ $\bar{p}$ ’. Entailment entails compatibility, then, only on the condition that the antecedent of the entailment is self-compatible, or that it is monadically possible.

DT-8 ∀x Y K Z Fx Fx Y Fx Gx Z Fx Gx (cf. **T-1** and **T-14**)

DT-9 ∀x Y K M Fx Y Fx Gx Z Fx Gx

The monadic and dyadic operators are related to each other through the notions of self-entailment and self-compatibility. A statement which is possible is self-compatible, but not vice versa, and a statement which is entailed by its own negation is necessary, but not vice versa.

DT-10 ∀x Y M Fx Z Fx Fx

DT-11 ∀x Y Y $\bar{F}$x Fx L Fx (cf. **T-24**)

The following theorems are related in straightforward ways to theorems of monadic modal logic given in Chapter III.

DT-12 ∀x Y M Fx A Z Fx Gx Z Fx $\bar{G}$x (cf. **T-2** and **DT-9**)
DT-13 ∀x Y Z Fx A Gx Jx A Z Fx Gx Z Fx Jx (cf. **T-3**)
DT-14 ∀x Y A Z Fx Gx Z Fx Jx Z Fx A Gx Jx (cf. **T-4**)
DT-15 ∀x Y K Y Fx Gx Y Fx Jx Y Fx K Gx Jx (cf. **T-5**)
DT-16 ∀x Y Y Fx K Gx Jx K Y Fx Gx Y Fx Jx (cf. **T-6**)
DT-17 ∀x Y Z Fx K Gx Jx K Z Fx Gx Z Fx Jx (cf. **T-7**)
DT-18 ∀x Y A Y Fx Gx Y Fx Jx Y Fx A Gx Jx (cf. **T-8**)
DT-19 ∀x Y Y Fx Gx C L Fx L Gx (cf. **T-9**)
DT-20 ∀x Y Y Fx Gx C M Fx M Gx (cf. **T-10**)

The result of replacing the occurrences of ‘ C ’ by occurrences of ‘ Y ’ in **DT-19** and **DT-20** are not theorems of the present system. They will be theorems of a later system where iterations are collapsed.

DT-21 ∀x Y K L Fx L Gx Z Fx Gx (cf. **T-12**)
DT-22 ∀x Y K L Fx M Gx Z Fx Gx (cf. **T-13**)
DT-23 ∀x Y Y K Fx Gx Jx Y Fx C Gx Jx (cf. **T-16**)

Exercise

Construct cancellation proofs of the following theorems of **CY1**: **DT-12**, **DT-19**, **DT-20**, **DT-22**.

3.2 Entailment, Logical Validity and Logical Equivalence. In Chapter II we discussed material implication, material validity of arguments and material equivalence. The test for each of these involved the cancellation of a formula of the form ‘ C $\alpha\beta$ ’. *Strict* implication, validity and equivalence involve the same tests, but the formula of the form ‘ C $\alpha\beta$ ’

is replaced by a formula of the form ' L C α β ', as described in Chapter III. We now define *logical* entailment, logical validity and logical equivalence using the ' Y ' operator.

A wff, α, *logically entails* another, β, if ' Y α β ' cancels to closed branches.

An argument is *logically valid* if the conjunction of its premises logically entails its conclusion.

Two wffs are *logically equivalent* if they logically entail each other.

Formulas which were given in the exercises for Chapter II as materially equivalent are also logically equivalent when the test is carried out in the system **CY1**. Argument forms which were given in the exercises for Chapter II as materially valid are also logically valid when the test is carried out in **CY1**. Certain of these argument forms will not be logically valid with respect to the second systems of entailment to follow.

Exercises

1. Determine whether or not every two necessary statements entail each other.

2. Demonstrate whether or not the following arguments are (a) strictly valid; (b) logically valid.

$$\frac{\overline{\mathrm{M}}\mathrm{p}}{\mathrm{LCpq}} \qquad \frac{\overline{\mathrm{M}}\mathrm{p}}{\mathrm{Ypq}} \qquad \frac{\begin{matrix}\mathrm{p}\\ \bar{\mathrm{p}}\end{matrix}}{\mathrm{Kpq}}$$

3.3 Interpretation of the CY Systems. We now make the appropriate additions to the model system to accommodate dyadic modal logic, using ' Sλμ ' as an abbreviation for " (Model set) λ is an S-alternative to (model set) μ." In addition to the conditions on the model system for truth-functional operators given at IV.2, those for monadic modal operators given at IV.3, and those for quantifiers given at V.3, we add the following conditions.

3.3.1 [Y α β] ∈ μ if and only if for every model set λ such that Sλμ, either [α] ∉ λ or [β] ∈ λ.

3.3.2 [Z α β] ∈ μ if and only if there is at least one model set λ such that Sλμ and both [α] ∈ λ and [β] ∈ λ.

3.3.3 For any model sets λ and μ, if Hλμ, then Sλμ.

3.3.4 The fixed model set, 0, is H-alternative to itself.

3.3.5 Every model set is S-alternative to itself.

3.3.6 Every model set which is H-alternative to a model set is H-alternative to itself.

The condition *Ex* is included in the conditions in IV.3 and does not need to be added here.

The expedient of interpreting entailment on the S-relation and the monadic operators on the H-relation provides a distinction between entailment and strict implication, given condition 3.3.3 above. [L C p q] is a member of a given model set μ if and only if [A $\bar{p}$ q] is a member of every H-alternative to μ. [Y p q] is a member of a given model set μ if and only if [A $\bar{p}$ q] is a member of every H-alternative *and* every S-alternative to μ. Some of the S-alternatives will also be H-alternatives; some may not be.

3.4 Additions to the Quantificational Analogue. To the vocabulary of the Quantificational Analogue given at IV.7.1 and V.4 we make the following addition.

> **S——** A predicate of degree 2 ("S-alternativeness"), taking model set variables or constants as arguments. 'S$\lambda\mu$' is read "(Model set) λ is S-alternative to (model set) μ."

3.4.1 Mapping Rules. The following rules, taken together with those given at V.6.1, allow the translation of wffs of dyadic modal logic into the Quantificational Analogue. As before, mapping rules may be regarded as rules for the removal of parentheses.

'$(\mathrm{Y}\,\alpha\beta)\boldsymbol{\mu}$' is translated to '$\mathbf{\forall\lambda\ C\ S\lambda\mu\ A}\ (\bar{\alpha})\boldsymbol{\lambda}\ (\beta)\boldsymbol{\lambda}$'.
'$(\mathrm{Z}\,\alpha\beta)\boldsymbol{\mu}$' is translated to '$\mathbf{\exists\lambda\ K\ S\lambda\mu\ K}(\alpha)\boldsymbol{\lambda}\ (\beta)\boldsymbol{\lambda}$'.
'$(\bar{\mathrm{Y}}\,\alpha\beta)\boldsymbol{\mu}$' is translated to '$\mathbf{\exists\lambda\ K\ S\lambda\mu\ K}\ (\alpha)\ \boldsymbol{\lambda}\ (\bar{\beta})\boldsymbol{\lambda}$'.
'$(\bar{\mathrm{Z}}\,\alpha\beta)\boldsymbol{\mu}$' is translated to '$\mathbf{\forall\lambda\ C\ S\lambda\mu\ A}\ (\bar{\alpha})\boldsymbol{\lambda}\ (\bar{\beta})\boldsymbol{\lambda}$'.

As before, we will drop the square brackets around atomic formulas as a notational convenience. (For the second systems of dyadic modal logic, we will be forced to retain the square brackets.) In Figure 7 is a step by step translation of **DT-3** into QA. Translation rules from Chapter V and from this section are cited as they are used. The First Interpretation of basic formulas is used for simplicity.

∀x Y M K Fx Gx Z Fx Gx	
(∀x Y M K Fx Gx Z Fx Gx)**0**	
∀x C Bx0(Y M K Fx Gx Z Fx Gx)**0**	V.6.1.8
∀x C Bx0 ∀λ C Sλ0 A ($\bar{\mathrm{M}}$ K Fx Gx)**λ** (Z Fx Gx)**λ**	3.3.1
∀x C Bx0 ∀λ C Sλ0 A ∀μ C Hμλ ($\bar{\mathrm{K}}$ Fx Gx)**μ** (Z Fx Gx)**λ**	V.6.1.13
∀x C Bx0 ∀λ C Sλ0 A ∀μ C Hμλ A $\bar{\mathbf{F}}$xμ $\bar{\mathbf{G}}$xμ (Z Fx Gx)**λ**	V.6.1.4
∀x C Bx0 ∀λ C Sλ0 A ∀μ C Hμλ A $\bar{\mathbf{F}}$xμ $\bar{\mathbf{G}}$xμ ∃μ K Sμλ K Fxμ Gxμ	3.3.1

FIGURE 7

3.4.2 The Antecedent Assumptions of **CY1**. To prevent the two alternativeness relations from collapsing into one, we must hedge the assumptions somewhat. The reflexivity assumptions are as follows.

Rf.S	$\forall\lambda\, S\lambda\lambda$
LRf	$\forall\lambda\, C\, \exists\mu\, H\lambda\mu\, H\lambda\lambda$
0*Rf*	$H00$
Ex	$\forall\lambda\, \exists\mu\, H\mu\lambda$

The antecedent assumption relating the two alternativeness relations we call '*H-S*'.

H-S $\forall\lambda\, \forall\mu\, C\, H\lambda\mu\, S\lambda\mu$

The remaining assumption is the assumption of singular existence.

SEx $\forall\lambda\, \exists x\, Bx\lambda$

The only assumption needed to close the cancellation of **QA: DT-3** above is *H-S* (see Figure 8).

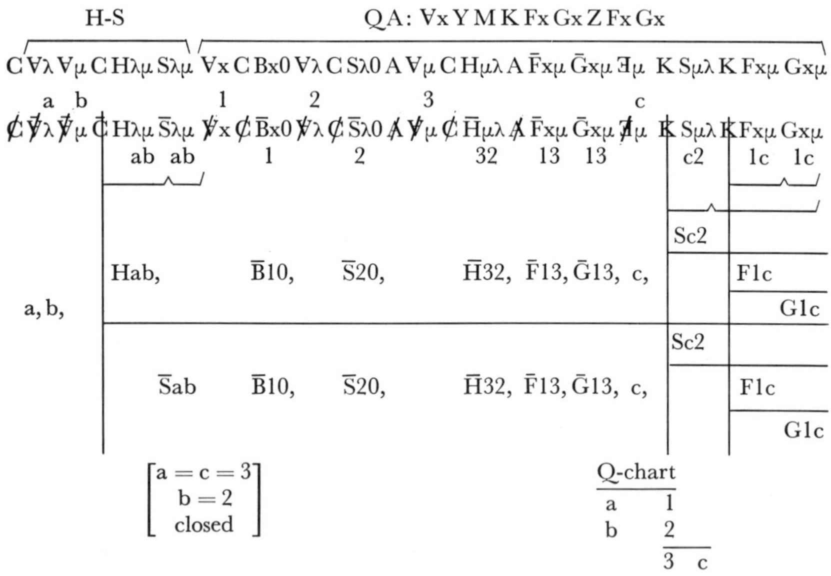

FIGURE 8

Exercise

Translate **DT-4** and **DT-5** into QA and cancel. Determine which of the permissible antecedent assumptions for **QA: CY1** are needed to close each cancellation.

3.5 Extensions of the System CY1. We will not pursue in detail the entailment analogues of **CS4** and **CS5** which are simple extensions of **CY1**. It is easy enough to see, however, the form that such extensions will take. For the system **CY2**, analogous to **CS4**, we add to the Quantificational Analogue the assumption that the S-alternativeness relation is transitive:

Tr.S $\forall\lambda\,\forall\mu\,\forall\nu\,\mathrm{C\,K\,S}\lambda\mu\,\mathrm{S}\mu\nu\,\mathrm{S}\lambda\nu$.

To the object language we add the closure condition analogous to **CS4** I.

CY2 I. *If two starred constants occur consecutively in a modal index, it is permissible to delete the upper of the two constants before replacing dummies by constants.*

For the system **CY3**, analogous to **CS5**, we add to the Quantificational Analogue the assumption that the S-alternativeness relation is symmetrical:

Sm.S $\forall\lambda\,\forall\mu\,\mathrm{C\,S}\lambda\mu\,\mathrm{S}\mu\lambda$.

To the object language, we add the closure condition analogous to **CS5** I.

CY3 I. *If all the entries in a modal index column are starred, all but the bottom entry may be deleted.*

If the **CS4** and **CS5** assumptions on H-alternativeness are added to the model system, the corresponding closure conditions may be added for unstarred dummies and constants. In modal indices which have starred and unstarred entries, the deletion of a starred entry causes the next lower entry to be starred. In the case of the **CS5** and **CY3** closure conditions, the bottom entry becomes starred if any of the deleted entries are starred.

4 THE SECOND SYSTEMS OF DYADIC MODAL LOGIC, THE SYSTEMS **CYC**

The systems **CYC** depart from the "conventional" systems of modal logic in that there are two distinct symbols for negation: the negation bar '‾', which we have been using right along, and the cancellation symbol '–', written before certain formulas as a result of the application of cancellation rules. We make the distinction by calling this latter symbol the *denial stroke*. The result of writing the denial stroke to the left of a wff is the *denial* of that wff; the result of writing the negation bar over the left-most sign of a wff is, as before, the *negation* of that wff.

Making a distinction between negation and denial is not without philosophical ramifications. In general, the distinction will always be between the negation of a predicate and the denial of a statement in which that predicate occurs. The formal results of making the distinction are respectable enough, and we will leave discussion of it for a later section. It may be reassuring to the dubious, however, to recall Aristotle's exploration of the various permutations of "Man (not-man) is (is-not) just (not-just)" in *De Interpretatione* (9b: 18–40), and his discussion of whether or not one of an affirmation or denial must be true and the other false (18a: 10, 28; 19b: 1–4). In some respects the present account seems to be more conservative than Aristotle's.

4.1 The System CYC1. A full set of cancellation rules for a formal system containing both negation and denial contains thirty-eight distinct rules. In order to keep the presentation of the **CYC** systems as simple as possible, we will impose certain limitations on iwffs, thereby reducing the number of cancellation rules needed to seventeen.

4.1.1 Vocabulary. The vocabulary of the **CYC** systems is the vocabulary of the system **CY1**, with the omission of the logical constant 'C' and the addition of the denial stroke '–'.

4.1.2 Formation. An initial well-formed formula of the **CYC** systems is one of the following.

(i) A predicate of degree *n* followed by *n* individual variables or constants.
(ii) The formula which results from prefixing one of 'A', 'K', 'Y' or 'Z' to a pair of iwffs.
(iii) The formula which results from writing one of '∃' or '∀', followed by an individual variable, to the left of an iwff.
(iv) The formula which results from writing one of 'M' or 'L' to the left of an iwff.
(v) The formula which results from placing the negation bar over the left-most sign of an iwff, *if that left-most sign is a predicate constant.*

The denial stroke does not occur in initial well-formed formulas under this limited set of formation rules.

Under this restricted definition of initial well-formedness, not every iwff of the **CY** systems is also an iwff of the **CYC** systems. But every iwff of the **CY** systems is *logically equivalent* to an iwff of the **CYC** systems in the sense of logical equivalence described in §3.2 above. '∀xCFxGx', for example, is an iwff of the **CY** systems, but it does not meet the conditions above for the **CYC** systems. '$\forall x A \bar{F}x Gx$', however, is logically equivalent to '$\forall x C Fx Gx$', and is an iwff of both systems.

In general, the formation rules above simply eliminate the operator 'C' and confine negation to atomic formulas.

4.1.3 Cancellation Rules. The complete set of cancellation rules for the **CYC** systems is given in Figure 9. The denial stroke '$-$' appears in these rules.

A *pair of opposites* in the **CYC** systems consists of a basic subformula and its *denial*. A pair like '$F_{01}, \bar{F}_{01}$' is no longer a pair of opposites, unless it can be converted to '$F_{01}, -F_{01}$' by the closure rules to follow.

4.1.4 **CYC1** *Closure.* The closure rules of the system **CYC1** are closure rules I–V of the system **CY1** above, plus the following.

CYC1 I. *Basic formulas which are of the form* '$-\bar{F}_{\substack{0\\mi}}$' (*where* 'm' *is a modal dummy or constant and* 'i' *is an individual dummy or constant*) *may be further cancelled to formulas of the form* '$F_{\substack{0\\mi}}$' provided that the bottom entry in the modal index column is unstarred.

CYC1 II. *Basic formulas which are of the form* '$\bar{F}_{\substack{0\\mi}}$' *may be changed to the form* '$-F_{\substack{0\\mi}}$', provided that the bottom entry in the modal index column is unstarred.

The closure rules will be applied in theorem proofs in the next section.

4.1.5 Important Theorems of **CYC1**. The theorems of **CYC1** will be labeled '**DCT**' to distinguish them from the theorems of the more conventional system of dyadic modal logic **CY1**.

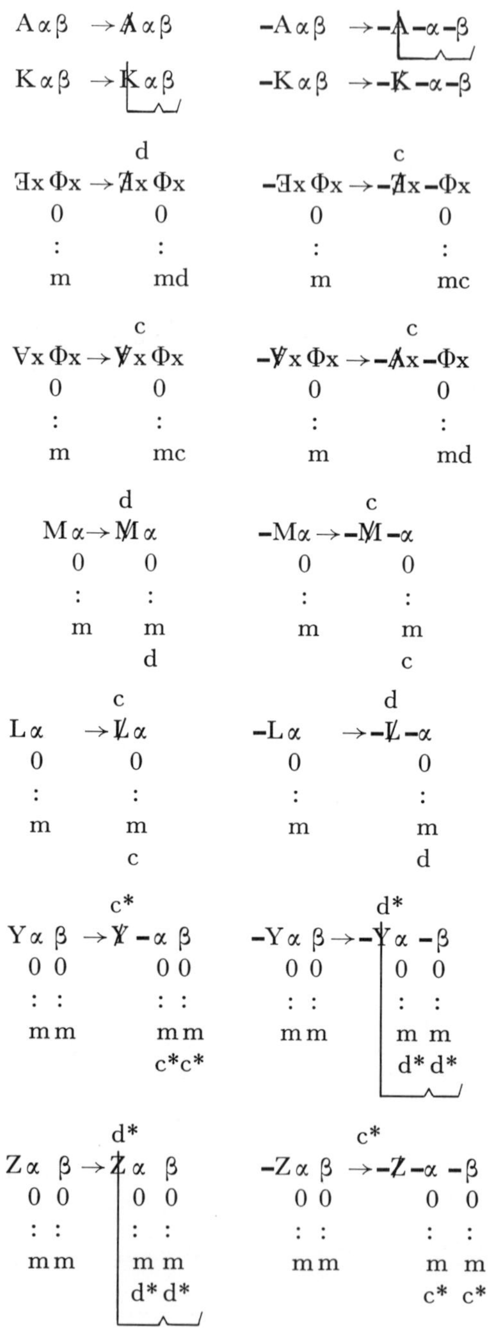

FIGURE 9

DCT-1 Vx Y Y Fx Gx L A $\bar{F}$x Gx

```
        1 2*  a*          3
        ∀x Y  –Y  Fx–Gx L A F̄x Gx
                  01  01    01 01
                  2*  2*    2* 2*
                  a*  a*    3  3
                ____________
                  F,        F̄, G        [ 'F̄' = '–F', CYC1 II ]
                  01        01 01        [ a* = 3, CY1 II      ]
                  2*        2* 2*        [ closed              ]
                  a*        3  3
                ____________
                     –G ,   F̄, G
                     01     01 01
                     2*     2* 2*
                     a*     3  3
```

FIGURE 10

All of the numbered theorems **DT-1** through **DT-23** of **CY1** are also theorems of **CYC1**, with the exception of the following:

DT-2 Vx Y Y Fx Gx A $\bar{F}$x Gx (Not a theorem of **CYC1**)
DT-5 Vx Y K Fx $\bar{F}$x Gx (Not a theorem of **CYC1**)
DT-23 Vx Y Y K Fx Gx Jx Y Fx A $\bar{G}$x Jx (Not a theorem of **CYC1**)

The paradox of consistency is, of course, **DT-5**, and it is precisely this formula that the system **CYC1** is designed to avoid. **DT-2** and **DT-23** may be regarded desirable as theorems, although it is possible to argue against their acceptability. We will shortly suggest an alteration in **CYC1** that will bring **DT-2** and **DT-23** into the system as theorems without also bringing in **DT-5**. The cancellation of **DT-5** is given in Figure 11. None

```
        Vx Y   K  Fx  F̄x Gx
         1 2*
        ∀x Y  –K –Fx –F̄x Gx
                   01  01 01
                   2*  2* 2*

                 –F , –F̄ , G
                   01   0   01
                   2*   2*  2*
```

FIGURE 11

of the closure rules will close the branch. The subformula '–$\bar{F}$ ' could be
01
2*

changed to ‘F’ only if the bottom entry were unstarred.

```
01
2*
```

CYC1 is not totally paradox-free. Although formulas of the form ‘$\text{YK}\alpha\bar{\alpha}\beta$’ are not provable for every α and β, formulas of this form are provable in some cases where α contains monadic modal operators (see Figure 12).

```
VxY   K  L  Fx  M  F̄x Gx
 1 2*      a       3
V̸x Y̸  –K̸–L̸–Fx–M̸–F̄x Gx
            01        01 01
            2*        2* 2*
            a         3
           –F,       –F̄, G        [ ‘–F̄’ = ‘F’, CYC1 I ]
            01        01 01        [ a = 3, CY1 III     ]
            2*        2* 2*        [ closed             ]
            a         3
```

FIGURE 12

This and other “higher-order” paradoxes could be eliminated from the system by eliminating the monadic modal operators, and perhaps defining new monadic operators in terms of the dyadic operators by strengthening **DT-10** and **DT-11** to logical equivalences. Restrictions on iterations of the ‘Y’ operators might also be in order.

As the system stands, however, it has some desirable properties. Perhaps the most important of these is the relationship between the theorems of **CYC1** and formulas which meet Binkley's Test for entailment (§1.1 above). *Where α and β do not themselves contain modal operators,* ‘$\text{Y}\alpha\beta$’ *is a theorem of* **CYC1** *if and only if α entails β according to Binkley's Test.*[4]

[4] *Proof*: Consider the cancellation of an iwff of the form ‘$\text{Y}\alpha\beta$’. If there are no modal operators in either of α or β, the modal index of every basic formula in α and β will consist of the constant ‘0’ plus a single starred constant, with the starred constant in the bottom of the modal index column. None of the **CYC** closure rules are applicable in such a case except those inherited from the **CY** systems. One member of a pair of opposites in a branch must contain an occurrence of the denial stroke. But the denial stroke does not occur in any iwff of the system, and in a formula of the form ‘$\text{Y}\alpha\beta$’ where ‘Y’ is the only modal operator, any occurrence of the denial stroke must originate with the cancellation of ‘Y’. The cancellation of ‘Y’ will introduce a denial stroke only before the antecedent, α. The cancellation rules given above and the limitation of iwffs guarantee that all and only the bsfs within α will contain occurrences of the denial stroke. Thus, the finished branch chart will contain bsfs containing occurrences of the denial stroke, all of which must have come from the antecedent, α, and will contain bsfs *not* containing occurrences of the denial stroke, all of which must have come from the consequent, β. The closure elements of each branch then fulfill the condition of Binkley's Test.

4.1.6 Some Odd Features of **CYC1**. Apart from the missing link between entailment and material implication reflected in the fact that **DT-2** has no counterpart theorem in **CYC1**, there is a remaining oddity in the system. The so-called "disjunctive syllogism" of elementary logic is not logically valid according to the **CYC** version of entailment. That is, the argument form

$$\frac{\begin{array}{c}\alpha \\ A\bar{\alpha}\beta\end{array}}{\beta}$$

if tested for logical validity in the usual way, by constructing the entailment formula '$YK\alpha A\bar{\alpha}\beta\beta$', will not cancel to closed branches. This purported entailment is also not a theorem of the Anderson-Belnap systems of entailment, nor does it meet Binkley's Test. While such a result may seem to provide grounds for rejecting **CYC**, Binkley's Test and the Anderson-Belnap systems, it can be defended. Anderson and Belnap argue that this purported entailment is fallacious because it commits a "fallacy of relevance" [VII.6, p. 19]. The thrust of the argument is this. The antecedent of the purported entailment is logically equivalent to '$AK\alpha\beta K\alpha\bar{\alpha}$'. An alternation entails another statement (and materially or strictly implies it) only if each member of the alternation entails that statement. The purported entailment is acceptable, then, only if both '$YK\alpha\beta\beta$' and '$YK\alpha\bar{\alpha}\beta$' are acceptable. The former is of course a true entailment, but the latter is just the paradox of consistency. Since the acceptability of the disjunctive syllogism rests upon the acceptance of the paradox, Anderson and Belnap reject the disjunctive syllogism.

One of Anderson and Belnap's suggestions concerning the general acceptability of the disjunctive syllogism is particularly interesting. They hold that the argument pattern is logically valid only when interpreted as justifying an inference from proven premises: If both α and $A\bar{\alpha}\beta$ are theorems, then so is β [VII.2, pp. 10–12]. It is interesting to note in this connection that while '$\forall x\, YKFxA\bar{F}xGxGx$' is not a theorem of **CYC1**, the formula '$\forall x\, YKLFxLA\bar{F}xGxLGx$' *is* a theorem. Monadic necessity and theoremhood are not one and the same, of course, but all formulas which are theorems are also monadically necessary.

The loss of the disjunctive syllogism does not rob the system of the valid argument form called *modus ponens*. 'p', taken together with "If p, then q," does entail 'q'. The if-then premise of the argument is properly symbolized in this system using entailment, not material implication (or the equivalent of material implication, '$A\bar{p}q$'). The formula

$$\forall x\, YKFxYFxGxGx$$

is a theorem of **CYC1**.

That entailment entails strict implication without entailing material implication is odd indeed. The **CYC** system does not, as it stands, have **DT-2**, which would read '∀x Y Y Fx Gx A F̄x Gx'. The problem here is partly a mechanical one, since the corresponding formulas with 'M' and 'L' inserted before the 'A' are theorems. What might be needed is an "actuality" operator, like that suggested by Nicholas Rescher for von Wright's dyadic systems [VII.27]. Such an operator would carry the sense "It is true that—," which would make it a non-modal assertion operator, further complicating an already very complicated system. Instead, we can add a closure condition to **CYC1** which takes the place of such an operator:

> *New Closure Condition. An occurrence of the special modal constant* '0' *may be added at the bottom of a modal index column, provided the following conditions are met:* (i) *the bottom position of the modal index column contains a constant before* '0' *is added;* (ii) *such an addition is made to only one of any pair of opposites.*

The occurrence of '0' may then be used to replace any dummies in the corresponding position of the opposing index.

A cancellation of **DCT-2**, using the new closure condition, is shown in Figure 13.

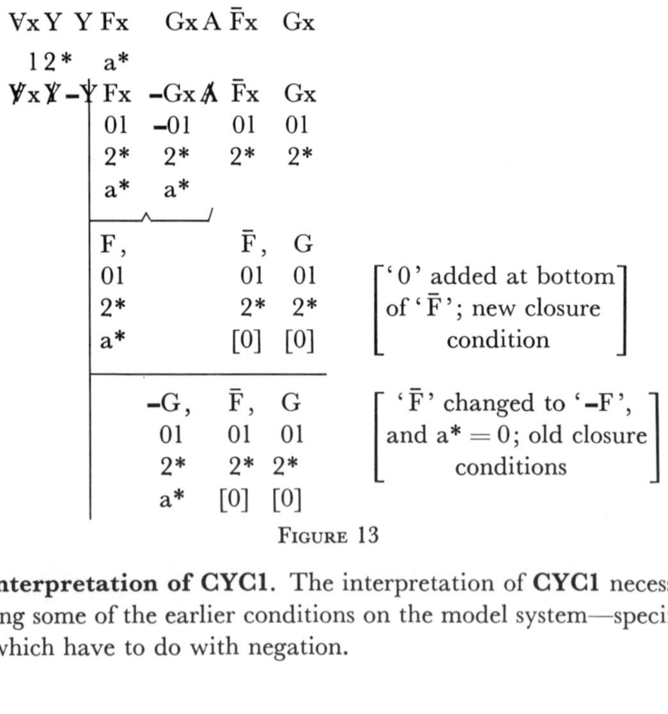

FIGURE 13

4.2 Interpretation of CYC1. The interpretation of **CYC1** necessitates changing some of the earlier conditions on the model system—specifically those which have to do with negation.

A model set μ in a model system Ω obeys the conditions at IV.2.1.2 through IV.2.1.4, at IV.3.1.1 through IV.3.1.4, and at 3.3.1 through 3.3.6 above. These are the conditions governing 'K', 'A', 'M', 'L', 'Y' and 'Z', plus the condition that no model set is empty, and the *Ex* and *Rf* assumptions taken for the H- and S- relations above. In addition, a model set obeys the conditions on quantifiers at V.3, and the following conditions.

4.2.1 If there is a model set λ such that $H\mu\lambda$, then $[\bar{\alpha}] \in \mu$ if and only if $[\alpha] \notin \mu$.

This condition requires consistency of those model sets that are H-alternatives to other model sets.

4.2.2 $[-\alpha] \in \mu$ if and only if $[\alpha] \notin \mu$.

It is now denial, not negation, that is given the general interpretation of "non-membership" in a model set. This condition amounts to saying that a sentence cannot be both a member and a non-member of a model set.

Taken together, the conditions 4.2.1 and 4.2.2 justify the **CYC1** closure conditions. Negation and denial *are* one and the same with respect to model sets that are H-alternative to other model sets, and with respect to such model sets, we might as well say that $[\bar{F}x]$ is a member as that $[-Fx]$ is a member. Basic subformulas are judged to be true if they are members of the model sets assigned the bottom entries in their modal index columns. Unstarred entries are assigned to model sets which are H-alternatives to the sets above them and starred entries are assigned to model sets which are S-alternatives to the sets above them. If the bottom entry in the modal index column under '$\bar{F}$' is unstarred, then it is both an H-alternative and an S-alternative to the entry above it, and 4.2.1 and 4.2.2 justify our changed '$\bar{F}$' to '$-F$'. If the bottom entry is starred, it is an S-alternative to the entry above it, but might not be an H-alternative, and that limiting condition on 4.2.1 removes our justification for changing a negation to a denial.

4.3 CYC1 in the Quantificational Analogue. In order to interpret **CYC1** in the Quantificational Analogue, we must, first of all, restore the square brackets around basic formulas. Heretofore, we have dropped the square brackets required by IV.7.2.5, shortening '$[Fx]\lambda$', for example, to '$Fx\lambda$'. Now we must restore the brackets in order to keep track of negations that are inside the brackets as distinct from negations that are outside. For notational convenience, we will use '$-[Fx]$' for negation outside brackets, and '$[\bar{F}x]$ for negations that are inside.

4.3.1 Mapping Rules. In translating iwffs of the **CYC** systems into QA, basic subformulas are enclosed in square brackets *as the first step in translation.*

The translation rules at V.6.1 are retained, as are the translation rules for 'Y' and 'X' given at 3.4.1 above.

Negations inside square brackets may not be brought outside, and negations outside square brackets may not be brought inside.

To get right to the difficult case, we translate the paradox formula into QA as follows:

$$\forall x\, Y\, K\, Fx\, \bar{F}x\, Gx$$

$$\forall x\, Y\, K\, [Fx]\, [\bar{F}x]\, [Gx] \qquad \text{(above)}$$

$$(\forall x\, Y\, K\, [Fx]\, [\bar{F}x]\, [Fx])\mathbf{0}$$

$$\boldsymbol{\forall x\, C\, Bx0}\, [\, Y\, K\, [Fx]\, [\bar{F}x]\, [Gx])\mathbf{0} \qquad \text{V.6.1.8}$$

$$\boldsymbol{\forall x\, C\, Bx0\, \forall\lambda\, C\, S\lambda 0\, A}\, (\bar{K}\, [Fx]\, [\bar{F}x]\boldsymbol{\lambda}\, [Gx]\boldsymbol{\lambda} \qquad \text{3.4.1}$$

$$\boldsymbol{\forall x\, C\, Bx0\, \forall\lambda\, C\, S\lambda 0\, A\, A - [Fx]\lambda - [\bar{F}x]\lambda\, [Gx}(\boldsymbol{\lambda}) \qquad \text{V.6.1.4}$$

Like the cancellation rules for the **CYC** systems, the QA translation rules would need to be considerably expanded if we did not confine negations in iwffs of the object language.

What the modifications of the mapping rules amount to is this: In constructing and cancelling formulas in QA, negation remains the same as before, except that the negations *acquired by* a formula in square brackets remain outside the brackets at this stage. An antecedent assumption will connect inside and outside negations for certain formulas.

4.3.2 Antecedent Assumptions of **CYC1**. All the antecedent assumptions of **CY1** given at §3.3.2 above may be taken for **CYC1**. In addition, the following assumptions may be taken for any predicate constant F.[5]

$$\boldsymbol{\forall x\, \forall\lambda\, \forall\mu\, C\, H\lambda\mu\, C\, Bx\lambda\, C\, [Fx]\lambda - [\bar{F}x]\lambda}$$

$$\boldsymbol{\forall x\, \forall\lambda\, \forall\mu\, C\, H\lambda\mu\, C\, Bx\lambda\, A\, [Fx]\lambda\, [\bar{F}x]\lambda}$$

The first assumption requires that a formula and its negation not both be members of a set which is H-alternative to another set; the second

[5] Given the extended primitive vocabulary for QA discussed in the notes at IV. 7.1 and V.4.1, "quantification over predicates"—quantification over those marks of the model system which are occurrences of predicate letters—can be introduced for these two assumptions. They will read as follows in the "condensed" QA of the text:

$$\forall x\ \forall f\ \forall\lambda\ \forall\mu\ C\ H\lambda\mu\ C\ Bx\lambda\ C\ [fx]\lambda - [\bar{f}x]\lambda$$

$$\forall x\ \forall f\ \forall\lambda\ \forall\mu\ C\ H\lambda\mu\ CBx\lambda\ A\ [fx]\lambda\ [\bar{f}x]\lambda$$

If this is to be done, a set of predicate dummies will be needed for QA cancellation. In the extended QA vocabulary, of course, all QA quantification is over individual variables and the universe of discourse is the marks and sets of marks in the model system. Denial is handled as the negation of the QA predicate 'E', introduced in the note at V.4.1.

requires that of a formula and its negation, one or the other must be a member of any given set which is H-alternative to another set. The two assumptions can be separated, although they are not considered separately with respect to the system **CYC1**.

5 QUANTIFICATION AND ENTAILMENT

With respect to the integration of quantifiers and modal operators, entailment is subject to the same general considerations as is monadic modal logic. The large number of variant systems of quantified modal logic that are considered in Chapter V can all be integrated with the operators 'Y' and 'Z'. We assume for the entailment systems that the Standard Interpretation of quantifiers is appropriate, and the cancellation step at V.6.5.1, which was omitted from the dyadic cancellation rules for the sake of simplifying cancellations, should be restored in the cancellation of any formulas involving more than the single quantifier "out front." As with monadic modality, the most appropriate quantificational variant for alethic systems of dyadic modality seems to be **Q1**. The systems **CY1Q1** and **CYC1Q1** will be those considered as "standard" systems of alethic dyadic modality.

6 TOWARD A STANDARD SYSTEM OF ENTAILMENT

Both the **CY** systems and the **CYC** systems provide articulations of the operators 'Y' and 'Z' that fulfill the conditions embodied in the DVT shown in §1.2.2 above. On no line of the DVT can we derive 'T' for 'Ypq' or 'F' for 'Zpq'. The operators in both systems are in this sense independent of the truth-value and modal status of component statements. The choice between the two systems must rest on the question of the paradox of consistency, 'YKp$\bar{p}$q'.

Do we or do we not want to avoid the entailment of every statement by a contradiction? If we go back to the locution of "supposing" as a prop to the intuitions, we might find some help. One can, and often does, "suppose" what is false. "What would happen if I hit the cue ball just so? Well suppose I did: it would knock the four ball into the seven ball, tipping the eight ball into the corner pocket. Better not." The use of counterfactuals may be interpreted as supposing what happens to be false. The lump of sugar on my table has never been immersed in a cup of hot coffee, but suppose it were: it would dissolve. One might even suppose what is theoretically impossible. Some physicists are currently considering what theoretically impossible "super-c" phenomena would be like: "supposing" that a body could travel faster than the speed of light (*c*) and considering what the consequences would be of motion at such a speed.

Can we also "suppose" what is logically impossible? More to the point, can we imagine that a blatant contradiction is true and consider what its genuine consequences would be? Probably not. The **CYC** systems do not seem to provide a formal model of "the way people think" in this sense. They are worth pursuing as a means of following through a well-articulated notion of "logical consequence" as given by Anderson and Belnap, or "relevance" as given in Binkley's Test, but they are probably somewhat extreme for analytic purposes. Given the way that formal relevance has been ignored by logicians until recently, it may be that extremism in the defense of relevance is no vice. But for most analytic purposes, the **CYC** systems are probably more trouble than they are worth. They are, perhaps, best relegated to the status of an interesting formal exercise. Whatever value lies in the **CYC** systems might lie in the fact that they can be constructed at all. The distinction between negation and denial is an interesting one, and worth pursuing. We will mention it in several places in the final chapter, as an aid to straightening out some confusions concerning existence and modality. It is comforting to know that a formal system can include both without being unsound. But Binkley's Test or no Binkley's Test, the more straightforward **CY** systems, in meeting the criterion of independence as laid out on the DVT, are preferable to the convoluted **CYC** systems.

In the following sections, then, we will understand "entailment" as referring to the articulation of 'Y' in the **CY** systems and not the **CYC** systems.

6.1 Entailment and Implication. A minimal requirement for the entailment of a statement 'q' by another, 'p', is that if the entailment holds, then it must be impossible for 'p' to be true and 'q' false.

$$\mathrm{Y\,Y\,p\,q\,\bar{M}\,K\,p\,\bar{q}}$$

must be a theorem of any acceptable system of dyadic modal logic. This formula is, of course, a theorem of **CY1**. It is equivalent to **DT-1**, 'YYpqLCpq'. The conjunction of 'p' and '$\bar{q}$' may be impossible for other reasons than the presence of a genuine entailment—for example, it is impossible if 'p' is impossible or if '$\bar{q}$' is—but in the presence of 'Ypq' it must be impossible.

Whereas material and strict implication can be trivially true—the cases where they produce a 'T' on the DVT—entailment is trivially true only when its antecedent is overtly contradictory, and in some cases involving nested entailments.

7 SPECIES OF ENTAILMENT

The entailment system **CY1** provides a reasonable articulation of the concept of *lawlikeness*: the connection can only be trivialized by taking an antecedent that is overtly contradictory. Lawlikeness comes in as many species as does alethic modality. The *logical laws* that result from the articulation of logical constants are reflected in the theorems of **CY1** already cited. There are also conceptual entailments and theoretical entailments, which depend upon an articulation of the non-logical constants in the vocabulary of a theory under analysis.

Systems of conceptual and theoretical entailment can be added to the system of logical entailment in precisely the same way that conceptual and theoretical modality were developed in Chapter VI. A set of predicate constants is articulated, and two new alternativeness relations are defined and ranked for strength. 'S_{log}', 'S_{con}' and 'S_{the}' relate to each other in precisely the same way that 'H_{log}', 'H_{con}' find 'H_{the}' were related at VI.2.2.3 and VI.2.3.1. The resultant systems of logical, conceptual and theoretical entailment have some interesting properties: all the desirable logical features of **CY1** are preserved in the three species, and in addition, the three sorts of "law" can be kept distinct. Such distinctions can be extremely important in the analysis of scientific theories.

In phylogenetic taxonomy, for example, there are several vocabularies that are interrelated, but not strictly by definition. There are names of taxa (species names, genus names, etc.), phenetic predicates (roughly, the observable properties of organisms) and phylogenetic predicates (roughly, the evolutionary properties and relations of taxa). In addition, there is a set of category names ("genus" , "family", "species", etc.) which are ordered in a systematic way so as to schematize the taxonomic hierarchy. For the evolutionary taxonomist, there are conceptual entailments among taxa names, phylogenetic predicates and phenetic predicates. A taxon name bears conceptual relations to the set of phylogenetic predicates whose presence or absence defines the taxon. But the criteria for membership in a taxon are phenetic, and although phenetic predicates cannot be used to define a taxon name, there are entailment relations between being a member of a certain taxon and having certain phenetic properties. Attempts to analyze these relations which employed elementary logic were failures, although they failed for interesting reasons. More recent attempts employing the more sophisticated logical machinery show promise of success.[6]

[6] The discussion is a fairly complicated one, but understanding it does not involve knowing a lot of biology. The argument generally takes place in the pages of the journal *Systematic Zoology*. See, in particular, [VII.11] and [VII.17].

In the analysis of a physical theory or of an explanation pattern within a physical theory, it is important that the formal resources employed in the analysis involve as few "inferential anomalies" as possible. Of the formal systems considered here, the system **CY1**, with 'Y' as the principal if-then operator, is the most reasonable analytic tool for such analysis.

8 ENTAILMENT AND COUNTERFACTUALS

In work in the philosophy of science, a good deal of attention has been given to the counterfactual conditional, "If p were the case, then q would be." The counterfactual is one of the few surviving subjunctive forms in contemporary English, and it functions often in reasoning based on what are taken to be lawlike connections. "If this lump of sugar were immersed in water, it would dissolve," carries a complex sense: first of all, that this bit of sugar has not been immersed in water, and secondly that if we did immerse it, it would dissolve: "Both not-p and if p, then q."

When the if-then connection in a counterfactual conditional is analyzed as material implication, a really paradoxical result occurs. '$K\bar{p}Cpq$' *is equivalent to simply* '$\bar{p}$'. It doesn't matter what 'q' is. So long as the bit of sugar in our example hasn't been immersed in water, we might as well say that if it were immersed, then it would prove Fermat's last theorem.

There are also "counterpossible" conditionals that figure in significant reasoning. One familiar example of this is Newton's First Law, in one of its more elementary formulations: "A body subject to no external forces maintains constant velocity." Since the rest of Newton's Laws entail that there are no such bodies, we end up with "It is impossible that there be bodies subject to no external forces, but if there should be bodies subject to no external forces they would maintain constant velocity." Analyzed in monadic modal logic, using strict implication for the "If—, then—", this produces a result similar to the result above for material implication. '$K\bar{M}pLCpq$' *is equivalent to simply* '$\bar{M}p$'. Again, it doesn't matter what 'q' is on the strict implication analysis. So long as it is impossible that there be bodies subject to no external forces, we can say "A body subject to no external forces maintains a velocity that varies inversely with the level of the tide at Jones Beach," if strict implication is the best we can do for "If, —then—".

Using the entailment relation as articulated in **CY1**, however, the counterfactual can be shown as '$K\bar{p}Ypq$' and the "counterpossible" as '$K\bar{M}pYpq$' without 'q' being merely a trivial appendage.

9 DEONTIC ENTAILMENT

In Chapter VI we took note of an unfortunate result of treating strict implication on the deontic interpretation. If a given act is morally forbidden, then it "morally implies" doing everything. '$C\bar{M}pLCpq$' is just the antecedent paradox of strict implication. The formula, in this mode as in the others, is paradoxical only when read in such a way as to suggest entailment; the equivalent formulas '$C\bar{M}pLA\bar{p}q$' and '$C\bar{M}p\bar{M}Kp\bar{q}$' are reasonable enough—if 'p' is forbidden, then it is forbidden in combination with anything else.

The entailment operator can be adapted to capture the sense of "moral commitment" or "moral obligation," if it is desirable to consider the way in which one act or state of affairs commits us to another. The adaptation should come as no surprise: reflexivity has to be dropped, as it was dropped for the system of deontic logic in VI. In **CY1** with reflexivity, 'Cpq' is entailed by 'Ypq' (**DT-2**). If this occurrence of 'Y' is interpreted as moral commitment, we have "p morally commits us to q" entailing "If p, then q." If reflexivity is dropped semantically, and the corresponding closure condition **CY1** I is dropped, then **DT-2** is no longer a theorem, nor is **DT-4**, which has 'Zpq' entailed by 'Kpq'. "Both p and q" should not entail "p and q are morally compatible." If Schulz both makes a promise (p) and breaks it (q), it does not follow that promise-making and promise-breaking are morally compatible.

The remaining theorems of **CY1**, which do not depend on the closure condition dropped, are acceptable on the "moral commitment" and "moral compatibility" interpretation of 'Y' and 'Z'. The "paradoxical" **DT-5** does no more harm on the moral intepretation than it does on the alethic theoretical interpretation. "If a contradiction happens, then you are obligated to do everything." Society won't crumble on that principle.

BIBLIOGRAPHY—CHAPTER VII

[VII.1] Anderson, Alan Ross. "Modalities in Ackermann's 'Rigorous Implication'," *Journal of Symbolic Logic* **24** (1959), pp. 107–111.

[VII.2] ———. "Some Open Problems Concerning the System *E* of Entailment," *Acta Philosophica Fennica* **16** (1963), pp. 7–18.

[VII.3] Anderson, Alan Ross, and Nuel D. Belnap, Jr. "A Simple Treatment of Truth Functions," *Journal of Symbolic Logic* **24** (1959), pp. 301–302.

[VII.4] ———. "Enthymemes," *Journal of Philosophy* **58** (1961), pp. 713–723.

[VII.5] ———. "The Pure Calculus of Entailment," *Journal of Symbolic Logic* **27** (1962), pp. 19–52.

[VII.6] ———. "Tautological Entailments," *Philosophical Studies* **13** (1962), pp. 9–24.

[VII.7] Angell, R. B. "A Propositional Logic with Subjunctive Conditionals," *Journal of Symbolic Logic* **27** (1962), pp. 327–343.

[VII.8] Belnap, Nuel D., Jr. "Entailment and Relevance," *Journal of Symbolic Logic* **25** (1960), pp. 144–146.

[VII.9] ———. "Pure Rigorous Implication as a 'Sequenzen-Kalkul'," *Journal of Symbolic Logic* **24** (1959), p. 282.

[VII.10] Bronstein, Daniel J. "The Meaning of Implication," *Mind* **45** (1936), pp. 157–180.

[VII.11] Buck, Roger, and David Hull. "The Logical Structure of the Linnaean Hierarchy," *Systematic Zoology* **15** (1966), pp. 97–111.

[VII.12] Curme, George O. *Syntax*. (D. C. Heath, Boston, Mass., 1931)

[VII.13] Emch, Arnold F. "Implication and Deducibility," *Journal of Symbolic Logic* **1** (1936), pp. 26–35. (Addendum, p. 58, same volume).

[VII.14] Geach, Peter. "Necessary Propositions and Entailment Statements," *Mind* **57** (1948), pp. 491–493.

[VII.15] Hampshire, Stuart N. "Mr Strawson on Necessary Propositions and Entailment Statements," *Mind* **57** (1948), pp. 354–357.

[VII.16] Hockney, D., and K. Wilson. "In Defense of a Relevance Condition," *Logique et Analyse* **8** (1965), pp. 211–220.

[VII.17] Hull, David L., and D. Paul Snyder. "Contemporary Logic and Evolutionary Taxonomy," *Systematic Zoology* **18** (1969), pp. 57–63.

[VII.18] Körner, Stefan. "On Entailment," *Aristotelian Society Proceedings* **47** (1946–47), pp. 143–162.

[VII.19] ———. "Entailment and the Meaning of Words," *Analysis*. (1949)

[VII.20] Kripke, Saul. "The Problem of Entailment," *Journal of Symbolic Logic* **24** (1959), p. 324.

[VII.21] Lewis, Clarence, I. "Emch's Calculus and Strict Implication," *Journal of Symbolic Logic* **1** (1936), p. 77.

[VII.22] Nelson, Everett J. "Intensional Relations, " *Mind* **39** (1930), pp. 440–453.

[VII.23] Pap, Arthur. "Strict Implication, Entailment, and Modal Iteration," *Philosophical Reivew* **64** (1955), pp. 604–613.

[VII.24] Prior, Arthur N. "Facts, Propositions, and Entailment," *Mind* **57** (1948), pp. 62–68.

[VII.25] Putnam, Hilary. Review of von Wright's Logical Studies, *Philosophical Review* **72** (1963), pp. 242–249.

[VII.26] Rescher, Nicholas. "A Contribution to Modal Logic," *Review of Metaphysics* **12** (1958), pp. 186–199.

[VII.27] ———. "A Factual Analysis of Counterfactual Conditionals," *Philosophical Studies* **11** (1960), pp. 49–54.

[VII.28] ———. *Hypothetical Reasoning*. (North Holland, Amsterdam, 1964)

[VII.29] Robinson, Richard. "Partial Entailment and the Causal Relation," *Mind* **70** (1961), pp. 526–533.

[VII.30] Sellars, Wilfrid. "Inference and Meaning," *Mind* **62** (1953), pp. 313–338.

[VII.31] Smiley, T. J. "Entailment and Deducibility," *Proceedings of the Aristotelian Society* **59** (1959), pp. 233–254.

[VII.32] ———. "Relative Necessity," *Journal of Symbolic Logic* **28** (1963), p. 113 ff.

[VII.33] Snyder, Donald Paul. *Causation, Predication and Modal Logic*. (University Microfilms, Ann Arbor, 1964) (Dissertation: Duke University)

[VII.34] ———. "Models for Logical Entailment," *Logique et Analyse* **9** (1966), pp. 344–359.

[VII.35] Strawson, Peter. "Necessary Propositions and Entailment Statements," *Mind* **57** (1948), pp. 184–200.

[VII.36] ———. *Introduction to Logical Theory*. (Methuen, London, 1952)

[VII.37] ———. Review of von Wright, *Logical Studies*, *Philosophical Quarterly* **8** (1958), pp. 372–376.

[VII.38] von Wright, G. H. "A New System of Modal Logic," in *Logical Studies*. (Routledge and Kegan Paul, London, 1957), pp. 89–126.

[VII.39] ———. "The Concept of Entailment," in *Logical Studies*. (Routledge and Kegan Paul, London, 1957), pp. 166–191.

[VII.40] ———. "A Note on Entailment," *Philosophical Quarterly* **9** (1959), pp. 363–365.

[VII.41] Woods, John. "Relevance," *Logique et Analyse* **7** (1964), pp. 130–137.

[VII.42] ———. "Relevance Revisited," *Logique et Analyse* **9** (1966), pp. 364–371.

Chapter VIII

MODALITY, REFERENCE AND IDENTITY

In Chapter V we made provision for dealing with initial well-formed formulas containing individual constants, but we did not deal with such formulas. We have left the discussion of individual reference until last because there is a cluster of problems having to do with reference, identity and modality. Some of these problems have relatively straightforward solutions. Others do not, and will demand the use of all the interpretative tools at our disposal.

It is worth noting at the outset that all theorems of the formal systems presented in the preceding chapters must be *initial* well-formed formulas of those systems. For the systems of quantified modal logic, initial well-formed formulas may contain individual constants, and they may contain individual variables bound by quantifiers. Initial well-formed formulas contain no variables that are not bound by quantifiers (cf. V.1.3.1).

1 QUANTIFIED MODALITY AND SINGULAR EXISTENCE

For the following considerations, cancellation proofs will be carried out in the formal system **CMQ1** under the Standard Interpretation, as described at V.6.5. With respect to the formulas to be considered in this section, there will be no difference among **CMQ1**, **CS4Q1** and **CS5Q1**. All the formulas considered here which are provable as material implications in **CMQ1** are also provable as entailments in **CY1Q1**.

1.1 Symbolizing with Proper Names. In symbolizing natural language statements containing proper names, we represent those names as individual constants. Any constant may be chosen for any proper name, so long as the choice is adhered to throughout a given formula. If the formula is to be cancelled, any constants introduced in the cancellation must be numerically later than the highest numbered individual constant already in the formula. It is advisable, then, to start choosing individual constants with '1' in symbolizing.

It is worth recalling at this point that the **Q1** systems allow model sets alternative to the fixed model set to have populations that differ from that of the fixed model set: individuals may *belong* to an alternative to the fixed model set without belonging to the fixed set itself. We must then allow ourselves to use individual constants to name "possible individuals" as well as real individuals, although some care is needed. Fictional individuals can be discussed, but not as if they existed. Fictional individuals do, however, provide a reasonable way of viewing "possible individuals." "Sherlock Holmes was a detective" is simply false. There was never any such person. "It is possible that Sherlock Holmes was a detective," however, is true. Conan Doyle did his job well, and Mr. Holmes is, among other things, a fine possible individual.

Following are some simple symbolizations involving individual constants, using this vocabulary:

Fx—x is a freshman.	
Gx—x is a girl.	1—Susie
Jxy—x is younger than y.	2—Charley

Susie is a freshman girl.	K F1 G1
Susie is a girl who is younger than Charley.	K G1 J12
Some girl is younger than Charley.	∃x K Gx Jx2
Susie is a girl who is younger than some freshman.	K G1 ∃x K Fx J1x
Susie is a girl who is possibly younger than Charley.	K G1 M J12
It is possible that someone is younger than Susie.	M ∃x Jx1
It is possible that some freshman is younger than Susie.	M ∃x K Fx Jx1
There is a freshman who is possibly younger than Susie.	∃x K Fx M Jx1

1.2 Cancellation with Proper Names. Cancellation of formulas containing individual constants, even with modal operators on the scene, can be approached without adding any new procedures. An additional closure condition will be added shortly. The cancellation rules for quantifiers require that we choose the first unused constant in the stock of constants in cancelling '∀' or '$\bar{\exists}$'. Constants which appear in the original iwff are used, so in cancelling '∀x C K Fx Gx Jx2', for example, the first constant that can be introduced is '3'.

In order to begin the treatment of individual constants with a minimum of complication, we *temporarily* extend the cancellation step at V.6.5.1 to individual constants already in an iwff, as well as to individual dummies and constants introduced by the cancellation of quantifiers. *For now, all individual dummies and constants, wherever they come from, are repeated under their original occurrence, in the bottom position of the index matrix.* This extension of the cancellation rule is not justified by the interpretation of bsfs, and it will be withdrawn in a later section. This gives us a way of getting constants from an iwff into the index matrix, to open the discussion.

Formulas which contain individual constants, but neither quantifiers nor modal operators, are simply formulas of statement logic. They are provable just in case the corresponding formulas of statement logic are provable. 'C K F1 G2 F1', for example, will cancel to closed branches, and might as well be written 'C K p q p'. Formulas which contain individual constants and modal operators, but no quantifiers, will be provable or not, according to whether or not the corresponding formulas of the modal systems of Chapter III are provable. 'C K F1 G1 M K F1 G1' cancels to closed branches, as does 'C K p q M K p q'. The first case where the presence of individual constants makes a difference comes when we have individual constants and quantifiers present in the same formula, but no modal operators. The following should, when symbolized, be theorems of standard elementary logic.

If Susie is a freshman, then someone is.	C F1 ∃x Fx
If everyone is a freshman, then Susie is.	C ∀x Fx F1

Both formulas cancel to closed branches under the rules of **CMQ1** (see Figure 1).

C F1 ∃x Fx				C ∀x Fx F1				
a				a				
Ȼ F̄1 ∄x Fx				Ȼ ∀̸x F̄x F1				
01 0a				0a 01				
F̄ , a, F		$\begin{bmatrix} a = 1 \\ \text{closed} \end{bmatrix}$		a, F̄ , F		$\begin{bmatrix} a = 1 \\ \text{closed} \end{bmatrix}$		
01 0a				0a 01				

FIGURE 1

Formulas containing individual constants, quantifiers and modal operators, *where the individual constants do not occur under modal operators*, are also straightforward. All of the following are intuitively correct, and when symbolized (see Figure 2), are provable in **CMQ1**.

If Susie is a freshman, then it is possible that someone is a freshman.	C F1 M ∃x Fx
If Susie is a freshman, then there is someone who is possibly a freshman.	C F1 ∃x M Fx
If everyone is necessarily a freshman, then Susie is a freshman.	C ∀x L Fx F1
If it is necessary that everyone be a freshman, then Susie is a freshman.	C L ∀x Fx F1

```
C F1 M ∃x Fx        C F1 ∃x M Fx        C ∀x L Fx F1        C L ∀x Fx F1

    a    b               a  b               a b                a b
Ȼ F̄1 M̸ ∄x Fx        Ȼ F̄1 ∄x M̸ Fx        Ȼ ∀̸x L̸ F̄x F1        Ȼ L̸ ∀̸x F̄x F1
  01       0          01        0a                0a 01              0  01
           ab                   ba             ba                 ab

  F̄ ,a,   b,F̄         F̄ ,  a,b,  F          a,b,F̄ ,F          a, b,  F̄ ,F
  01       0          01        0a                0a 01              0  01
  01       ab         01        ba                ba 01              ab 01
 ┌CM I  ┐           ┌CM I  ┐           ┌CM I  ┐           ┌CM I  ┐
 │a = 0 │           │a = 1 │           │a = 1 │           │a = 0 │
 │b = 1 │           │b = 0 │           │b = 0 │           │b = 1 │
 └closed┘           └closed┘           └closed┘           └closed┘
```

FIGURE 2

Where individual constants appear in formulas falling under modal operators, as in 'M F1', the situation is not so straightforward. The difference between "It is possible that Susie is a freshman" and "It is possible that Sherlock Holmes lived at 221b Baker Street" lies not in logical form but in the tacit assumption that Susie is a real girl, while Holmes is a fictional detective. The two statements will be symbolized in the same way, and any inference that depends upon the assumption that Susie is something more than a "possible individual" will fail *unless we make the tacit assumption explicit.*

Without any assumption that Susie exists, the following is correct:

If it is possible that Susie is a freshman,
then it is possible that someone is a freshman. C M F1 M ∃x Fx

```
C M F1 M ∃x Fx

   2    a  b
Ȼ M̸ F̄1 M̸ ∄x Fx
    0        0      ┌a = 2 ┐
    21       ab     │b = 1 │
                    └closed┘
```

Neither 'M F1' nor 'M ∃x Fx' commits us to the "real world" existence of anything (the position of individual dummies and constants in the cancelled formula makes this clear). '1' could represent a purely "possible individual" so far as either of these formulas is concerned. The following, however, is correct only given the assumption that Susie exists.

If it is possible that Susie is a freshman, then there is someone who is possibly a freshman. $C\,M\,F1\,\exists x\,M\,Fx$

$$\begin{array}{cccccc} C & M & F1 & \exists x & M & Fx \\ & 2 & & a & b & \\ \not{C} & \not{\bar{M}} & \bar{F}1 & \not\exists x & \not{M} & Fx \\ & & 0 & & & 0a \\ & & 21 & & & ba \end{array} \qquad \text{[no closure]}$$

The occurrence of ‘a’ next to the “real world” constant has no replacement.

The problem is not as dramatic as we are making it appear. If the natural language statement above proved ambiguous, we might clarify it by rephrasing the antecedent from “It is possible that Susie is a freshman” to “Susie is a girl who is possibly a freshman,” to emphasize that Susie really *is* something, and is also *possibly* a freshman. With this new element in the antecedent, the formula should be provable; we will shortly introduce a closure condition to make it so. The formula cancellation is given in Figure 3. We have in this branch chart a situation that could not

$$\begin{array}{cccccccc} C & K & G1 & M & F1 & \exists x & M & Fx \\ & & & 2 & & a & b & \\ \not{C} & \not{\bar{K}} & \bar{G}1 & \not{\bar{M}} & \bar{F}1 & \not\exists x & \not{M} & Fx \\ & & 01 & & 0 & & & 0a \\ & & & & 21 & & & ba \\ & & \bar{G}, & & \bar{F}, & a, b, & & F \\ & & 01 & & 0 & & & 0a \\ & & & & 21 & & & ba \end{array} \qquad \text{[no closure]}$$

FIGURE 3

arise in a system where individual constants do not appear in iwffs. The individual constant ‘1’ appears in different places in the matrices of ‘$\bar{G}$’ and ‘$\bar{F}$’. Such a situation cannot occur when an individual constant is brought into the branch chart by the cancellation of a quantifier, because at the point where a quantifier is cancelled, all the basic formulas within its scope have precisely the same modal indices, and the variable of that quantifier will therefore be placed in the same position in each index in which it occurs. To meet this new situation, we must introduce a new closure condition, and it is reasonable enough: the element ‘$\bar{G}_{01}$’ in the branch chart in Figure 3 “places” the individual named by ‘1’ in the model set (world description) named by '0', and we should be allowed to write '1' next to '0' in the matrix of '$\bar{F}$' also, thereby closing the branch. We state the new closure condition in a general way, even though the rules assure that it can only apply in cases where an individual constant occurred in the original iwff. Examples are given in Figure 4.

```
F ,G,J
01 0  0
   a1 a12
   b1
```

'1' may be written next to '0' in the matrices of 'G' and 'J'

```
F ,G,J
01 01 01
   a1 a12
   b1
```

```
F ,G,J
0  01 0
a1    b23
```

'1' may be written next to '0' in the matrix of 'F' (but not of 'J')

```
F ,G,J
01 01 0
a1    b23
```

FIGURE 4

1.2.1 When an individual constant appears in the index matrix of a basic subformula on the same horizontal line with a modal dummy or constant, it may be written next to that modal dummy or constant *in any basic subformula of the same branch where it has an occurrence*, in the same vertical line as that original occurrence.

To return to the cancellation of "Susie is a girl who is possibly a freshman," we can now apply the new closure rule and show the formula to be provable. The branch chart, when last seen, appeared thus:

```
Ḡ, F̄ , a, b, F
01 0        0a
   21       ba
```

The closure condition at 1.2.1 allows the following change:

```
Ḡ, F̄ , a, b, F
01 01       0a
   21       ba
  ⎡a = 1 ⎤
  ⎢b = 2 ⎥
  ⎣closed⎦
```

The new closure condition requires no new semantic conditions on the model system for **CMQ1** and no new antecedent assumptions in QA. It is justified by the translation rules given in Chapter V, even though the formulas treated in Chapter V were of such a nature that the new condition could not have been applicable to them.

The inference from "It is possible that so-and-so is an F" to "There is something that is possibly an F" is not, in general, correct. The inference is valid *only when* the "so-and-so" may be assumed to exist. "Sherlock Holmes exists and possibly lives at 221b Baker Street" implies "There is someone who possibly lives at 221b Baker Street." The existence assumption could as well be supplied by the introduction of the existence predicate 'Q' discussed in Chapter V, but for the present we will not have to resort to direct existence assertions. *It is sufficient to establish the existence of an individual that something can be predicated of it outside a modal operator.* The new closure condition makes this explicit.

The two remaining formulas to be considered are equivalent to the last two considered, and will produce corresponding results.

If everything is necessarily a freshman, then Susie is necessarily a freshman.	C∀xLFxLF1
If it is necessary that everything is a freshman, then it is necessary that Susie is a freshman.	CL∀xFxLF1

The first formula is provable only if we can be sure that '1' names an existing individual; the second is provable without any further assumption.

In Chapter V we mentioned the ancient distinction between modality *de dicto* and modality *de re* as it was reflected in the formulas 'M∃xFx' and '∃xMFx' respectively. What is at stake in the question of existential commitment connected with formulas like 'MF1' is precisely this distinction. Modality *de dicto* is modality as attached to statements, as in "*It is possible that* Susie is a freshman." Modality *de dicto* does not in itself commit one to the existence of mentioned individuals. Modality *de re*, on the other hand, has to do with the "manner of predication"—the way in which an individual does or does not have a property. Modality *de re* is generally phrased in such a way that the individual is named *outside* the modal qualifier, making it clear that we are talking about an existing individual, as in "Susie is possibly a freshman." In the usual notation of modern logic, it is impossible to make the distinction between modality *de dicto* and modality *de re* when a proper name is involved. The ploy of establishing existence by asserting an actual property of the individual concerned is only partially satisfactory, and in the following section a modification of the formal system will be suggested to make the distinction in a more satisfactory way.

1.3 Fictional Individuals. We claimed above that the statement "Sherlock Holmes was a detective" is simply false. It is false, but not *simply* so. The sense in which it is false may be compared with the sense in which "John XXIII was a Catholic" is true. John XXIII was a real individual; Sherlock Holmes a fictional one. If we compare "Sherlock Holmes was a detective" with "David Copperfield was a detective," however, the Holmes statement might be said to be true and the Copperfield statement false. The difference is not especially subtle: in the one case we are speaking of Holmes and John XXIII "in the same breath," as if they were both real people; in the other, we are clearly speaking of two fictional individuals, and referring to the roles they played in their respective works of fiction. In the latter case, there is a tacit "suspension" of the usual mode of referring to individuals. The tacit suspension can be made explicit.

At the end of Chapter V, it was suggested that writing a work of fiction is rather like constructing a model set. The more structural features that model set has in common with the set of sentences which describe the real world, the more realistic the fiction. Science fiction in the 1930's often went beyond what was theoretically possible, and is sometimes called "fantasy" for this reason—the "model set" of the story obeyed logical and conceptual conditions but not theoretical ones. More recent science fiction, and most ordinary fiction, is more realistic: what is claimed is logically, conceptually and theoretically possible. The more *basic sentences* (contingent statements) a fictional "model set" has in common with the set of sentences describing the real world, the more plausible the fiction. There are, both in the Holmes stories and in the real world, a Scotland Yard, a 221b Baker Street, moors in Scotland, and so on. The populations of the two model sets—the individual names which "belong" to them—are different. Holmes does not appear in the fixed model set, and whoever really lived at the Baker Street house does not appear in the set of sentences describing Holmes' world. But the populations are not totally different: Queen Victoria, at least, belongs to both worlds.

The point of the comparison is simply this: the tacit "suspension" of existence presuppositions in making statements about Sherlock Holmes and David Copperfield is very much like the sort of suspension involved in making statements under a possibility operator. To say that (fictionally) Holmes was a detective may be interpreted as saying that there is a set of statements which is an alternative to the set of statements describing the world, in which "Holmes was a detective" appears. A "fiction" operator has no place in the general scheme of modality, but "It is fictional that—" may be construed as entailing "It is theoretically possible that—." Statements about possible individuals do not allow us to derive existentially quantified statements, as we saw above: from "It is theoretically possible that Holmes was a detective," we cannot infer "There is something such that it is theoretically possible that it is a detective."

The **Q1** systems under the Standard Interpretation are more or less satisfactory with respect to their treatment of fictional individuals. We can never justify reasoning from a true statement about Sherlock Holmes to a false statement. 'M F1' does not, for example, imply or entail '∃x M Fx'. But there is a little fancy footwork involved in justifying 'C ∀x Fx F1' if '1' is allowed to represent a fictional individual. One line of argument that has been taken with respect to such situations is this (we state it without subscribing to it): '∀x Fx' does indeed imply 'F1', as the provable material implication shows (and strictly implies it and entails it, as the appropriate tests would show); BUT (watch closely now) even if '1' names a merely possible individual, this will never justify arguing from a true statement to a false one, because there are no true statements of the form '∀x Fx' where 'Fx' is atomic. If 'F1' is something other than an atomic formula, say 'C G1 J1', then it involves no commitment to the existence of anything, since existence (like belonging to a model set) follows only from being mentioned in a true atomic formula. On certain linguistic views of what atomic statements are, this argument might have some plausibility. Having avoided taking any position on such matters thus far, however, it is a little late in the game to adopt such a view here.

Instead, we offer two alternative suggestions: a restriction with respect to symbolizing formulas containing individual constants and, if this seems odious, a different way of interpreting formulas containing individual constants. The first suggestion reflects a common practice in elementary logic. The second suggestion will be adopted here.

1.3.1 An Ad Hoc Restriction. In symbolizing statements involving proper names, choose a constant for each proper name involved, as before. Constants may be chosen to represent fictional individuals *provided that* in the symbolization such a constant is used only within the scope of a modal operator. The restriction is not quite as *ad hoc* as it sounds. Consider the following dialogue: "If everything is a non-dragon, then Sherlock Holmes is a non-dragon." "Wait a minute, you can't say that; Sherlock Holmes isn't anything at all, not even a non-dragon." There is at least a twinge of common-sense objection to even considering statements like "Sherlock Holmes is a non-dragon," and the restriction reflects this twinge.

1.3.2 An Interpretation of Unquantified Well-formed Formulas. The restriction just suggested is not very satisfactory. It puts the problem outside the formal system, when it belongs inside. Let's take stock of the formal situation. The formulas that are bothersome are 'C ∀x Fx F1' and

'CF1∃xFx', in cases where '1' does not represent the name of an existing individual. If the '1' were picked up by a quantifier in front of the whole formula there would be no problem: '∀yC∀xFxFy' doesn't involve any non-existent individual. It is interpreted in QA as '**∀yCBy0**(C∀xFxFy)**0**'. Now consider the interpretation of 'C∀xFxF1', moving right to the crucial bit, the 'F1'. The first part of the formula will be interpreted with negations confined, but there will be no negation on '(F1)**0**' when we get to it in the QA translation. It comes across as '**CB10F10**', which doesn't place '1' in the fixed model set at all. The conditional form in QA that is essential to the proper interpretation of atomic subformulas containing individual *variables* makes atomic formulas containing individual *constants* misleading. There is, of course, a way out. We will interpret occurrences of constants differently from the way in which we interpret occurrences of variables of quantification. Subformulas like 'Jx2' will have to work out properly too, of course, and they will, if the two translation rules are applied in order.

Basic subformulas are left as before when they contain individual variables. Where '**x**' is an individual variable,

'(Ψx)**μ**' is translated to '**CBxμΨxμ**';
'(Ψ̄x)**μ**' is translated to '**CBxμΨ̄xμ**'.

These are the same conditions given for the Standard (Fifth) Interpretation at V.6.2.4. But bsfs containing individual *constants* are interpreted as follows; where '**i**' is an individual constant,

'(Ψi)**μ**' is translated to '**Ψiμ**';
'(Ψ̄i)**μ**' is translated to '**Ψ̄iμ**'.

The cancellation rule at V.6.5.1 is no longer extended to individual constants that are not introduced by quantifiers. As stated, the rule applies only to dummies and constants that are in the index matrix. The extension of the rule in §1.2 above to constants that are not in the index matrix is now withdrawn. Neither of the following cancellations can be closed without the extension of the rule:

CF1∃xFx	C∀xFxF1
a	a
C̸F̄1∄xFx	C̸∀̸̄xF̄xF1
0 0a	0a 0

This time we will be more careful about bringing constants from the original iwff into the index matrix. First, however, we should reassure ourselves that the following two cancellations still close:

$$\begin{array}{ccc} \begin{array}{l} \forall y\ C\ Fy\ \exists x\ Fx \\ \ 1 \qquad\quad\ \ a \\ \not\forall y\ \not C\ \bar{F}y\ \not\exists x\ Fx \\ \qquad\ \ 01 \qquad 0a \end{array} & \qquad & \begin{array}{l} \forall y\ C\ \forall x\ Fx\ Fy \\ \ 1 \qquad a \\ \not\forall y\ \not C\ \bar{\not\forall} x\ \bar{F}x\ Fy \\ \qquad\qquad\quad\ 0a\ 01 \end{array} \\ \begin{bmatrix} a = 1 \\ \text{closed} \end{bmatrix} & & \begin{bmatrix} a = 1 \\ \text{closed} \end{bmatrix} \end{array}$$

The quantifier "out front" places the occurrence of '1' in the index matrix.

There are two ways of assuring ourselves that the unclosed cancellations above are closed only when '1' represents an existing individual. We can introduce a "quasi-quantifier" or revive the existence predicate 'Q' that was considered briefly in Chapter V. Whichever we choose, the QA reflection will have to be the same, and 'Q' is a little easier to handle in the object language until we arrive at a final formulation of existence.

Formally, then, we introduce the new constant 'Q' as a predicate (or "quasi-predicate") of degree one, with the following formation rule.

'Q' followed by an individual constant is an iwff.

We add the following translation rules for the new constant, so that it can be treated in QA.

'$(Qi)\boldsymbol{\mu}$' is translated to '$\mathbf{Bi}\boldsymbol{\mu}$'.
'$(\bar{Q}i)\boldsymbol{\mu}$' is translated to '$\bar{\mathbf{B}}\mathbf{i}\boldsymbol{\mu}$'.

The new symbol is treated like any predicate in cancellation, and acquires a modal index in the process. It is not a predicate in the usual sense, of course, as its QA translation shows: it is interpreted solely on the hardware of QA, which is not the case with ordinary predicate constants. 'Q' gets special treatment, as befits its special interpretation. The cancellation rule at V.6.5.1 *does* apply to '$\bar{Q}$': any constant written next to '$\bar{Q}$' in an iwff is written next to the bottom entry in the modal index column of '$\bar{Q}$' as part of the cancellation procedure. The closure rule at 1.2.1 may be applied as before.[1]

[1] The variety of manipulations of the cancellation procedure make it appear that one can do almost anything he likes. This is not quite the case, and the reader may reassure himself by translating '$C\forall x\,F\,x\,F1$' into QA as would have been done prior to this section (with negations confined, of course) and translating '$C\forall x\,Fx\,C\,Q1\,F1$' into QA according to the present rules. The translations are precisely the same. The 'Q' predicate is Hintikka's, and like his it can be eliminated from the system once identity is introduced. See [VIII.5].

With the addition of 'Q' to the object language vocabulary, it is possible to make a neat distinction between "If everything is an F, then Sherlock Holmes is an F"—'C∀xFxF1'—which is not provable, and "If everything is an F, then if Sherlock Holmes exists, Sherlock Holmes is an F"—'C∀xFxCQ1F1'—which is provable. The corresponding distinction for the existential quantifier is between 'CF1∃xFx' and 'CKQ1F1∃xFx'. The cancellations are as follows.

$$\begin{array}{lcl} \mathrm{C\,K\,Q1\,F1\,\exists x\,Fx} & \qquad & \mathrm{C\,\forall x\,Fx\,C\,Q1\,F1} \\ \qquad\qquad\quad \mathrm{a} & & \quad\mathrm{a} \\ \not{\mathrm{C}}\,\not{\mathrm{K}}\,\bar{\mathrm{Q}}1\,\bar{\mathrm{F}}1\,\not{\exists}\mathrm{x\,Fx} & & \not{\mathrm{C}}\not{\forall}\mathrm{x}\,\bar{\mathrm{F}}\mathrm{x}\,\not{\mathrm{C}}\,\bar{\mathrm{Q}}1\,\mathrm{F1} \\ \quad 01\ \ 0 \qquad 0\mathrm{a} & & \qquad\quad 0\mathrm{a}\quad 01\ \ 0 \end{array}$$

The closure rule at 1.2.1 above allows us to place '1' in the index matrices under its original occurrences:

$$\begin{array}{lcl} \bar{\mathrm{Q}},\ \bar{\mathrm{F}}\ ,\ \mathrm{a},\ \mathrm{F} & \qquad & \mathrm{a},\ \bar{\mathrm{F}}\ ,\ \bar{\mathrm{Q}},\ \mathrm{F} \\ 01\ \ 01 \quad\ 0\mathrm{a} & & 0\mathrm{a}\ \ 01\ \ 01 \end{array}$$

$$\begin{bmatrix} \mathrm{a} = 1 \\ \text{closed} \end{bmatrix} \qquad \begin{bmatrix} \mathrm{a} = 1 \\ \text{closed} \end{bmatrix}$$

The examples of modality *de dicto* and *de re* that were considered at the end of §1.2 can now be distinguished formally using 'Q':

It is possible that Susie is a freshman.	MKQ1F1
Susie is possibly a freshman.	KQ1MF1

Exercise

Using the translation rules for bsfs containing individual constants, which are given above, and the Standard Interpretation translation rules for bsfs containing individual variables (which are repeated above), translate the following wffs into QA and cancel.

CF1∃xFx C∀xFxF1 ∀yCFy∃xFx ∀yC∀xFxFy

Then, using the translation rule for 'Q' given above, translate the following into QA and cancel.

CKQ1F1∃xFx C∀xFxCQ1F1

2 IDENTITY

In the use of "miniature universes" to generate examples of quantified formulas and to interpret the various species of alethic modality, we used a predicate constant 'I' to represent "same individual." "x is the same individual as y," written 'Ixy', was a necessary ingredient of the vocabulary if we were to be able to say, for example, "There are two dots"—'∃x ∃y K K Dx Dy $\bar{\mathrm{I}}$xy'. The incorporation of the identity predicate into quantified modal logic raises problems. It is now time to turn to these problems and lay them out, if not solve them.

First, we introduce identity to the system of elementary logic described in Chapter III. No new formation rule is required, since 'I' is treated like a predicate constant of degree two, and 'I' followed by two individual variables or constants constitutes an iwff in the same way that any predicate of degree two does. What must be added to elementary logic are two new closure conditions for identity. Examples are given below in Figure 5.

Id. I. *Where* 'i' *is an individual constant, a branch is closed if it contains a formula of the form* 'Iii'.

Id. II. *Where* 'i' *and* 'j' *are individual constants, and a branch contains a formula of the form* '$\bar{\mathrm{I}}$ij', *the branch may be expanded by adding, for each bsf in the branch of the form* 'Ψi', *a bsf of the form* 'Ψj' [*and for each bsf in the branch of the form* 'Ψj' *a bsf of the form* 'Ψi'].

∀x Ixx	∀x ∀y C K Fx Ixy Fy
1	1 2
∀̸x Ixx	∀̸x ∀̸y C̸ K̸̄ $\bar{\mathrm{F}}$x $\bar{\mathrm{I}}$xy Fy
11 [Id. I	1 12 2 [Id. II
I11 closed]	$\bar{\mathrm{F}}$1, $\bar{\mathrm{I}}$12, F2, $\bar{\mathrm{F}}$2, $\bar{\mathrm{I}}$21, F1 closed]

FIGURE 5

The closure condition Id. II can also take another form, which is somewhat easier with pencil and paper than with type. Instead of expanding the branch, simply strike occurrences of the one constant and replace them with occurrences of the other *ad libitum*.

Two additional theorems for identity, which cancel easily to closed branches using Id. II, are '∀x ∀y ∀z C K Ixy Iyz Ixz' (identity is transitive) and '∀x Ay C Ixy Iyx' (identity is symmetrical).

Exercise

Cancel the two formulas in the last paragraph of §2. Show the replacements needed for closure, and cite the appropriate Identity closure conditions.

2.1 Identity and Modality. In giving the special closure rules for identity in systems of modal logic, we have to take into account the position of individual dummies and constants with respect to modal dummies and constants. For Id. II, this is straightforward enough. We need only assure ourselves that the interchange of individual constants take place *wherever the identity holds*, and that it not take place where the identity does not hold. We require, then, that the modal index columns of 'Iij', 'Ψi' and 'Ψj' be the same, and that the interchange take place only on the line of the index matrix where 'i' and 'j' occur under 'I'. Before laying down the closure condition formally, let us take a look at the examples given in Figure 6. In the first example, the antecedent 'L Ixy'

∀x ∀y C L Ixy C L Fx L Fy	∀x ∀y C Ixy C L Fx L Fy
1 2 a b 3	1 2 a 3
$\not\forall$x $\not\forall$y $\not C$ $\bar{\not L}$ $\bar{I}$xy $\not C$ $\bar{\not L}$ $\bar{F}$x $\not L$ Fy	$\not\forall$x $\not\forall$y $\not C$ $\bar{I}$xy $\not C$ $\bar{\not L}$ $\bar{F}$x $\not L$ Fy
012 01 02	012 01 02
a12 b1 32	a1 32
[a = b = 3]	[a = 3]

FIGURE 6

assures that the identity holds in all possible worlds, so if x is an F in all possible worlds—'L Fx'—then so is y—'L Fy'. The interchange of '1' and '2' is permitted on both horizontal lines of the index matrix.

In the second example, the antecedent 'Ixy' only assures us that the identity holds in the real world. 'x' and 'y' might end up representing distinct individuals in some other possible state of affairs. The interchange is allowed on the horizontal line headed by '0', but there is no basis for making it on any lower lines in the matrix. The cancellation cannot be closed.

We can state the modal version of Id. II as follows:

MId. II. *Where 'm' is a modal constant, 'i' and 'j' are individual constants, and a branch contains a formula of the form '$\bar{I}_{mij}$', the branch may be expanded by adding, for each bsf in the branch of the form 'Ψ_{mi}', a bsf of the form 'Ψ_{mj}'.*

The "vice versa" part of the rule Id. II is not really needed.

An important, and at times controversial, question has to do with whether or not there are "contingent identities." Can two individuals be identical in one state of affairs without being identical in all? Perhaps the use of "two individuals" in phrasing the question is misleading, and it generally comes up with respect to individuals designated by description rather than by name. Teddy Roosevelt is identical with Teddy Roosevelt, and it is difficult to suggest a reasonable way to construct a state of affairs in which Teddy Roosevelt is not identical with Teddy Roosevelt. But Cicero is identical with Tully, and we might do something with that, since there are two distinct referring expressions for one individual. More to the point, however, consider the following two miniature universes:

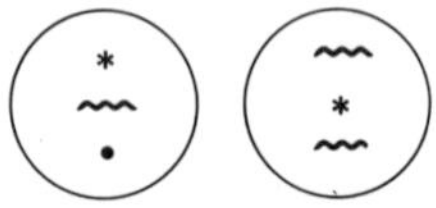

In the left-hand world, *the star* is identical with *the topmost object*. In the right-hand world, it is not. If we are to permit descriptions in our vocabulary at all, we clearly cannot have all identities necessary. Richard M. Nixon is identical with the man who was elected President in 1968. But surely there is an alternative world in which this is not the case.

The modal version of Id. I is as follows, with examples given in Figure 7.

MId. I. *Where* 'i' *is an individual constant and* 'm' *a modal constant, a branch is closed if it contains a formula of the form* 'I_{mii}', *provided that this configuration occurs on every horizontal line of the index matrix where* 'i' *has an occurrence.*

I		I	
012	[No closure]	011	[MId. I closed]
311		211	

FIGURE 7

In the **Q1** systems, '∀x∀y C Ixy L Ixy' does not cancel to closed branches under the Standard Interpretation. Its converse, of course, does, in **CMQ1** and subsequent systems (see Figure 8).

2.1.1 Interpretation of Identity in QA. There is no special translation rule needed for translating 'I' into the Quantificational Analogue. There will, of course, be assumptions and closure conditions in QA to cover the closure conditions for 'I'. Where 'x' and 'y' are individual variables, '(Ixy)**μ**' is translated, like any other formula, to '**C Bxμ C Byμ Ixyμ**'. Negated identities come across in the corresponding way; '($\bar{\text{I}}$xy)**μ**' is translated to '**C Bxμ C Byμ $\bar{\text{I}}$xyμ**'.

The rule for translating formulas containing individual constants is the same as that for any predicate. Where 'i' and 'j' are individual constants, '(Iij)**μ**' becomes simply '**Iijμ**', and '($\bar{\text{I}}$ij)**μ**' becomes '**$\bar{\text{I}}$ijμ**'.

```
 1   2         3               1   2   a
∀̸x ∀̸y C̸ Īxy L̸ Ixy           ∀̸x ∀̸y C̸ L̸̄ Īxy Ixy
          012   012                        012 012
                312                        a12
          Ī  ,  I                       a, Ī  , I
          012   012                        012 012
                312                        a12
     [No closure]                  [No closure]
-------------------------      -------------------------
          Ī  ,  I  , I                  a, Ī  , I
          012   012 011                    012 012
                312 312                    a12 012
     [ MId. II    ]                 [ CM I   ]
     [ No closure ]                 [ a = 0  ]
                                    [ closed ]
```

FIGURE 8

An antecedent assumption is needed in QA to justify Id. I. It is called the Principle of Self-Identity (*SI*), and is as follows:

SI **∀x V λ C Bxλ Ixxλ**

Note that the condition *does not require* that every individual be self-identical in every model set, but only that *if* an individual belongs to a model set then it is self-identical in that model set. This is tantamount to requiring that a model set contain no sentences of the form 'Īii' (cf. [VIII.5, p. 57]).

The justification of MId. II requires a new closure condition for QA, and cannot be covered by an antecedent assumption. It is worthwhile recalling here that QA is just elementary logic, with appropriate abbreviations added to cover constructions like 'Vλ' (cf. Chapter IV, footnote 6). Since Id. II requires a closure condition in elementary logic, MId. II will require a closure condition in that version of elementary logic that we call QA. The closure condition, in its QA version, is called *Expansion*, and is as follows.

Expansion (QA). *Where* 'i' *and* 'j' *are individual constants and* 'm' *is a modal constant, and a branch contains a formula of the form* '**Īijm**', *the branch may be expanded by adding, for each bsf in the branch of the form* '**Ψim**', *a bsf of the form* '**Ψjm**'.

Exercises

1. Translate '∀x ∀y C L Ixy Ixy' into QA and show that it cancels to closed branches in **QA: CMQ1**.

2. Translate '∀x ∀y C Ixy L Ixy' into QA.

2.2 The Elimination of 'Q'. The existence predicate 'Q' is an unnecessary appendage when identity is added to the system. *SI* can and should be strengthened to a biconditional. Once this is done, the effect of 'Q' can be gotten by using 'I'. First, we give the new principle in QA. It is called the *Principle of the Existence of Self-Identicals* (*ESI*).

ESI **∀x∀λ C Ixxλ Bxλ**

The assumptions *SI* and *ESI*, taken together, yield a QA equivalence between '**Bxλ**' and '**Ixxλ**'. Since the 'Q' predicate was interpreted simply on '**B——**', it can now be eliminated from the object language. We can show this by an object language definition, and drop 'Q' from the object language vocabulary altogether.

$$\mathrm{Qi} =_{\mathrm{df}} \exists x\, \mathrm{Ixi}$$

Constants occurring in iwffs are now brought into the cancellation index matrix by means of 'I', rather than by means of 'Q', in the following way: When '$\bar{\mathrm{I}}$' occurs and one of the positions in its index matrix is unfilled, the individual constant from the iwff may be brought down to fill the position. We give an example in Figure 9 and the formal statement of the rule below. There is no closure until '1' is brought down into the branch chart to fill the empty position in the index matrix of '$\bar{\mathrm{I}}$'. When this is done, each branch chart contains a formula of the from '$\underset{021}{\bar{\mathrm{I}}}$', which allows '1' to be entered next to '0' under its original occurrence in any bsf, by 1.2.1 above.

$$
\begin{array}{llllllllllllll}
\mathrm{C} & \mathrm{K} & \exists x & \mathrm{Ix1} & \mathrm{F1} & \exists x & \mathrm{Fx} & \quad & \mathrm{C} & \forall x & \mathrm{Fx} & \mathrm{C} & \exists x & \mathrm{Ix1}\ \mathrm{F1} \\
 & & 2 & & & \mathrm{a} & & & & \mathrm{a} & & & 2 & \\
\not{\mathrm{C}} & \not{\mathrm{K}} & \not{\bar{\exists}} x & \bar{\mathrm{I}}\mathrm{x1} & \bar{\mathrm{F}}1 & \not{\bar{\exists}} x & \mathrm{Fx} & & \not{\mathrm{C}} & \not{\bar{\forall}} x & \bar{\mathrm{F}}\mathrm{x} & \not{\mathrm{C}} & \not{\bar{\exists}} x & \bar{\mathrm{I}}\mathrm{x1}\ \mathrm{F1} \\
 & & & 02 & 0 & & 0\mathrm{a} & & & & 0\mathrm{a} & & & 02\ \ 0 \\
 & & & \bar{\mathrm{I}}\ 1, & \bar{\mathrm{F}}1, & \mathrm{a}, & \mathrm{F} & & & & \mathrm{a}, \bar{\mathrm{F}}\ , & & 2, & \bar{\mathrm{I}}\ 1, \mathrm{F1} \\
 & & & 02 & 0 & & 0\mathrm{a} & & & & 0\mathrm{a} & & & 02\ \ 0
\end{array}
$$

FIGURE 9

The formation and cancellation rules assure that a fragment like '$\underset{02}{\bar{\mathrm{I}}}$' cannot occur in a branch *unless* the remaining position next to 'I' contained an individual constant in the original iwff. The closure rule, then, is as follows.

MId. III. *Where* 'i' *and* 'j' *are individual constants and* 'm' *is a modal constant, and a branch contains a formula of the form* 'Ī j' *or* 'Ī j ',
mi– m–i
and 'j' *is the individual constant in the original formula which occurs above the open position of the index matrix,* 'j' *may be written under its original occurrence on any horizontal line of the index matrix where* 'i' *appears.*

The two branch charts in Figure 9 can now be closed, as shown in Figure 10. Since the circumstances covered by 1.2.1 can only happen

Ī 1, F̄1, a, F	a, F̄ , Ī 1, F1
02 0 0a	0a 02 0
[No closure]	[No closure]
----------	----------
Ī , F̄1, a, F	a, F̄ , Ī , F1
021 0 0a	0a 021 0
[MId. III]	[MId. III]
----------	----------
Ī , F̄ , a, F	a, F̄ , Ī , F
021 01 0a	0a 021 01
[VIII.1.2.1 a = 1 closed]	[VIII.1.2.1 a = 1 closed]

FIGURE 10

with respect to individual constants from an iwff, and since these come into the branch chart only under MId. III, we combine the two rules henceforth, adding to MId. III the following sentence:

'j' *may then be written next to* 'm' *in the index matrix of any basic formula where it has an original occurrence, under that original occurrence.*

Exercise

Translate and cancel in QA: 'C∀x Fx C∃x Ix1 F1', with the antecedent assumption of *ESI*.

2.3 Modality De Dicto and Modality De Re Revisited. Written with '∃x Ix1' in place of 'Q1', modality *de dicto* and modality *de re* appear as follows.

It is possible that Susie is a freshman.	(*de dicto*)	M K ∃x Ix1 F1
Susie is possibly a freshman.	(*de re*)	K ∃x Ix1 M F1

Not only can *de dicto* and *de re* modalities be distinguished in the notation, it can now be shown (Figure 11) that possibility *de re* implies

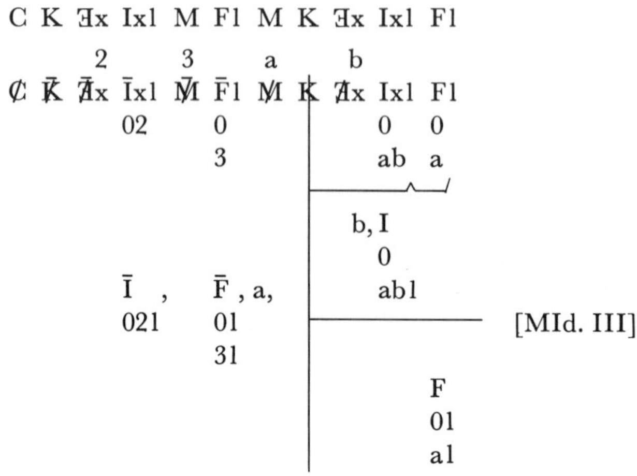

FIGURE 11

possibility *de dicto*, and that possibility *de dicto* does not imply possibility *de re* (cf. V.5.1). '1' is brought into the index matrices by MId. III. Since it occurs next to the modal constant '0' in the index of 'Ī', it is written next to '0' under its original occurrence in every other basic formula in the same branch(es) as 'Ī', and then, like any other dummy or constant, repeated in the bottom position. Both branches are closed with the replacements a = 3, b = 1 (the upper branch closes by MId. I, not by a pair of opposites 'Ī, I').

The corresponding cancellation of 'C M K ∃x Ixl Fl K ∃x Ixl M Fl' cannot be closed.

3 THE DEFINITE DESCRIPTION

In the last section we made passing appeal to the expressions "the star" and "the topmost object." Such expressions are called *definite descriptions*, and they are intended to pick out and refer to unique individuals in somewhat the same way that proper names do. Definite descriptions are usually represented by the use of the sign 'ι' (iota), sometimes written upside down like the 'V' of quantification. "The star" is represented 'ιx Sx', for example, and "The star is over all dots," in the miniature universe vocabulary, would be represented 'Vz C Dz O(ιx Sx)z'—"For all z, if z is a dot then the star is over z."

The sign 'ι' is not a new primitive, but is defined in context in the following way:

$$\Phi\,(\iota x\,\Psi x) =_{\mathrm{df}} \exists x\, K\, K\, \Psi x\, \forall y\, C\, \Psi y\, Iyx\, \Phi x.$$

"The Ψ is a Φ" is equivalent by definition to "There is an **x** such that **x** is Ψ, and anything that is Ψ is the same individual as **x**, and **x** is Φ." The formula above, representing "The star is above all dots," comes into primitive notation as

$$\exists x\, K\, K\, Sxy\, \forall\, C\, Sy\, Iyx\, \forall z\, C\, Dz\, Oxz.$$

"There is an x such that x is a star, and anything that is a star is the same individual as x, and for all z, if z is a dot, then x is over z." An important feature of such statements is that they are falsifiable in several ways. In the following "worlds," for example, "The star is above all dots" is true in the first and false in both the second and the third.

The formula is false in the second world, not because there is some dot over a star, but because there is more than one star. The part of the translation that requires uniqueness, "—anything that is a star is the same individual as x—" is false, and the whole statement is therefore false.

In many ways, expressions of the form '$\iota x\,\Psi x$' may be used like proper names (individual constants). But care must be taken with the use of such expressions in any formal system. Even more care is needed in elementary logic than is needed in a system like **CMQ1**, where existence assumptions are made explicit. A common mistake in elementary logic results from the use of the definite description as if it were an individual constant. In elementary logic, it is usually assumed that "everything exists." In the QA interpretation, '**∀x Bx0**'. Since there are no modal operators, and no need to interpret any part of the system on an alternativeness relation, there is usually no place in the system for the notion of a "possible individual." On the assumption that everything exists, formulas like 'C ∀x Fx F1' are provable without any further existence condition. The *ad hoc* restriction suggested at 1.3.1 is not *ad hoc* at all in elementary logic, but is common practice. Most important, '∀x Ixx' is a theorem of elementary logic, as it is of **CMQ1**, and with the overall existence assumption, so are 'I11', 'I22', 'I33', and so on. If the definite description is used as if it were an individual constant, then the fallacious inference from '∀x Ixx' to 'I (ιx Fx) (ιx Fx)' might be committed. This formula should not, of course, be a theorem of any formal system. That is, in primitive notation, it should never be *provable* that

$$\exists x\, K\, K\, Fx\, \forall y\, C\, Fy\, Iyx\, \exists z\, K\, K\, Fz\, \forall y\, C\, Fy\, Iyx\, Ixz.$$

Such a mistake is pernicious enough in elementary logic, where it seems to make *provable* the statement that something is an F. It is worse yet in a system of modal logic, where provable statements are also necessary.

In general, any formulas provable in the **Q1** systems under Standard Interpretation that contain segments of the forms '$K\exists x\, Ixi\, Gi$' or '$C\exists x\, Ixi\, Gi$'—that is, "i exists and is a G" or "If i exists, then i is a G," where 'i' is an individual constant—are provable also where 'i' is replaced by an expression of the form '$\iota x\, \Psi x$'. But *a formula containing a definite description is provable if and only if the result of replacing the description according to the definition above is provable.* The "shorthand" use of 'ι' requires some care.[2]

4 NEGATION, DENIAL AND EXISTENCE PRESUPPOSITIONS

In the entailment systems **CYC** of the preceding chapter, we introduced two distinct negative operators, called *negation* and *denial.* These may be viewed as negation *de dicto* and negation *de re*, in much the same sense that modality can be viewed as either *de dicto* or *de re.* Denial, as presented in VII, is negation *de dicto*: "*It is not the case that* Susie is a freshman." Negation is *de re*: "Susie is a non-freshman." The four interpretations of basic formulas considered in V provide a way of distinguishing the two semantically.

Negation *de re*, like modality *de re*, has to do with the question whether or not an (existing) individual has a certain property. The statements "Susie is a freshman" and "Susie is a non-freshman" are symbolized and interpreted as follows:

Susie is a freshman.	(F1)**0**	**KB10**[**F1**]**0**
Susie is a non-freshman.	($\bar{F}$1)**0**	**KB10**[$\bar{\mathbf{F}}$**1**]**0**

Both the assertion that Susie is a freshman and the *de re* negation are interpreted in such a way that Susie is understood to be an existing individual.

Negation *de dicto*, like modality *de dicto*, has to do with the question whether or not a given statement is true. The statements "It is not the case that Susie is a freshman" and "It is not the case that Susie is a non-freshman" are not semantically equivalent to the *de re* negation formulas just symbolized:

[2] And often doesn't get it. For further observations on this subject, see N. L. Wilson [VIII.23], especially pp. 111-112, and [VIII.24].

Not: Susie is a freshman.	$-(\mathrm{F1})\mathbf{0}$	$\overline{\mathbf{K}}\,\mathbf{B10}\,[\mathbf{F1}]\mathbf{0}$
Not: Susie is a non-freshman.	$-(\overline{\mathrm{F}}1)\mathbf{0}$	$\overline{\mathbf{K}}\,\mathbf{B10}\,[\overline{\mathbf{F}}\mathbf{1}]\mathbf{0}$

Consider now the case where Susie does not exist. Then the *denials* of *both* "Susie is a freshman" *and* "Susie is a non-freshman" are true. That is, given '$\overline{\mathbf{B}}\mathbf{10}$', it follows that '$\overline{\mathbf{K}}\,\mathbf{B10}\,[\mathbf{F1}]\mathbf{0}$' and that '$\overline{\mathbf{K}}\,\mathbf{B10}\,[\overline{\mathbf{F}}\mathbf{1}]\mathbf{0}$' are true. If Susie does exist, then one or the other of "Susie is a freshman" or "Susie is a non-freshman" must be true. The relations among the four statements can be schematized as follows.

A statement and its denial (*de dicto*) are *contradictories*: precisely one of them is true, and the other false. "Susie is a freshman" and "It is not the case that Susie is a freshman" are contradictories, and "Susie is a non-freshman" and "It is not the case that Susie is a non-freshman" are contradictories.

A statement and its negation (*de re*) are *contraries*: at least one of them is false, and both of them may be false. "Susie is a freshman" and "Susie is a non-freshman" are *contraries.*

The denials (*de dicto*) of a pair of contraries are *subcontraries*: at least one of them is true, and both of them may be true. "It is not the case that Susie is a freshman" and "It is not the case that Susie is a non-freshman" are subcontraries. Both of them are true when Susie does not exist.

Finally, in a "normal" world—in a model set which is an H-alternative to another model set, in the vocabulary of the model system—a statement entails the denial (*de dicto*) of its negation (*de re*): "Susie is a freshman" entails "It is not the case that Susie is a non-freshman," and "Susie is a non-freshman" entails "It is not the case that Susie is a freshman." The four statements and their Quantificational Analogue interpretations appear in the diagram in Figure 12. Each of the upper statements entails the statement diagonally opposite it.

Susie is a freshman. $\mathbf{K}\,\mathbf{B10}\,[\mathbf{F1}]\mathbf{0}$	CONTRARIES (At least one false)	Susie is a non-freshman. $\mathbf{K}\,\mathbf{B10}\,[\overline{\mathbf{F}}\mathbf{1}]\mathbf{0}$
CONTRADICTORIES (One true; one false)		CONTRADICTORIES (One true; one false)
It is not the case that Susie is a freshman. $\overline{\mathbf{K}}\,\mathbf{B10}\,[\mathbf{F1}]\mathbf{0}$	SUBCONTRARIES (At least one true)	It is not the case that Susie is a non-freshman. $\overline{\mathbf{K}}\,\mathbf{B10}\,[\overline{\mathbf{F}}\mathbf{1}]\mathbf{0}$

FIGURE 12

4.1 Existence Presuppositions. It should be noted emphatically, before we go any further, that it is only in the abnormally complicated **CYS** systems that the distinction between negation and denial can be reflected formally. The negation-denial distinction can be adopted without taking the further step of adopting the 'Y' operator of **CYC**, however, and both modal and non-modal logics can be developed in which the distinction is made.

A logic in which the negation-denial distinction is made would allow us to interpret as *false* all statements about individuals which do not exist, unless those statements fall under modal operators. "Sherlock Holmes was a detective" is false. "It is possible that Sherlock Holmes was a detective" is true.[3]

But the familiar (non-modal) logic of quantification does not include a negation-denial distinction, nor do most of the systems of modal logic that we have considered. The interpretations of basic formulas containing individual variables and those containing individual constants cannot reflect all four of the formulas on the Square of Opposition in Figure 12. The interpretations of basic formulas containing individual variables that were considered in Chapter V reflected the different combinations of these four formulas. The Standard Interpretation—the only one of the four that preserved the desired equivalences between modal operators and between quantifiers—treated '$(Fx)\mu$' and '$(\bar{F}x)\mu$' not as contradictories but as subcontraries, with the quantifier interpretations supplying the extra ingredients to make pairs of formulas like '∃x Fx' and '$\bar{\exists}$x Fx' turn out to be contradictory. On the Standard Interpretation, the two formulas are translated as follows:

∃x Fx	**∃x K Bx0 C Bx0 Fx0**
$\bar{\exists}$x Fx	**∀x C Bx0 C Bx0 $\bar{\text{F}}$x0.**

With the redundancies eliminated, these become

∃x Fx	**∃x K Bx0 Fx0**
$\bar{\exists}$x Fx	**∀x C Bx0 $\bar{\text{F}}$x0.**

The fragment '**C Bx–**' ceases to be redundant when a modal operator

[3] A famous and oft-reprinted philosophical exchange between Bertrand Russell and P. F. Strawson centers on precisely how to treat statements when the subject term does not refer to anything. The example has to do with statements about "The present King of France," and raises some special problems about referring by means of description rather than by means of proper name. But the central issue is worth considering in the light of the present distinction. See Russell, "On Denoting," *Mind* 14 (1905); Strawson, "On Referring," *Mind* 59 (1950); and Russell, "Mr. Strawson on Referring," *Mind* 66 (1957).

intervenes between the quantifier and the basic formula.

$$\exists x\, M\, Fx \qquad \mathbf{\exists x\, K\, Bx0\, \exists\lambda\, K\, H\lambda 0\, C\, Bx\lambda\, Fx\lambda}$$
$$\bar{\exists} x\, M\, Fx \qquad \mathbf{\forall x\, C\, Bx0\, \forall\lambda\, C\, H\lambda 0\, C\, Bx\lambda\, \bar{F}x\lambda}$$

It is not therefore eliminable from the interpretation of basic formulas that contain individual variables. Its function is reflected in the cancellation procedure for the Standard Interpretation.

A familiar deductive principle of elementary logic is called *Existential Generalization.* It allows the inference from a formula of the form '$\Phi 1$' to a formula of the form '$\exists x \Phi x$'. A corresponding principle is *Universal Instantiation,* which allows the inference from a formula of the form '$\forall x \Phi x$' to a formula of the form '$\Phi 1$'. Both of these principles have been considered in elementary form in earlier sections of this chapter, when we considered whether or not '$\exists x\, Fx$' was implied by '$F1$' and whether '$F1$' was implied by '$\forall x\, Fx$'. Under the new translation rules for formulas containing individual constants given at 1.3.2 above, neither of these principles is valid with respect to the **Q1** systems. This can be seen easily enough by translating the appropriate material implication formulas into QA.

$$C\, F1\, \exists x\, Fx \qquad \mathbf{A\, \bar{F}10\, \exists x\, K\, Bx0\, Fx0} \quad \text{(Existential Generalization)}$$
$$C\, \forall x\, Fx\, F1 \qquad \mathbf{A\, \exists x\, K\, Bx0\, \bar{F}x0\, F10} \quad \text{(Universal Instantiation)}$$

The reason is obvious enough, given our earlier discussion of existence. It needs to be established in one way or another that the individual designated by '1' belongs to **0** in order for the implication to hold. In elementary (non-modal) logic, this is usually covered by an *existence presupposition.* The assumption *PUP*—$\mathbf{\forall x\, \forall\lambda\, Bx\lambda}$—which is taken for the **Q2** systems is such a presupposition. If it is presupposed that every individual exists in every possible state of affairs, then these two principles are valid. A weaker assumption that covers the ground for elementary logic, where the only model set involved is 0, is the assumption *everything exists*:

EE $\mathbf{\forall x\, Bx0}$

Either *EE* or *PUP* is sufficient to close the QA cancellations of the two formulas above. That is, either of these two presuppositions will justify the principles of Existential Generalization and Universal Instantiation.

For the systems of modal logic, however, these assumptions are not in general appropriate. Recently, a number of logicians have considered systems of elementary logic without existence presuppositions.[4] Such

[4] cf. Hintikka [VIII.5] and Hugues Leblanc and T. Hailperin [VIII.9].

systems employ a way of specifying existence in the object language, either by means of an existence predicate like our discarded 'Q', or by means of the quantifier-identity combination used in §2.2 above, or by means of a more complicated object language formula that allows a semantic interpretation corresponding to our '**B10**'.

In the absence of the existence presuppositions *EE* or *PUP*—specifically, in the desirable **Q1** systems—the principles Existential Generalization and Universal Instantiation are valid only when supplemented by an existence specification, as we found in the discussion of fictional individuals:

C K x∃ Ixl Fl ∃x Fx	"If l exists and is an F, then something F's."
C ∀x Fx C ∃x Ixl Fl	"If everything F's, then if l exists, l F's."

4.2 Negation De Dicto and De Re in the Presupposition-Free Object Language. The four permutations of "Susie is a freshman" considered above can be reflected in the object language without existence presuppositions; that is, in the **Q1** systems under the new Standard Interpretation of basic formulas containing individual constants, without introducing a denial operator in addition to the usual negation operator. The QA formula '**B10**' which was used earlier to make the distinction clear is, assuming the principles of Self-Identity and the Existence of Self-Identicals, adequately reflected in the object language by '∃x Ixl'. The four statements appear as follows.

Susie is a freshman.	K ∃x Ixl Fl
Susie is a non-freshman.	K ∃x Ixl $\bar{\text{F}}$l
It is not the case that Susie is a freshman.	$\bar{\text{K}}$ ∃x Ixl Fl
It is not the case that Susie is a non-freshman.	$\bar{\text{K}}$ ∃x Ixl $\bar{\text{F}}$l

5 CONCLUSIONS CONCERNING MODALITY AND EXISTENCE

Many of the objections to modal logic that have been raised by philosophers and logicians have to do with problems of reference to individuals. What has developed in this chapter is the following result: If the existence presuppositions that are assumed for most formulations of elementary logic are assumed for modal logic, two important distinctions become unaccessible to the formal system: the distinction between modality *de dicto* and modality *de re* in unquantified formulas, and the distinction between negation *de dicto* and negation *de re*. In a logic free of the existence presuppositions *PUP* and *EE*, these distinctions can be preserved in the object language of modal logic *and* the distinction between negation *de dicto* and negation *de re*, which has to do with the problem of naming non-existent individuals, can be preserved in elementary logic as well.

It is not our business here to argue that a presupposition-free logic is the sole correct system of either modal or elementary logic. At the beginning of the Introduction we took the position that there is no such thing as the sole correct system of logic, and there is no reason to abandon this position now. But the mechanics of building formal systems developed here using the Quantificational Analogue make it clear that in elementary logic as well as modal logic the business of choosing a formal system for a given task must be carried out with an awareness of the presuppositions —antecedent assumptions—that one wants to accept and those that one wants to reject. This awareness comes not from accepting an object language system and treating it as primitive, but from interpreting the object language system in such a way that all its assumptions and presuppositions are out in the open.

BIBLIOGRAPHY—CHAPTER VIII

[VIII.1] Chisholm, Roderick M. "Identity Through Possible Worlds: Some Questions," *Nous* **1** (1967), pp. 1–8.

[VIII.2] Donnellan, Keith S. "Reference and Definite Descriptions," *Philosophical Review* **75** (1966), pp. 281–304.

[VIII.3] Fitch, F. B. "Morning Star and Evening Star," *Philosophy of Science* **16** (1949), pp. 137–141.

[VIII.4] Hintikka, Jaakko. "Existential Presuppositions and Existential Commitments," *Journal of Philosophy* **56** (1959), pp. 125–137.

[VIII.5] ———. "Studies in the Logic of Existence and Necessity," *The Monist* **50** (1966), pp. 55–76.

[VIII.6] ———. "Individuals, Possible Worlds and Epistemic Logic," *Nous* **1** (1967), pp. 33–62.

[VIII.7] Kemeny, John G. Review of Quine's "Reference and Modality," *Journal of Symbolic Logic* **19** (1954), pp. 137–138.

[VIII.8] Kripke, Saul. "Semantical Considerations in Modal Logic," *Acta Philosophica Fennica* **16** (1963), pp. 83–94.

[VIII.9] Leblanc, Hugues, and T. Hailperin. "Nondesignating Singular Terms," *Philosophical Review* **68** (1959), pp. 239–243.

[VIII.10] Marcus, Ruth Barcan. Review of Smullyan's "Modality and Description," *Journal of Symbolic Logic* **13** (1948), pp. 149–150.

[VIII.11] ———. "Extensionality," *Mind* **69** (1960), pp. 55–62.

[VIII.12] Quine, Willard V. "Notes on Existence and Necessity," *Journal of Philosophy* **40** (1943), pp. 113–127.

[VIII.13] ———. "The Problem of Interpreting Modal Logic," *Journal of Symbolic Logic* **12** (1947), pp. 43–48.

[VIII.14] ———. "Reference and Modality," in *From a Logical Point of View.* (Harvard University Press, Cambridge, 1953)

[VIII.15] Russell, Bertrand. "On Denoting," *Mind* **14** (1905),

[VIII.16] ———. "Mr. Strawson on Referring," *Mind* **66** (1957), pp. 385–389.

[VIII.17] Searle, John R. "Proper Names," *Mind* **67** (1958), pp. 166–173.

[VIII.18] Smullyan, Arthur. Review of Quine's "The Problem of Interpreting Modal Logic," *Journal of Symbolic Logic* **12** (1947) pp. 139–141.

[VIII.19] ———. "Modality and Description," *Journal of Symbolic Logic* **13** (1948), pp. 31–37.

[VIII.20] Strawson, Peter. "On Referring," *Mind* **59** (1950), pp. 320–344.

[VIII.21] Vorsteg, Robert. "Definite Descriptions and Existential Entailment," *The Monist* **51** (1967), pp. 136–150.

[VIII.22] Wilson, N. L. "The Trouble with Meanings," *Dialogue* **3** (1964), pp. 52–64.

[VIII.23] ———. "Modality and Identity, A Defense," *Journal of Philosophy* **62** (1965), pp. 471–477.

[VIII.24] ———. Review of P. A. Schilpp (Ed.) "The Philosophy of Rudolf Carnap," *Dialogue* **4** (1965), pp. 102–112.

Appendix A

CANCELLATION RULES AND CLOSURE CONDITIONS

1 ELEMENTARY LOGIC (cf. Chapter II, §4 and §5)

$$
\begin{array}{ll}
A\alpha\beta \rightarrow \not{A}\alpha\beta & \bar{A}\alpha\beta \rightarrow \dagger\!\!\bar{A}\,\bar{\alpha}\bar{\beta} \\
K\alpha\beta \rightarrow \dagger\!\!K\,\alpha\beta & \bar{K}\alpha\beta \rightarrow \not{\bar{K}}\,\bar{\alpha}\bar{\beta} \\
C\alpha\beta \rightarrow \not{C}\,\bar{\alpha}\beta & \bar{C}\alpha\beta \rightarrow \dagger\!\!\bar{C}\,\alpha\bar{\beta}
\end{array}
$$

(In the three rules written with $\dagger$, the connective is crossed by a vertical stroke, with a bracket drawn under the formula.)

$$\bar{\bar{\alpha}} \rightarrow \not{\bar{\bar{\alpha}}}$$

$$
\begin{array}{ll}
\exists x\,\Phi x \rightarrow \not{\exists} x\,\Phi^{d} x_{d} & \bar{\exists} x\,\Phi x \rightarrow \not{\bar{\exists}}^{\,c} x\,\bar{\Phi} x_{c} \\
\forall x\,\Phi x \rightarrow \not{\forall} x\,\Phi^{c} x_{c} & \bar{\forall} x\,\Phi x \rightarrow \not{\bar{\forall}} x\,\bar{\Phi}^{d} x_{d}
\end{array}
$$

Q-closure I. *Dummy constants in the branch chart may be replaced by numerical constants from the branch chart within the restrictions of the Uniformity Requirement and the Order Requirement.*

Q-closure II. *When two dummies appear in a pair of opposing formulas and no constant is present to replace them, a new numerical constant may be introduced into the branch chart to replace the two dummies.*

2 MONADIC MODAL LOGIC

Rules for 'A', 'K', 'C' and their negations, and '⁼', are as above, except that an occurrence of '0' is written below every basic subformula before cancellation begins (cf. Chapter III, §2).

$$\begin{array}{cccc} & \text{d} & & \text{c} \\ M\alpha \rightarrow & \not{M}\alpha & \bar{M}\alpha \rightarrow & \bar{\not{M}}\bar{\alpha} \\ 0 & 0 & 0 & 0 \\ & \text{d} & & \text{c} \end{array}$$

$$\begin{array}{cccc} & \text{c} & & \text{d} \\ L\alpha \rightarrow & \not{L}\alpha & \bar{L}\alpha \rightarrow & \bar{\not{L}}\bar{\alpha} \\ 0 & 0 & 0 & 0 \\ & \text{c} & & \text{d} \end{array}$$

CMn-_closure_ I. _Dummy constants in a modal index column may be replaced by numerical constants from the branch chart within the restrictions of the Uniformity Requirement and the Order Requirement._

CMn-_closure_ II. _When two dummies appear in a pair of opposing formulas and no constant is present to replace them, a new numerical constant may be introduced into the cancellation to replace the two dummies._

CM-_closure_ I. _A dummy constant may be deleted uniformly in the indices of basic subformulas in any branch._

OR

CM-_closure_ I (alternate). _A numerical constant may be repeated immediately below its original occurrence in any modal index column._

CS4-_closure_ I. _If two constants (excluding zero) occur consecutively in a cancellation index, it is permissible to delete the upper constant before making replacements of dummies by constants._ (_This rule may be applied as often as needed to a given index._)

CS5-_closure_ I. _It is permissible to delete all entries in a modal index column except the bottom entry._

3 QUANTIFIED MONADIC MODAL LOGIC (Standard Interpretation)

The rules given in §2 apply, in addition to the following.

$$\begin{array}{cccc} & \text{d} & & \text{c} \\ \exists x\,\Phi x \rightarrow & \not{\exists} x\,\Phi x & \bar{\exists} x\,\Phi x \rightarrow & \bar{\not{\exists}} x\,\bar{\Phi} x \\ 0 & 0 & 0 & 0 \\ \vdots & \vdots & \vdots & \vdots \\ \text{m} & \text{md} & \text{m} & \text{mc} \end{array}$$

$$\begin{array}{cccc} & \text{c} & & \text{d} \\ \forall x\,\Phi x \rightarrow & \not{\forall} x\,\Phi x & \bar{\forall} x\,\Phi x \rightarrow & \bar{\not{\forall}} x\,\bar{\Phi} x \\ 0 & 0 & 0 & 0 \\ \vdots & \vdots & \vdots & \vdots \\ \text{m} & \text{mc} & \text{m} & \text{md} \end{array}$$

As the last step in the cancellation procedure (Standard Interpretation) all dummies and constants in the index matrix of a basic subformula are repeated under their original occurrence in the index matrix, on the bottom line of the matrix (omitted for First Interpretation).

The two Q-closure conditions in §1 above apply in cancellation of quantified modal formulas. Q-variants are as follows.

Q1-*closure. No further manipulation is permitted.*
Q2-*closure. An individual constant occurring in an index matrix may be repeated on every horizontal line above and below its original occurrence.*
Q3-*closure. An individual constant occurring in an index matrix may be repeated on every horizontal line below its original occurrence.*
Q4-*closure. An individual constant occurring in an index matrix may be repeated on every horizontal line above its original occurrence.*

4 QUANTIFICATION WITH IDENTITY

Two versions of the closure conditions for identity are given; the first for elementary logic with existence presuppositions, the second for quantified modal logic and elementary logic without existence presuppositions. The latter is best treated using the modal device of the index matrix, placing dummies and constants next to the occurrence of '0' that heads the matrix, even though '0' will be the only modal index in the matrix.

4.1 Identity Closure in Elementary Logic

Id. *closure* I. *Where* 'i' *is an individual constant, a branch is closed if it contains a formula of the form* 'Iii'.

Id. *closure* II. *Where* 'i' *and* 'j' *are individual constants, and a branch contains a formula of the form* '$\bar{\text{I}}$ij', *the branch may be expanded by adding, for each basic subformula in the branch of the form* 'Ψi', *a basic subformula of the form* 'Ψj'.

4.2 Identity Closure in Modal Logic

MId. *closure* I. *Where* 'i' *is an individual constant and* 'm' *a modal constant, a branch is closed if it contains a formula of the form* 'I_{mii}', *provided that this configuration occurs on every horizontal line of the index matrix where* 'i' *has an occurrence.*

MId. *closure* II. *Where* 'm' *is a modal constant,* 'i' *and* 'j' *are modal constants, and a branch contains a formula of the form* '$\bar{\text{I}}_{\text{mij}}$', *the branch may be expanded by adding, for each bsf in the branch of the form* 'Ψ_{mi}', *a bsf of the form* 'Ψ_{mj}'.

MId. *closure* III. *Where* 'm' *is a modal constant, and* 'i' *and* 'j' *are individual constants, and a branch contains a formula of the form* '$\bar{I}\,j_{mi-}$' *or* '$\bar{I}\,j_{m-i}$', and 'j' *is the individual constant in the original formula which appears above the open position in the index matrix,* 'j' *may be written under its original occurrence on any horizontal line of the index matrix where* 'i' *appears.* 'j' *may then be written under its original occurrence in any basic formula, next to* 'm'.

5 DYADIC MODAL LOGIC: **CY** SYSTEMS

The rules given in §2 and §3 apply, in addition to the following.

$$\begin{array}{ccc} Y\alpha & \beta \\ 0 & 0 \\ \vdots & \vdots \\ m & m \end{array} \rightarrow \begin{array}{ccc} \overset{c*}{\not{Y}} & \bar{\alpha} & \beta \\ & 0 & 0 \\ & \vdots & \vdots \\ & m & m \\ & c* & c* \end{array} \qquad \begin{array}{ccc} \bar{Y}\alpha & \beta \\ 0 & 0 \\ \vdots & \vdots \\ m & m \end{array} \rightarrow \begin{array}{ccc} \overset{d*}{\bar{\not{Y}}} & \alpha & \bar{\beta} \\ & 0 & 0 \\ & \vdots & \vdots \\ & m & m \\ & d* & d* \end{array}$$

$$\begin{array}{ccc} Z\alpha & \beta \\ 0 & 0 \\ \vdots & \vdots \\ m & m \end{array} \rightarrow \begin{array}{ccc} \overset{d*}{\not{Z}} & \alpha & \beta \\ & 0 & 0 \\ & \vdots & \vdots \\ & m & m \\ & d* & d* \end{array} \qquad \begin{array}{ccc} \bar{Z}\alpha & \beta \\ 0 & 0 \\ \vdots & \vdots \\ m & m \end{array} \rightarrow \begin{array}{ccc} \overset{c*}{\bar{\not{Z}}} & \bar{\alpha} & \bar{\beta} \\ & 0 & 0 \\ & \vdots & \vdots \\ & m & m \\ & c* & c* \end{array}$$

CY1 I. *Any constant in a modal index column may be repeated immediately below the position in which it first occurs. All individual dummies and constants to the right of the modal index column are also repeated.*

CY1 II. *Starred dummies in a modal index column may be replaced by any modal constant in the branch chart.*

CY1 III. *Unstarred dummies in a modal index column may be replaced by unstarred constants* only.

CY1 IV. *A new constant may be introduced to replace two dummies in otherwise opposing formulas. If one of the dummies is unstarred, the new constant may be unstarred.*

CY1 V. *All replacements of dummies by constants are subject to the Uniformity Requirement and the Order Requirement.* (*The Order Requirement may be ignored.*)

CY2 I. *If two starred constants occur consecutively in a modal index, it is permissible to delete the upper of the two constants before making replacement of dummies by constants.*

CY3 I. *If all the entries in a modal index column are starred, all but the bottom entry may be deleted.*

6 ORDER FOR APPLYING CANCELLATION AND CLOSURE PROCEDURES

(1) Apply the cancellation rules to the original formula, moving from left to right through the formula until all cancellable elements have been attended to.

(2) Construct the branch chart below the cancelled formula, guided by the split markers. If no split markers occurred in the cancellation, this step is not really necessary. The cancelled formula can serve as its own branch chart.

(3) Apply the *quantifier* closure conditions appropriate to the formal system being used (V.5.4.1).

(4) Apply the *modal* closure conditions appropriate to the formal system being used (Chapter III and V.5.4.2).

(5) Replace dummies by constants as permitted within the Uniformity and Order Requirements.

Appendix B

QUANTIFICATIONAL ANALOGUE

1 VOCABULARY

To the vocabulary of elementary logic (Chapter II, §2 and §5) we add the following.

- **λ, μ, ν** variables ranging over model sets.
- **0** model set constant; the name of the fixed model set.
- **H——** a predicate of degree 2 (H-alternativeness), taking model set variables or constants as arguments. '**Hλμ**' is read "(Model set) λ is H-alternative to (model set) μ."
- **[—]—** a predicate of degree 2, taking an iwff of the object language for the first argument and a model set variable or constant for the second. '**[α]μ**' is read "[α] is a member of μ."
- **B——** a predicate of degree 2 ("belongs"), taking an individual variable or constant for the first argument and a model set variable or constant for the second. '**Bxμ**' is read "x belongs to μ."
- **S——** a predicate of degree 2 (S-alternativeness), taking model set variables or constants as arguments. '**Sλμ**' is read "λ is S-alternative to μ."

Quantification over model sets can be eliminated in favor of quantification that employs only individual variables and constants, as described in the note in Chapter II, §7.1.

2 TRANSLATION RULES

The first step in translation of an object language formula into QA is to enclose the formula in parentheses and write the modal constant '0' to the right of the parentheses. The translation rules are then rules for the removal of parentheses. The rules are shown in the form given at V.6, with negation confined.

2.1 ‘$(K\alpha\beta)\mu$’ is translated to ‘$\mathbf{K}\,(\alpha)\mu\,(\beta)\mu$’.
2.2 ‘$(A\alpha\beta)\mu$’ is translated to ‘$\mathbf{A}\,(\alpha)\mu\,(\beta)\mu$’.
2.3 ‘$(C\alpha\beta)\mu$’ is translated to ‘$\mathbf{A}\,(\bar{\alpha})\mu\,(\beta)\mu$.
2.4 ‘$(\bar{K}\alpha\beta)\mu$’ is translated to ‘$\mathbf{A}\,(\bar{\alpha})\mu\,(\bar{\beta})\mu$’.
2.5 ‘$(\bar{A}\alpha\beta)\mu$’ is translated to ‘$\mathbf{K}\,(\bar{\alpha})\mu\,(\bar{\beta})\mu$’.
2.6 ‘$(\bar{C}\alpha\beta)\mu$’ is translated to ‘$\mathbf{K}\,(\alpha)\mu\,(\bar{\beta})\mu$’.
2.7 ‘$(\exists x\,\Phi x)\mu$’ is translated to ‘$\mathbf{\exists x\,K\,Bx\mu}\,(\Phi x)\mu$’.
2.8 ‘$(\forall x\,\Phi x)\mu$’ is translated to ‘$\mathbf{\forall x\,C\,Bx\mu}\,(\Phi x)\mu$’.
2.9 ‘$(\bar{\exists} x\,\Phi x)\mu$’ is translated to ‘$\mathbf{\forall x\,C\,Bx\mu}\,(\bar{\Phi} x)\mu$’.
2.10 ‘$(\bar{\forall} x\,\Phi x)\mu$’ is translated to ‘$\mathbf{\exists x\,K\,Bx\mu}\,(\bar{\Phi} x)\mu$’.
2.11 ‘$(M\alpha)\mu$’ is translated to ‘$\mathbf{\exists\lambda\,K\,H\lambda\mu}\,(\alpha)\boldsymbol{\lambda}$’.
2.12 ‘$(L\alpha)\mu$’ is translated to ‘$\mathbf{\forall\lambda\,C\,H\lambda\mu}\,(\alpha)\boldsymbol{\lambda}$’.
2.13 ‘$(\bar{M}\alpha)\mu$’ is translated to ‘$\mathbf{\forall\lambda\,C\,H\lambda\mu}\,(\bar{\alpha})\boldsymbol{\lambda}$’.
2.14 ‘$(\bar{L}\alpha)\mu$’ is translated to ‘$\mathbf{\exists\lambda\,K\,H\lambda\mu}\,(\bar{\alpha})\boldsymbol{\lambda}$’.
2.15 ‘$(Y\alpha\beta)\mu$’ is translated to ‘$\mathbf{\forall\lambda\,C\,S\lambda\mu\,A}\,(\bar{\alpha})\boldsymbol{\lambda}\,(\beta)\boldsymbol{\lambda}$’.
2.16 ‘$(Z\alpha\beta)\mu$’ is translated to ‘$\mathbf{\exists\lambda\,K\,S\lambda\mu\,K}\,(\alpha)\boldsymbol{\lambda}\,(\beta)\boldsymbol{\lambda}$’.
2.17 ‘$(\bar{Y}\alpha\beta)\mu$’ is translated to ‘$\mathbf{\exists\lambda\,K\,S\lambda\mu\,K}\,(\alpha)\boldsymbol{\lambda}\,(\bar{\beta})\boldsymbol{\lambda}$’.
2.18 ‘$(\bar{Z}\alpha\beta)\mu$’ is translated to ‘$\mathbf{\forall\lambda\,C\,S\lambda\mu\,A}\,(\bar{\alpha})\boldsymbol{\lambda}\,(\bar{\beta})\boldsymbol{\lambda}$’.

2.19 First Interpretation of Basic Formulas. This is not the Standard Interpretation. ‘Ψy’ represents an atomic formula, and ‘y’ may be either an individual variable or an individual constant.

2.19.1 ‘$(\Psi y)\mu$’ is translated to ‘$\mathbf{\Psi y\mu}$’.
2.19.2 ‘$(\bar{\Psi} y)\mu$’ is translated to ‘$\mathbf{\bar{\Psi} y\mu}$’.

2.20 Standard (Fifth) Interpretation of Basic Formulas. (‘Ψx’ represents an atomic formula; ‘x’ represents an individual variable; ‘i’ represents an individual constant.)

2.20.1 ‘$(\Psi x)\mu$’ is translated to ‘$\mathbf{C\,Bx\mu\,\Psi x\mu}$’.
2.20.2 ‘$(\bar{\Psi} x)\mu$’ is translated to ‘$\mathbf{C\,Bx\mu\,\bar{\Psi} x\mu}$’.
2.20.3 ‘$(\bar{\Psi} i)\mu$’ is translated to ‘$\mathbf{\Psi i\mu}$’.
2.20.4 ‘$(\bar{\Psi} i)\mu$’ is translated to ‘$\mathbf{\bar{\Psi} i\mu}$’.

3 ANTECEDENT ASSUMPTIONS

3.1 *Ex* (Existence) $\mathbf{\forall\lambda\,\exists\mu\,H\mu\lambda}$
3.2 *Rf* (Reflexivity) $\mathbf{\forall\lambda\,H\lambda\lambda}$

3.3	*Tr* (Transitivity)	**∀λ∀μ∀ν C K Hλμ Hμν Hλν**
3.4	*Sm* (Symmetry)	**∀λ∀μ C Hλμ Hμλ**
3.5	*Qe* (Quasi-equivalence)	**∀λ∀μ∀ν C K Hλμ Hνμ Hλν**
3.6	*LRf* (Limited Reflexivity)	**∀λ C ∃μ Hλμ Hλλ**
3.7	*LSm* (Limited Symmetry)	**∀λ∀μ C K Hμμ Hλμ Hμλ**
3.8	*Pop* (Population)	**∀x ∃λ Bxλ**
3.9	*SEx* (Singular Existence)	**∀λ ∃x Bxλ**
3.10	*PUP* (Uniform Population)	**∀x ∀λ Bxλ**
3.11	*PHEP* (Historical Expansion)	**∀x∀λ∀μ C K Bxλ Hμλ Bxμ**
3.12	*PIEP* (Imaginary Expansion)	**∀x∀λ∀μ C K Bxλ Hλμ Bxμ**
3.13	*Rf.S* (S-Reflexivity)	**∀λ Sλλ**
3.14	0*Rf* (Fixed Set Reflexivity)	**H00**
3.15	*H-S*	**∀λ∀μ C Hλμ Sλμ**
3.16	*Tr.S*	**∀λ∀μ∀ν C K Sλμ Sμν Sλν**
3.17	*Sm.S*	**∀λ∀μ C Sλμ Sμλ**
3.18	*EE* (Everything Exists)	**∀x Bx0**
3.19	*SI* (Self-Identity)	**∀x∀λ C Bxλ Ixxλ**
3.20	*ESI* (Existence of Self-Identicals)	**∀x∀λ C Ixxλ Bxλ**

4 QUANTIFICATIONAL ANALOGUES FOR FORMAL SYSTEMS

To determine whether or not a given formula is a truth of logic with respect to the model for a given formal system, first translate the formula into the Quantificational Analogue using the translation rules in §1 above. For each formal system discussed, there is a set of antecedent assumptions that may be used in QA, selected from the list in §2 above. For each formula in a given formal system, there is a formula of the Quantificational Analogue, constructed in the following way: where α is a formula of a given formal system, the corresponding formula of the Quantificational Analogue is $\mathbf{C}\boldsymbol{\beta\gamma}$, where $\boldsymbol{\beta}$ is the conjunction of the antecedent assumptions for that formal system and $\boldsymbol{\gamma}$ is the Quantificational Analogue translation of α. We speak of the Quantificational Analogue of a given formula, α, in a given formal system, **S**, as **QAS:**$\boldsymbol{\alpha}$. The formal system **CMn**, for example, has the antecedent assumption *Ex*—**∀λ∃μ Hμλ**. The Quantificational Analogue of 'C L p M p' in **CMn**, then, is called

QACMn: C L p M p,

and appears as follows:

Ex ‾‾‾ C L p M p

C ∀λ ∃μ Hμλ A ∃λ K Hλ0 P̄λ ∃λ K Hλ0 Pλ

The antecedent assumptions for each formal system discussed in the text are given below. Standard Interpretation is assumed.

4.1 Antecedent Assumptions for Elementary Logic. For elementary (non-modal) logic, there is a choice of assumption, depending upon whether or not individual constants occur in iwffs. If no individual constants appear in iwffs, the assumption *SEx* (3.9) is sufficient. If individual constants do appear, the assumption *EE* or *PUP* may be taken (3.18 or 3.10), for "standard" systems with existence presuppositions. For systems without existence presuppositions, see §5 below.

4.2 Antecedent Assumption for CMn (without quantification). *Ex* (3.1).

4.3 Antecedent Assumption for CM. *Rf* (3.2).

4.4 Antecedent Assumptions for CS4. *Rf* and *Tr* (3.2 and 3.3).

4.5 Antecedent Assumptions for CS5. *Rf*, *Tr* and *Sm* (3.2, 3.3 and 3.4).

4.6 Antecedent Assumptions for Brouwer System. *Rf* and *Sm* (3.2 and 3.4).

4.7 Antecedent Assumptions for Deontic System. *Ex*, *LRf*, *Tr*, *Qe* and *LSm* (3.1, 3.6, 3.3, 3.5 and 3.7).

4.8 Antecedent Assumptions for Epistemic System. (Same as **CS4**)

4.9 Antecedent Assumptions for Doxastic System. *Ex*, *LRf* and *Tr*. (3.1, 3.6 and 3.3).

4.10 Antecedent Assumptions for CY1. *Rf.S*, *LRf*, *0Rf*, *Ex*, *H-S* (3.13, 3.6, 3.14, 3.1, 3.15).

4.11 Antecedent Assumptions for CY2. Add to **CY1** *Tr.S* (3.16).

4.12 Antecedent Assumptions for CY3. Add to **CY2** *Sm.S* (3.17).

4.13 Antecedent Assumptions for Q1 Systems of Quantified Modal Logic. Add to the assumptions for the appropriate modal system the assumptions *Pop* and *SEx* (3.8 and 3.9). The system **CMnQ1**, for example, has the assumptions for **CMn** plus the **Q1** assumptions.

4.14 Antecedent Assumptions for Q2 Systems. Add to the assumptions for the basic modal system the assumption *PUP* (3.10).

4.15 Antecedent Assumptions for Q3 Systems. Add to the assumptions for the basic modal system the assumptions *Pop*, *SEx* and *PHEP* (3.8, 3.9 and 3.11).

4.16 Antecedent Assumptions for Q4 Systems. Add to the assumptions for the basic modal system the assumptions *Pop*, *SEx* and *PIEP* (3.8, 3.9 and 3.12).

5 ANTECEDENT ASSUMPTIONS AND CLOSURE CONDITIONS FOR IDENTITY IN QA

Any system that includes identity in its vocabulary has the antecedent assumptions *SI* and *ESI* (3.19 and 3.20) in the Quantificational Analogue. In addition, all such systems have the following QA closure condition:

> *Expansion* (QA). *Where* 'i' *and* 'j' *are individual constants and* 'm' *is a modal constant, and a branch contains a formula of the form* '$\bar{\text{I}}$ijm', *the branch may be expanded by adding, for each bsf in the branch of the form* 'Ψim' *a bsf of the form* 'Ψjm'.

Systems of quantified modal logic or non-modal logic for which the assumptions *EE* or *PUP* are taken are called systems with *Existential Presupposition.* That is, they presuppose the existence of every individual designated by an individual constant in the vocabulary. The **Q1** systems do not have this presupposition. The existence of an individual 'i' must be specified by a formula of the form '∃x Ixi'. The identity closure condition above and the antecedent assumptions for identity are the same in either case. The differentiation is made by the choice of assumption for the appropriate **Q**-variant system (4.13 through 4.16 above).

Appendix C

DEDUCTIVE TREATMENTS OF MONADIC MODALITY

1 EQUIVALENCE

The systems **CMn**, **CM**, **CS4** and **CS5** are equivalent to the deductive formal systems **Mn**, **M**, **M′** and **M″** of G. H. von Wright. This equivalence will be demonstrated in Appendix E. The von Wright presentation of **M** is, in turn, equivalent to the system **T** of Feys and Gödel, and the systems **M′** and **M″** are equivalent to the Lewis systems **S4** and **S5**. The following deductive presentations are von Wright's, adapted to the present notation. The sequence of theorems in Chapter III can be proved in the order given, using these deductive formalizations. The proofs are somewhat more straightforward if the logical constant 'E' (biconditional) is restored to the formal system and the pairs of conditionals given as **T-3** and **T-4**, **T-5** and **T-6**, **T-23** and **T-24**, **T-25** and **T-26**, and the **CS4** and **CS5** equivalences are proved as biconditionals.

2 VOCABULARY AND FORMATION

The vocabulary of the system **Mn** is the vocabulary of iwffs of the system **CMn** (III.2.1), with the addition of the logical constant 'E', which we introduce by the following definition.

$$\mathrm{E}\alpha\beta =_{\mathrm{df}} \mathrm{K\,C}\alpha\beta\,\mathrm{C}\beta\alpha$$

The formation rules for the system **Mn** are those given for iwffs of **CMn** (III.2.1.1), with 'E' added to 'K', 'A' and 'C' in formation rule (ii). It is best (but not essential) to limit the definition of well-formedness for the system **CMn** (only) by adding to the formation rules the requirement that all formulas must, in addition to meeting the conditions in III.2.1.1, be **Mn** formulas as defined in III.3—formulas in which each atomic subformula falls under the same number of modal operators.

The following definition is assumed:

$$L\alpha =_{df} \bar{M}\bar{\alpha}.$$

3 AXIOMS AND RULES OF INFERENCE

The axioms of the deductive systems of modal logic are given as axiom *schemata*. There are therefore an infinite number of axioms for each formal system: all formulas of a given form, say, '$CL\alpha M\alpha$', which is the first axiom of **Mn** below. Von Wright's axiomatization calls for an appropriate set of axioms and rules for truth-functional logic, following the usual custom. In general, this means that any tautology of truth-functional logic may be cited as a theorem of the systems of modal logic, and that any formula which results from uniformly substituting wffs of modal logic for statement letters in a truth-functional tautology may be cited as a theorem. For example, 'Cpp' is a truth-functional tautology. 'CMpMp' is a result of uniform substitution in 'Cpp', and is a theorem of all systems of modal logic. The truth-functional rules and equivalences given in the second and third sets of exercises for Chapter II may be used freely, or another set of rules for truth-functional logic substituted.

3.1 Axioms and Rules of Mn

3.1.1 Axioms.

Group A. (A set of axiom schemata for truth-functional logic)

Group B. (1) $CL\alpha M\alpha$

(2) $EMA\alpha\beta AM\alpha M\beta$

Axiom B(2) is the biconditional formed from **T-3** and **T-4** of **CMn.** Axiom B(1) is **T-1.**

3.1.2 Rules of Inference.

Group A. (An appropriate set of rules for truth-functional logic)

Group B. (1) If $E\alpha\beta$ is provable, then $EM\alpha M\beta$ is provable. (The M-Rule of Extensionality)

(2) If α is provable, then $L\alpha$ is provable. (The L-Rule of Tautology)

To shorten proofs, the following rule of inference may be added, which is equivalent to the M-Rule of Extensionality:

Rule B (1′) If $E\alpha\beta$ is provable then $EL\alpha L\beta$ is provable.

3.1.3 Theoremhood in **Mn**. The theorems of the systems **Mn** are all formulas which satisfy one or more of the following conditions:

(i) The formula is an axiom of **Mn**.

(ii) The formula can be derived from an axiom or previously proved theorem of **Mn** by one of the rules of inference of **Mn**.

(iii) The formula can be obtained from a tautology of truth-functional logic by substituting (uniformly) wffs of the system **Mn** for its atomic subformulas, so long as the resulting formula is an **Mn**-formula.

3.2 Axioms and Rules of M. The wffs of **M** are those of **Mn**, plus those formulas which result from deleting the restriction that all wffs must be **Mn**-formulas. Included in the wffs of the system **M**, then, are formulas where atomic subformulas fall under different numbers of modal operators.

3.2.1 Axioms. The axioms of **M** are the axioms of **Mn**, with axiom schema B(1) replaced by the following axiom schema:

Axiom B(1′) $C\alpha M\alpha$.

3.2.2 Rules. The rules of inference for **M** are those for **Mn**.

3.2.3 Theoremhood in **M**. The theorems of **M** are conditions (i) through (iii) of the system **Mn** (3.1.3), with "**Mn**" replaced everywhere in these conditions by "**M**."

3.3 Axioms and Rules of M′ (S4). The wffs of **M′** are those of **M**. The axioms and rules of inference of **M′** are those of **M**, with the addition of the following axiom schema:

Axiom C(1) $CMM\alpha M\alpha$.

3.4 Axioms and Rules of M″ (S5). The wffs of **M″** are those of **M**. The axioms and rules of inference of **M″** are those of **M**, with the addition of the following axiom schema:

Axiom C(2) $CML\alpha L\alpha$.

Axiom C(1) may be derived from Axiom C(2) and the theorems of **M**.

Appendix D

RECLAMATION OF UNCANCELLED FORMULAS

Once one has developed some facility with cancellation procedures, it becomes fairly easy to "read" the structure of the original uncancelled formula from a branch chart. In using the Quantificational Analogue, for example, it has by now become easy to move from the discovery that a given bsf will close a branch to the decision that a given iwff, taken as an antecedent assumption, will yield just such a bsf on cancellation. Although such moves come easily with experience, laying down a formal procedure for the reclamation of uncancelled formulas turns out to be rather complicated. The demonstrations of the next two appendixes are dependent upon there being such a procedure, however, and it is in general worth laying out.

It is easy enough to expand the cancellation rules in such a way that they become reduction rules of the sort used at the beginning of Chapter II, and then, by simply reversing the rules, turn the reductive procedure into a deductive one designed to recapture a sequent of one well-formed member from a set of terminal sequents. Such a technique for reclamation has the advantage of clarity, but the decided disadvantage that it is extremely long and complicated. The reclamation procedures to be described here will be designed to move rather directly from the branch chart to a single iwff that either is, or is equivalent to, the original uncancelled formula.

We divide the reclamation procedure into three parts: truth-functional reclamation, modal reclamation and quantifier reclamation.

1 TRUTH-FUNCTIONAL RECLAMATION

The reclamation of the truth-functional structure of a formula from its branch chart is straightforward enough, if we resign ourselves to the fact that no reclamation procedure will guarantee that we recapture the precise formula cancelled, but can only guarantee that we recapture an equivalent formula that may differ from the original in several ways:

- The reclaimed iwff will have negation confined to atomic formulas.
- The reclaimed iwff will contain no occurrences of 'C', but only 'A' and 'K'.
- Long alternations and conjunctions may be associated differently on reclamation. Depending on the method used, an iwff of the form 'AαA$\beta\gamma$' may be reclaimed to the equivalent 'AA$\alpha\beta\gamma$', or vice versa.

1.1 A Simple Reclamation Procedure. The simplest reclamation procedure will recapture a formula that is equivalent to the original, but, when split markers occur in the branch chart, is more complex than the original. A more complicated procedure to be described below will recapture a formula that is both equivalent to, and not more complex than, the original. We present the simpler procedure first, in the form of two rules. (See Figure 1 for an example.)

Rule 1. List the bsfs of each branch of the branch chart. If there are *n* basic formulas in a branch, write *n*-1 occurrences of 'A' to the left of the branch and delete all commas.

Rule 2. Form the conjunction of all such alternations reclaimed from a given branch chart.

Original formula: C K p C p q K p q

Branch chart:

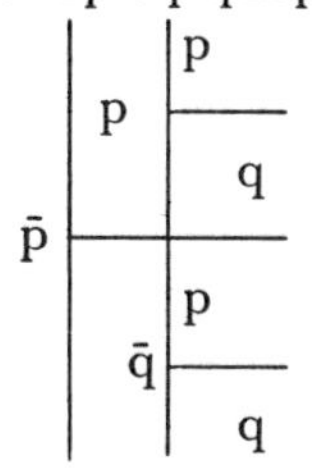

Rule 1. Top branch: A A $\bar{p}$ p p
Second branch: A A $\bar{p}$ p q
Third branch: A A $\bar{p}$ $\bar{q}$ p
Fourth branch: A A $\bar{p}$ $\bar{q}$ q
Rule 2. K K K A A $\bar{p}$ p p A A $\bar{p}$ p q A A $\bar{p}$ $\bar{q}$ p A A $\bar{p}$ $\bar{q}$ q

FIGURE 1

For a branch chart where there are several split markers, this procedure is clearly not satisfactory. The length of the resultant iwff can be reduced by the application of several equivalence rules (distribution and associativity; cf. Chapter II, 3.5, problems 10–24), but such simplification is not part of the reclamation procedure.

1.2 Full Truth-Functional Reclamation. There are several ways to shorten the results of reclamation, and several intermediate procedures that strike a balance between their own complexity and that of their result. We leave these to the reader and move on to a truth-functional reclamation procedure that is rather complicated, but that recaptures the original formula as closely as it can be recaptured on reclamation. We call this procedure *full reclamation.* The result of the full reclamation of a branch chart will be an iwff that is equivalent to the iwff whose cancellation produced the branch chart, and is of the same length as that iwff, and has the bsf occurring in it in the same order, but which may differ from the original iwff in the ways noted in §1 above.

We begin with two definitions. A *branch space segment* (BSS) is that area of a branch space between two vertical lines (from split markers), or between a vertical line and the right- or left-hand end of the branch chart. We say that two BSSs are *paired* when they are separated by a horizontal line that terminates at the left-hand boundary of the two segments. In the example given in Figure 2, there are six BSSs. The BSSs

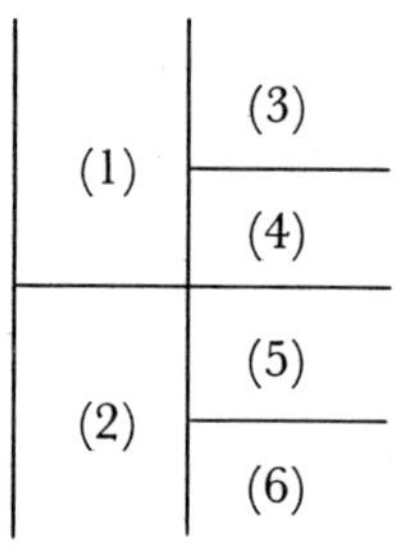

FIGURE 2

numbered (3) and (4) are paired, as are (5) and (6). Segments (4) and (5) are not paired, since the horizontal line that separates them does not terminate at their left-hand boundary.

In full reclamation, there is one procedure that is necessary in certain cases to prepare the branch chart for reclamation, and four truth-functional reclamation rules. The rules group bsfs in a branch chart and, when they are grouped, insert appropriate truth-functional operators. Each time a rule is applied, we cycle back through all four rules, applying them in order. The rules are applied to BSSs in a branch chart working from top to bottom, and from right to left. When no more moves can be made to the right of a given vertical line, we move to the left of that vertical line and begin again, always cycling through all rules to be sure that no more rules are applicable to the right of our position before we move further left. A long example is reclaimed step by step following the preparation and the four rules.

Preparation of the Branch Chart. In most cases, no preparation is necessary, and reclamation can begin with rule TF-1, following. In certain circumstances, however, preparation is needed to ensure that the reclaimed formula will be of the same length as the original. *When one or more bsfs occur in two BSSs that are adjacent and not all bsfs in those BSSs are the same, draw a new vertical line to separate the repeated bsfs from those that are not repeated.* Schematically:

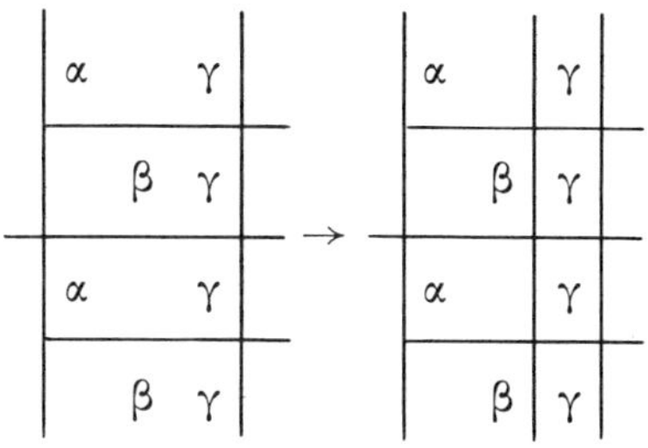

Reclamation Rule TF-1. Working from right to left within each BSS, delete commas and construct the alternation of each pair of wffs until the end of the BSS is reached.

$$\alpha, \beta \rightarrow A\alpha\beta$$

Reclamation Rule TF-2. When all the BSSs to the right of a given vertical line contain one wff each, form the conjunction of the wffs in each set of paired BSSs, and delete the horizontal line that separates the paired BSSs.

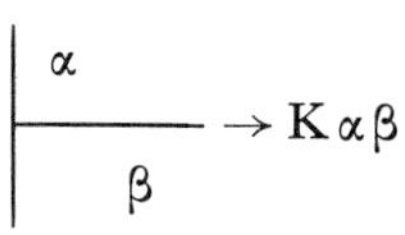

Reclamation Rule TF-3. When two paired BSSs contain precisely the same wff, the horizontal line which separates them is deleted and the two occurrences of the wff are reduced to one occurrence.

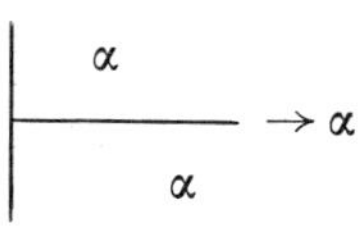

Reclamation Rule TF-4. When two horizontally adjacent BSSs are separated by a single vertical line, and no horizontal line terminates at the left-hand boundary of either, the vertical line which separates them is deleted and Rule TF-1 is applied.

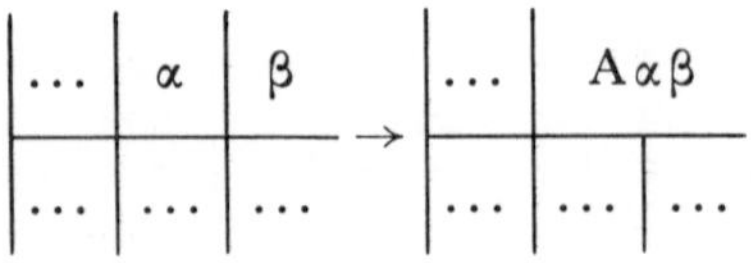

Example. We will now reclaim, step by step, the cancellation of a fairly long truth-functional formula, 'CApqCApKrsKpAqArs'. Since the reclaimed formula will not contain occurrences of 'C' and will have all negations confined to atomic formulas, it is helpful for the sake of the example to use the equivalences of Chapter I to construct the confined-negation equivalent of the formula for reference:

$$\text{`AK}\bar{p}\bar{q}\text{AK}\bar{p}\text{A}\bar{r}\bar{s}\text{KpAqArs'.}$$

The cancellation rules assure that the original conditional formula and its confined-negation equivalent will produce precisely the same branch

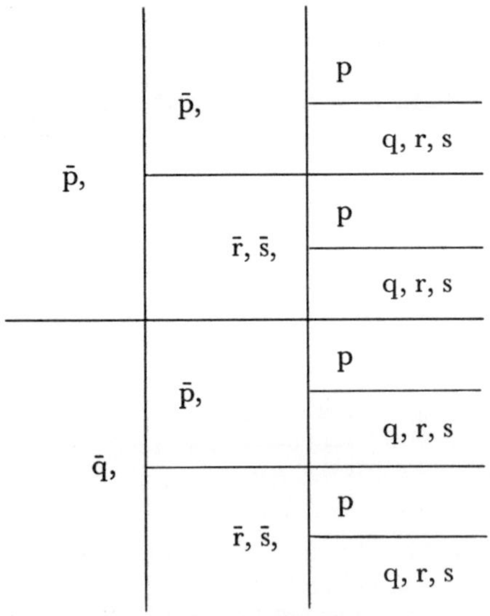

FIGURE 3

chart on cancellation (see Figure 3). No preparation of the branch chart is called for in this example.

Step 1. Rule TF-1 is applied to the right-most BSSs.

<table>
<tr><td rowspan="4">$\bar{p}$,</td><td rowspan="2">$\bar{p}$,</td><td>p</td></tr>
<tr><td>AqArs</td></tr>
<tr><td rowspan="2">$\bar{r}$, $\bar{s}$,</td><td>p</td></tr>
<tr><td>AqArs</td></tr>
<tr><td rowspan="4">$\bar{q}$,</td><td rowspan="2">$\bar{p}$,</td><td>p</td></tr>
<tr><td>AqArs</td></tr>
<tr><td rowspan="2">$\bar{r}$, $\bar{s}$,</td><td>p</td></tr>
<tr><td>AqArs</td></tr>
</table>

Step 2. Rule TF-2 is applied to paired BSSs to the right of the right-most vertical line.

<table>
<tr><td rowspan="2">$\bar{p}$,</td><td>$\bar{p}$,</td><td>KpAqArs</td></tr>
<tr><td>$\bar{r}$, $\bar{s}$,</td><td>KpAqArs</td></tr>
<tr><td rowspan="2">$\bar{q}$,</td><td>$\bar{p}$,</td><td>KpAqArs</td></tr>
<tr><td>$\bar{r}$, $\bar{s}$,</td><td>KpAqArs</td></tr>
</table>

Step 3. Rule TF-3 cannot be applied to the four conjunctions, since none of them are paired. No further moves can be made to the right of the vertical line, so we move left and apply TF-1.

<table>
<tr><td rowspan="2">$\bar{p}$,</td><td>$\bar{p}$,</td><td>KpAqArs</td></tr>
<tr><td>A$\bar{r}\bar{s}$,</td><td>KpAqArs</td></tr>
<tr><td rowspan="2">$\bar{q}$,</td><td>$\bar{p}$,</td><td>KpAqArs</td></tr>
<tr><td>A$\bar{r}\bar{s}$,</td><td>KpAqArs</td></tr>
</table>

Step 4. Rule TF-2 is applied twice.

<table>
<tr><td rowspan="2">$\bar{p}$,</td><td rowspan="2">$K\bar{p}A\bar{r}\bar{s}$,</td><td>$KpAqArs$</td></tr>
<tr><td>$KpAqArs$</td></tr>
<tr><td rowspan="2">$\bar{q}$,</td><td rowspan="2">$K\bar{p}A\bar{r}\bar{s}$,</td><td>$KpAqArs$</td></tr>
<tr><td>$KpAqArs$</td></tr>
</table>

Step 5. Rule TF-3 is applied twice to the right of the vertical line, where the branches are now paired.

<table>
<tr><td>$\bar{p}$,</td><td>$K\bar{p}A\bar{r}\bar{s}$,</td><td>$KpAqArs$</td></tr>
<tr><td>$\bar{q}$,</td><td>$K\bar{p}A\bar{r}\bar{s}$,</td><td>$KpAqArs$</td></tr>
</table>

Step 6. Rule TF-4 is applied to both branches.

<table>
<tr><td>$\bar{p}$,</td><td>$AK\bar{p}A\bar{r}\bar{s}KpAqArs$</td></tr>
<tr><td>$\bar{q}$,</td><td>$AK\bar{p}A\bar{r}\bar{s}KpAqArs$</td></tr>
</table>

Step 7. The two BSSs to the right of the vertical line are not paired, and no further rules can be applied to them. We move to the left, applying Rule TF-2 to the next group of BSSs, which are paired.

<table>
<tr><td rowspan="2">$K\bar{p}\bar{q}$,</td><td>$AK\bar{p}A\bar{r}\bar{s}KpAqArs$</td></tr>
<tr><td>$AK\bar{p}A\bar{r}\bar{s}KpAqArs$</td></tr>
</table>

Step 8. Cycling back, the pair of BSSs to the right of the vertical line are now paired, and Rule TF-3 is applied.

<table>
<tr><td>$K\bar{p}\bar{q}$,</td><td>$AK\bar{p}A\bar{r}\bar{s}KpAqArs$</td></tr>
</table>

Step 9. Finally, Rule TF-4 is applied.

$$AK\bar{p}\bar{q}AK\bar{p}A\bar{r}\bar{s}KpAqArs$$

The reclaimed formula is equivalent to the original. It differs from the confined-negation equivalent of the original only in that the last two alternations are associated differently.

2 MODAL RECLAMATION

To keep each aspect of the reclamation procedure as simple as possible, we separate modal reclamation from truth-functional reclamation. We assume in the following that truth-functional reclamation has been carried out, and that the result is a complex formula containing truth-functional operators ranging over basic formulas, with modal index columns occurring below the basic formulas.

As was the case with truth-functional reclamation, modal reclamation can guarantee only to recapture a wff that is equivalent to the original, with negation confined. '$\bar{L}$p' will be reclaimed to 'M$\bar{p}$', and so on. In addition, in cases where cancellation was carried out using the permissive rule which allows the same constant to be introduced on both sides of a split marker (cf. III.2.2), the formulas 'KLpLq' and 'LKpq', after cancellation and reclamation, will both appear as 'LKpq'. Since the two formulas are equivalent in all the modal systems under consideration, the guarantee that equivalent formulas will be reclaimed is not violated.

The rules for reclaiming modal formulas involve two sorts of moves: grouping subformulas according to the entries in their modal index columns, and reclaiming modal operators one at a time when all the subformulas having a given index have been grouped. As was the case with truth-functional reclamation, an abbreviated procedure could be adopted which would involve fewer reclamation steps, but the resulting formula would be more complex and more difficult to see as an equivalent to the original.

2.1 Full Modal Reclamation. The way in which modal operators are cancelled, and the way in which the dummies and constants resulting from modal cancellation fall in the index matrix, accomplish much of the organization of index matrices in advance. The three modal reclamation rules that follow are applied repeatedly, in order, until no index remains in the entire reclaimed formula except a single occurrence of '0'. The last step in reclamation is to delete the occurrence of '0'.

Rule Mod-1. Basic subformulas which have precisely the same modal index columns are grouped (by association) so that there is a well-formed complex subformula that contains all and only formulas with identical index columns. (See Figure 4 for an example.)

```
A p A q r → A A p q r
  0   0 0       0 0 0
  1   1 1       1 1 1
  a   a 2       a a 2
```

FIGURE 4

Rule Mod-2. Where two indexed wffs have precisely the same modal index, the index is assigned to the truth-functional operator that joins them. The entire wff formed by that operator is then regarded as having the index (Figure 5).

A	A	p	q	r	→	A	A	p	q	r
		0	0	0			0			0
		1	1	1			1			1
		a	a	2			a			2

FIGURE 5

Rule Mod-3. A dummy or constant in a modal index column is reclaimed only when (i) it occurs in precisely one modal index column in a formula and (ii) it occurs in the bottom position of that column. The dummy or constant is deleted and a modal operator is written to the left of the sign to which the index column is appended. The modal operator thus introduced then acquires the remainder of the index column. If the entry to be reclaimed is a dummy, the modal operator introduced is 'M'; if the entry is a constant, the modal operator is 'L'. (The rule is applied twice in Figure 6.)

A	A	p	q	r	→	A	M	A	p	q	L	r
	0			0			0				0	
	1			1			1				1	
	a			2								

FIGURE 6

In general, Rule Mod-1 will be applied only once, and the remainder of the procedure will consist of alternate applications of Mod-2 and Mod-3. To illustrate the applications of the three modal reclamation rules, we will use the same formula that was reclaimed by the truth-functional rules, but with a number of modal operators inserted. The original formula is as follows:

$$LCLA\bar{p}\bar{q}LCLA\bar{p}MK\bar{r}\bar{s}MKLLAMAqrMs.$$

Written with negation confined, the formula takes on the following form:

$$LAMK\bar{p}\bar{q}LAMK\bar{p}MA\bar{r}\bar{s}MKpLLAMAqrMs.$$

The branch chart resulting from the cancellation of either of the formulas above will have the same configuration as that shown in the section on truth-functional reclamation, except that every bsf will have a modal index column appended to it. Truth-functional reclamation is carried out, preserving the modal index columns, and when this is done, the formula appears as in Figure 7.

```
A K p̄ q̄ A K p̄ A r̄ s̄ K p A q A r s
    0 0     0   0 0   0   0   0 0
    1 1     1   1 1   1   1   1 1
    a a     2   2 2   2   2   2 2
            b   b b   d   d   d d
                c c       3   3 3
                          4   4 4
                          e   e f
```

FIGURE 7

Step 1. Rule Mod-1 is applied, regrouping the alternations.

```
A K p̄ q̄ A K p̄ A r̄ s̄ K p A A q r s
    0 0     0   0 0   0     0 0 0
    1 1     1   1 1   1     1 1 1
    a a     2   2 2   2     2 2 2
            b   b b   d     d d d
                c c         3 3 3
                            4 4 4
                            e e f
```

Step 2. Rule Mod-2 is applied three times.

```
A K p̄ q̄ A K p̄ A r̄ s̄ K p A A q r s
  0         0 0       0   0     0
  1         1 1       1   1     1
  a         2 2       2   2     2
            b b       d   d     d
              c           3     3
                          4     4
                          e     f
```

Step 3. Rule Mod-3 is applied four times.

```
A M K p̄ q̄ A K p̄ M A r̄ s̄ K p A M A q r M s
  0           0 0         0   0       0
  1           1 1         1   1       1
              2 2         2   2       2
              b b         d   d       d
                              3       3
                              4       4
```

Step 4. Rule Mod-2 is applied twice.

```
A M K p̄ q̄ A K p̄ M A r̄ s̄ K p A M A q r M s
  0         0             0 0
  1         1             1 1
            2             2 2
            b             d d
                            3
                            4
```

Step 5. Rule Mod-3 is applied three times (twice in succession for '3' and '4').

```
A M K p̄ q̄ A M K p̄ M A r̄ s̄ K p L L A M A q r M s
  0         0                 0 0
  1         1                 1 1
            2                 2 2
                              d d
```

Step 6. Rule Mod-2 applied once.

```
A M K p̄ q̄ A M K p̄ M A r̄ s̄ K p L L A M A q r M s
  0         0             0
  1         1             1
            2             2
                          d
```

Step 7. Rule Mod-3 applied once.

```
A M K p̄ q̄ A M K p̄ M A r̄ s̄ M K p L L A M A q r M s
  0         0             0
  1         1             1
            2             2
```

Step 8. Mod-2.

```
A M K p̄ q̄ A M K p̄ M A r̄ s̄ M K p L L A M A q r M s
  0       0
  1       1
          2
```

Step 9. Mod-3.

```
A M K p̄ q̄ L A M K p̄ M A r̄ s̄ M K p L L A M A q r M s
  0       0
  1       1
```

Step 10. Mod-2.

$$\underset{\substack{0\\1}}{A}\,M\,K\,\bar{p}\,\bar{q}\,L\,A\,M\,K\,\bar{p}\,M\,A\,\bar{r}\,\bar{s}\,M\,K\,p\,L\,L\,A\,M\,A\,q\,r\,M\,s$$

Step 11. Mod-3. The occurrence of '0' under the left-most sign is then deleted, and reclamation is finished.

$$\underset{\emptyset}{L}\,A\,M\,K\,\bar{p}\,\bar{q}\,L\,A\,M\,K\,\bar{p}\,M\,A\,\bar{r}\,\bar{s}\,M\,K\,p\,L\,L\,A\,M\,A\,q\,r\,M\,s$$

2.2 Modal Reclamation Without Grouping. In reclaiming modal indices when truth-functional reclamation has been carried out by the method of §1.1, it is necessary to retain modal dummies as subscripts under occurrences of 'M' when they are reclaimed. *Within each alternation*, the procedures of the preceding section are carried out. By retaining dummy subscripts below occurrences of 'M', the associativity and distribution equivalences of truth-functional logic are applied repeatedly until all subscripts are removed in one of the following ways:

(i) A dummy subscript below an occurrence of 'M' is deleted when it has precisely one occurrence in the entire formula.

(ii) When the configuration '$K M_d \alpha M_d \beta$' occurs (where the two occurrences of 'd' represent occurrences of the same dummy), it is reclaimed to '$M_d K \alpha \beta$', and the dummy is dropped if condition (i) is met.

3 RECLAMATION OF QUANTIFIED FORMULAS

The reclamation of quantified formulas is somewhat simpler than that of modal formulas, since the influence of a quantifier is determined both by the wff over which it has scope (as is the case with modal operators) and by the presence or absence of the variable of that quantifier in a given subformula. Since the truth-functional reclamation of a branch chart results in formulas where negation is confined to atomic subformulas and where 'A' and 'K' are the only binary truth-functional operators, we can take advantage of certain equivalences of elementary logic in quantifier reclamation. These are called *confinement equivalences*, and the relevant ones are as follows. 'α', in this context, represents a wff containing no occurrences of 'x'.

$K\alpha\forall x\Phi x$	is equivalent to	$\forall x K\alpha\Phi x$
$A\alpha\forall x\Phi x$	is equivalent to	$\forall x A\alpha\Phi x$
$K\alpha\exists x\Phi x$	is equivalent to	$\exists x K\alpha\Phi x$
$A\alpha\exists x\Phi x$	is equivalent to	$\exists x A\alpha\Phi x$

We are free, by virtue of these equivalences, to reclaim the branch 'Fa, Gb' to either 'A ∃x Fx ∃y Gy' or the equivalent '∃x ∃y A Fx Gy'.

In general, we will confine the scope of quantifiers to the smallest wff in which their variables occur. Quantifiers, like modal operators, are reclaimed in the reverse of the order in which they were introduced. Where predicates of degree greater than one are involved, it will be necessary to refer to the Q-chart in order to assure that the proper order of reclamation is followed.

It is assumed that truth-functional reclamation has been carried out, and that the result is a complex wff containing occurrences of individual dummies and constants within bsfs. The quantifier reclamation rules are as follows.

Rule Quant-1. Basic subformulas containing occurrences of the same dummy or constant are grouped (by association), so far as is possible, so that the occurrences of a given dummy or constant are limited to the smallest complex wff possible. (This step is not strictly necessary, given the equivalences above. It often simplifies the results of reclamation, however.) 'A Fa A Ga H1' is regrouped to 'A A Fa Ga H1'.

Rule Quant-2. Quantifiers are reclaimed in reverse order. A higher constant is reclaimed before a lower one, and an alphabetically later dummy is reclaimed before an alphabetically earlier one. In addition, an entry lower in a column of a Q-chart is reclaimed before an entry above it. The following Q-chart provides all that is needed for an example:

$$\begin{array}{ccc} & 1 & \\ \hline 2 & 3 & 5 \\ a & 4 & \\ & b & \end{array}.$$

The order of reclamation might be 5, b, 4, 3, a, 2, 1. This order meets the two conditions, but so do others, which may be used equally well; for example, b, a, 5, 4, 3, 2, 1.

Rule Quant-3. Occurrences of dummies and constants are reclaimed by placing quantifiers to the left of the smallest wff which contains all occurrences of the dummy or constant in the entire complex wff. A dummy is reclaimed by introducing an existential quantifier and replacing all occurrences of the dummy by occurrences of the variable of that quantifier. A constant is replaced in the same fashion by a universal quantifier. When the smallest wff containing all occurrences of a given dummy or constant is immediately preceded by a quantifier, the new quantifier resulting from the reclamation step is placed to the left of the already reclaimed quantifier.

Alternative Rule Quant-3. Given the equivalences above, all quantifiers may be brought "out front"—i.e., to the left of the entire reclaimed formula, in what is called "prenex form." In this case, Rule 1 may be omitted. The order of reclamation is that given in Rule 2, and each new quantifier reclaimed is placed to the left of the result of the preceding reclamation step. Dummies reclaim to existential quantifiers, and all occurrences of the dummy are replaced with occurrences of the variable of quantification. Constants reclaim to universal quantifiers in the same manner.

Appendix E

EQUIVALENCE OF CANCELLATION SYSTEMS AND AXIOMATIC SYSTEMS

1 EQUIVALENCE

We turn now to outlining a demonstration that the cancellation systems of monadic modal logic **CMn**, **CM**, **CS4** and **CS5** are equivalent to their "classical" axiomatic-deductive counterparts described in Appendix C. In several ways, this is already clear. The systems **CM** and **M** have the same model, for example, and a proof of the equivalence of the two systems can be based on this fact. With respect to the formulas discussed in the text, the two sets of formal systems share the same theorems and non-theorems. But to demonstrate that the systems are equivalent in the sense defined in II.1, we need to show that a formula is a theorem of a given cancellation system if and only if it is also a theorem of the axiomatic-deductive counterpart of that system. Such a proof involves recalling what it takes for a formula to be a theorem of each sort of formal system.

An iwff is a theorem of a given cancellation system, **CS**, if and only if it cancels to closed branches under the cancellation and closure rules of **CS**.

An iwff is a theorem of a given axiomatic-deductive system, **S**, if and only if it either is an axiom of **S**, or is derived from a theorem of S by a rule of inference or definition of **S**.

A demonstration that a cancellation system **CS** is equivalent to an axiomatic-deductive system **S** will involve the following sub-proofs.

1.1 *If an iwff* α *is a theorem of* **S**, *then* α *cancels to closed branches in* **CS**. This result may be shown by showing that all axioms of **S** cancel to closed branches in **CS**, and that if an iwff α cancels to closed branches in **CS**, and a second formula, β, can be derived from α by one of the rules of inference or definitions of **S**, then β also cancels to closed branches in **CS**.

1.2 *If an iwff* α *cancels to closed branches in* **CS**, *then* α *is a theorem of* **S**. This result may be shown for each pair of formal systems **S** and **CS** by showing that the reclamation (cf. Appendix D) of a branch that is closed in **CS** is a theorem of **S**.

1.3 Strategies and Definitions. Much of the wherewithal for the demonstration is already at hand, having been discussed in the text in one context or another. It is fairly easy to show that cancellation procedures and deductive procedures produce equivalent formal systems up to the point where modal closure conditions or modal axioms are brought into proofs, and the strategy for showing the equivalence beyond that point can be laid out in a straightforward way. There will be no attempt here to follow out all details of the demonstration. Certain parts of the proof are worthwhile exercises, however, and others are obvious enough to need no formal demonstration.

By a *modal instance of a truth-functional formula*, we mean a formula in which wffs of modal logic as defined in Chapter III are substituted uniformly for statement letters in a truth-functional formula. ‘$\mathrm{AMq\bar{M}q}$’ and ‘$\mathrm{ALKpq\bar{L}Kpq}$’ are both instances of the truth-functional tautology ‘$\mathrm{Ap\bar{p}}$’, for example.

By *opposites*, we mean, as before, any wff and its negation. We speak, for example, of ‘Mp’ as the opposite of ‘$\mathrm{\bar{M}p}$’ and vice versa. In addition, we will speak of truth-functional and modal operators as having opposites: ‘K’ is the opposite of ‘$\mathrm{\bar{K}}$’ and vice versa; ‘$\mathrm{\bar{L}}$’ is the opposite of ‘L’, and so on.

When two formulas are equivalent by definition, we will freely speak of them as having the same properties. The definitions involved are given in Chapter I and in Appendix C. In general, this will come up when we discuss a given formula and its confined-negation equivalent, like ‘$\mathrm{\bar{M}p}$’ and ‘$\mathrm{L\bar{p}}$’.

2 LEMMAS

The following statements are taken as lemmas and will not be demonstrated. Each of them is clearly true from an examination of the cancellation rules or a simple but perhaps lengthy induction on the cancellation rules.

Lemma I. *The succession of formal systems from statement logic through* **Mn**, **M**, **M′**(**S4**) *and* **M″**(**S5**) *is cumulative in both the cancellation and the axiomatic-deductive versions. That is, a theorem of statement logic is a theorem of all systems mentioned; a theorem of* **Mn** *in either version is a theorem of* **M**, **S4** *and* **S5**, *a theorem of* **M** *is a theorem of* **S4** *and* **S5**, *and a theorem of* **S4** *is a theorem of* **S5**.

Lemma II. *If the cancellation of a modal operator introduces a dummy into the modal index columns of bsfs within its scope, then the cancellation of the opposite of that modal operator introduces a constant, and vice versa.*

Lemma III. *A modal instance of a truth-functional tautology is a truth-functional tautology in the sense that it requires no modal axioms in its deductive proof and no modal closure conditions other than* **CMn** I *in its cancellation proof.*

Lemma IV. *A formula of the form* '$\mathrm{AM}\alpha\mathrm{L}\bar{\alpha}$' *is a modal instance of a truth-functional tautology. It is equivalent by definition to a formula of the form* '$\mathrm{AM}\alpha\bar{\mathrm{M}}\alpha$'.

Lemma V. *The cancellation of a wff whose left-most sign is an occurrence of a binary truth-functional operator yields a branch containing the elements* α, β *if and only if the cancellation of the opposite of that formula yields two branches, one of which contains the opposite of* α *and the other of which contains the opposite of* β.

Lemma VI. *If two formulas are equivalent by definition, they cancel in such a way as to produce either precisely the same branch charts or branch charts containing branches with precisely the same members, perhaps in different order.*

3 DEDUCTIVE THEOREMS ARE CANCELLATION THEOREMS

If an iwff α *is a theorem of an axiomatic-deductive system,* **S**, *then* α *cancels to closed branches in* **CS**.

We treat the formal systems under discussion as a cumulative sequence of formal systems (see Lemma I above). For each axiomatic-deductive formal system **S**, the theorems of **S** are the following (see Appendix C):

1. Tautologies of truth-functional logic
2. Modal instances of truth-functional tautologies
3. Axioms of **S**
4. The result of applying one of the rules of inference of **S** to a theorem of **S**.

For each of the four cases in which a formula may be said to be a theorem of **S**, we must show that that formula cancels to closed branches in **CS**.

3.1 *A tautology of truth-functional logic, of whatever complexity, cancels to closed branches.* Any tautology of truth-functional logic is, or is equivalent to, a formula of the form '$AA\alpha\bar{\alpha}\beta$', where '$\alpha$' is either an atomic or complex wff and β is (a) empty, (b) a basic formula or (c) an alternation or conjunction of basic or complex formulas. No matter how complex α and β might be, such a formula will cancel to closed branches. If α contains n binary truth-functional operators, Lemma V above is invoked n times in the argument.

3.2 *A modal instance of a truth-functional tautology, of whatever complexity, will cancel to closed branches.* Lemmas II, III, IV and VI above are invoked in the argument.

3.3 *Any axiom of a formal system* **S** *cancels to closed branches in* **CS**. That this holds for the simplest instances of each axiom has been shown in Chapter III. The **Mn** axioms are **T-1**, **T-3** and **T-4** of **CMn**; the **M** axiom is **T-29** of **CM**; the **M′**(**S4**) axiom is **T-45** of **CS4** and the **M″**(**S5**) axiom is **T-51** of **CS5**. That this result holds for complex instances of the axioms if it holds for the simplest instances may be shown by appeal to the same lemmas used in the two preceding sections.

3.4 *If α cancels to closed branches in* **CS**, *and β is derived from α by one of the rules of inference of* **S**, *then β cancels to closed branches in* **CS**. Aside from the truth-functional rules of inference, which we may safely regard as accounted for by the lemmas, each of the axiomatic systems of modal logic has two rules of inference, the Rule of Extensionality and the Rule of Tautology (cf. Appendix C). This result has been shown for each of these rules in Chapter III, §3.2.1.

4 CANCELLATION THEOREMS ARE DEDUCTIVE THEOREMS

If an iwff α cancels to closed branches in a cancellation system **CS**, *then α is a theorem of the corresponding axiomatic-deductive system* **S**. It is sufficient to show that the reclamation of any branch that is closed in **CS** is also a theorem of **S**. Given this result, if a cancellation branch chart contains more than one branch, and every branch is closed in **CS**, then the full reclamation will be a conjunction of wffs, each of which is a theorem of **S**. By the truth-functional rules which are common to all the formal systems under discussion, a conjunction of theorems of a formal system **S** is itself a theorem of **S**.

It is necessary, for this result, to account for every sort of situation where a branch is closed, which involves an induction of forbidding length. What will be outlined here is what is needed for the induction, and enough of each of the steps of the induction that the reader can carry out the details. We will need, first of all, a vocabulary for discussing sequents, parts of sequents and wffs formed from the members of sequents.

The *sequent alternation* related to a sequent, X, of *n* members, is the alternation formed by writing *n*-1 occurrences of 'A' to the left of the wffs which are the members of X. We represent the sequent alternation of a sequent, X, as 'A(X)'. In the following, the sequent variable 'Z' will be used to represent a sub-sequent containing some, but not necessarily all, of the members of the sequent designated by 'X', and the sequent alternation of Z will be designated 'A(Z)'. Finally, the sub-sequent composed of all the members of X which are not also members of Z will be designated 'X − Z', and the sequent alternation of this sub-sequent will be designated 'A(X − Z)'. Suppose 'X' represents the sequent 'p, q, r, $\bar{p}$, s, $\bar{q}$'. Then A(X) is 'AAAAApqr$\bar{p}$s$\bar{q}$'. The sub-sequent Z might be 'q, r, $\bar{p}$', in which case A(Z) is 'AAqr$\bar{p}$'. The residual sub-sequent X − Z is 'p, s, $\bar{q}$', and A(X − Z) is 'AAps$\bar{q}$'. In the following, we must allow for the cases where any or all of X, Z or X − Z contain precisely one well-formed member or no members. If a sequent, X, has precisely one member, α, then A(X) is just α. If a sequent, X, has no members, then there is no sequent alternation, and when we represent a formula such as 'AβA(X)', where β is a wff and X is empty, then the formula is simply β.

4.1 *The truth-functional reclamation of a closed branch is a tautology of truth-functional logic.* The reclamation of a closed branch is always a formula of the form 'AA$\alpha\bar{\alpha}$A(X)', where α is an atomic formula and A(X) is as described above. Any such formula is a truth-functional tautology, and a theorem of all the axiomatic-deductive systems under consideration.

4.2 *If a branch is closed by the application of a* **CMn** *closure condition, then the wff resulting from the modal reclamation of that branch is a theorem of* **Mn**. This can be shown in a straightforward way for the case where a closure condition is applied to the entry in the bottom position in a modal index column, and then the result is extended to positions above the bottom position. The possible configuration of dummies and constants in the bottom positions of modal index columns of the closure elements of a branch that is closed in **CMn** are given in Table 20. 'c' and 'd' represent constants and dummies, and 'd′' represents a dummy different from 'd'. For each of the four cases, it must be shown that the formula resulting from the modal reclamation of the bottom entry in the modal index columns of the bsfs in the branch is a theorem of axiomatic **Mn**, whether or not the constants and dummies in the indices of α and $\bar{\alpha}$ occur also in

TABLE 20

	α	$\bar{\alpha}$	Truth-Functional Reclamation	Closure Condition
Case 1	c	c	$A\,A\,\underset{c}{\alpha}\,\underset{c}{\bar{\alpha}}\,A(X)$	none
Case 2	c	d	$A\,A\,\underset{c}{\alpha}\,\underset{d}{\bar{\alpha}}\,A(X)$	**CMn I**
Case 3	d	d	$A\,A\,\underset{d}{\alpha}\,\underset{d}{\bar{\alpha}}\,A(X)$	**CMn II**
Case 4	d	d′	$A\,A\,\underset{d}{\alpha}\,\underset{d'}{\bar{\alpha}}\,A(X)$	**CMn II**

the indices of bsfs within X. Case 1 will be considered, even though no closure rules are required. For each case, a formula will be cited which is a theorem of axiomatic **Mn**. These may be proven as exercises in deductive proof.

Case 1a. $AA\underset{c}{\alpha}\underset{c}{\bar{\alpha}}A(X)$, where 'c' is '0'. Then the modal reclamation consists of dropping the occurrences of '0', and the case is that described in 4.1 above.

Case 1b. $AA\underset{c}{\alpha}\underset{c}{\bar{\alpha}}A(X)$, where 'c' is a constant other than '0' and does not occur in the indices of bsfs in X. The modal reclamation is $ALA\alpha\bar{\alpha}A(X)$, which is a theorem of **Mn**. Prove '$ALApp\bar{p}q$' from '$App\bar{p}$' by the Rule of Tautology and addition.

Case 1c. $AA\underset{c}{\alpha}\underset{c}{\bar{\alpha}}A(X)$, where 'c' occurs in the indices of some bsfs in X, say those in the sub-sequent Z (see §4). The modal reclamation is $ALAA\alpha\bar{\alpha}A(Z)\,A(X-Z)$, which is a theorem of **Mn**. Prove

$$ALAAp\bar{p}qr$$

from '$Ap\bar{p}$' by addition and the Rule of Tautology.

Case 2a. $AA\underset{c}{\alpha}\underset{d}{\bar{\alpha}}A(X)$, where neither 'c' nor 'd' occurs in the indices of any bsf in X. The modal reclamation is $AAL\alpha M\bar{\alpha}A(X)$, which is a theorem of **Mn**. Prove '$AALpM\bar{p}q$' from '$ALpM\bar{p}$' by addition. ('$ALpM\bar{p}$' is a modal instance of a truth-functional tautology.)

Case 2b. 'c' occurs in the index of some bsfs in X, say those in Z. The modal reclamation is $\mathrm{AALA}\alpha\mathrm{A(Z)\,M}\bar{\alpha}\mathrm{A}(X-Z)$, which is a theorem of **Mn**. Prove 'AALApqM$\bar{\mathrm{p}}$r'. (This follows easily from **T-19** by addition. The proof of **T-19** is tricky in von Wright's system, involving several truth-functional equivalences and several applications of the Rule of Extensionality and, of course, Axiom B(2).)

Case 2c. 'd' occurs in the index of some bsfs in X, say those in Z. The modal reclamation is $\mathrm{AAL}\alpha\mathrm{MA}\bar{\alpha}\mathrm{A(Z)\,A}(X-Z)$, which is a theorem of **Mn**. Prove 'AALpMA$\bar{\mathrm{p}}$qr' (from Axiom B(2)).

Case 3a. $\mathrm{AA}\underset{d\;\;d}{\alpha\bar{\alpha}}\mathrm{A(X)}$, where 'd' occurs in no index in X. The modal reclamation is $\mathrm{AMA}\alpha\bar{\alpha}\mathrm{A(X)}$, which is a theorem of **Mn**. Prove AMAp$\bar{\mathrm{p}}$q.

Case 3b. 'd' occurs in some index in X, say the indices of Z. The modal reclamation is $\mathrm{AMAA}\alpha\bar{\alpha}\mathrm{A(Z)\,A}(X-Z)$. Prove 'AMAAp$\bar{\mathrm{p}}$qr'.

Case 4a. $\mathrm{AA}\underset{d\;\;d'}{\alpha\bar{\alpha}}\mathrm{A(X)}$, where neither 'd' nor 'd′' occurs in any index in X. The modal reclamation is $\mathrm{AAM}\alpha\mathrm{M}\bar{\alpha}\mathrm{A(X)}$. Prove 'AAMpM$\bar{\mathrm{p}}$q' from Axiom B(1) by addition.

Case 4b. $\mathrm{AA}\underset{d\;\;d'}{\alpha\bar{\alpha}}\mathrm{A(X)}$, where one of the dummies occurs in some index in X, say those of the formulas in Z. The modal reclamation is $\mathrm{AAM}\alpha\mathrm{MA}\bar{\alpha}\mathrm{A(Z)\,A}(X-Z)$. Prove 'AAMpMA$\bar{\mathrm{p}}$qr' from Axioms B(1) and B(2), using addition as needed.

We have shown, then, that the wff resulting from the reclamation of the bottom positions of the modal index columns of any branch which is closed in **CMn** is a theorem of **Mn**, and thus also of **M**, **S4** and **S5**. We now show that if this holds for the *n*th position in the modal index columns of a closed branch, it holds for the $(n-1)$th position, and thus holds for all. In the interest of brevity we will suppress the passive elements in the branch which have been represented by 'X', 'Z' and '$X-Z$'. It is clear by now that if the result holds for the closure elements it can be shown to hold for alternations of the closure elements and those in X by applications of the addition rule of truth-functional logic at appropriate points in the proof.

We will use 'O', with primes as needed, to represent any modal operators written before the entire formula or before the closure elements α and $\bar{\alpha}$ as the result of the reclamation of operators from indices, working backward from the bottom position to the *n*th position (allowing for the case where the bottom position is the nth position). At this intermediate stage in reclamation, we will have a partially reclaimed formula which is by hypothesis a theorem of **Mn**, and where the next position in the modal index columns to be reclaimed is the $(n-1)$th position. The formula will have one of two forms: $AO\alpha O'\bar{\alpha}$, or $OAO'\alpha O''\bar{\alpha}$. The latter form could only have occurred if the nth position of the indices of α and $\bar{\alpha}$ contained occurrences of the same dummy or constant. The order of cancellation and writing of indices assures that this can be the case only when the $(n-1)$th positions of both indices also contained occurrences of the same dummy or constant.

We show, then, for any combination of dummies or constants which could occur in the $(n-1)$th position of index columns of the closure elements, that reclaiming these dummies and constants yields a theorem of **Mn** if reclaiming the dummies and constants in the nth position does. The possible combinations of dummies and constants are as before.

Case 1. $OAO'\alpha O''\bar{\alpha}$. Where 'c' is '0', the index is simply dropped,
c c
and the formula is provable in **Mn** by hypothesis. Where 'c' is some constant other than '0', the modal reclamation is $LOAO'\alpha O''\bar{\alpha}$. Prove from $OAO'\alpha O''\bar{\alpha}$ by the Rule of Tautology.

Case 2. $AO\alpha O'\bar{\alpha}$. The modal reclamation is $ALO\alpha MO'\bar{\alpha}$. Prove
c d
from $AO\alpha O'\bar{\alpha}$ by the Rule of Tautology and **T-10** of **Mn**.

Case 3. $OAO'\alpha O''\bar{\alpha}$. The modal reclamation is $MOAO'\alpha O''\bar{\alpha}$. Prove
d d
from $OAO'\alpha O''\bar{\alpha}$ by the Rule of Tautology and Axiom B(1).

Case 4. $AO'\alpha O''\bar{\alpha}$. The modal reclamation is $AMO'\alpha MO''\bar{\alpha}$. Prove
d d′
from $AO'\alpha O''\bar{\alpha}$ by the Rule of Tautology, Axiom B(1) and Axiom B(2).

We have shown that the reclamation of the bottom position of the modal index columns in a closed branch of **CMn** is reclaimed to a theorem of **Mn**, and that if any given position in a modal index column reclaims to a theorem of **Mn** then so does the next position above. Thus the full reclamation of a branch chart that is closed in **CMn** yields a theorem of **Mn**.

4.3 *If a branch is closed by the application of the* **CM** *closure condition, then the wff resulting from the modal reclamation of that branch is a theorem of* **M.** We will deal with the **CM** closure condition in the dummy-deletion version. To deal with the constant-repetition version, introduce a reclamation rule to cover the case where two occurrences of the same constant occur consecutively in a given index column. They simply reclaim as if they were one occurrence.

The **CM** closure condition is effective in closing branches only when all positions in the modal index column to which the closure condition is *not* applied are identical in both closure elements, or can be made identical by applications of the **CMn** closure conditions, which have already been accounted for. A branch can be closed by the **CM** condition only if, *after* the **CM** condition is applied, the branch is closed in **CMn**. What needs to be shown, then, is that if a branch is closed in **CMn** after application of the **CM** closure condition—i.e., given the preceding section, if the reclamation of that branch is a theorem of **Mn**—then the reclamation of the branch before application of the **CM** closure condition must be a theorem of **M**.

The configuration which produces closure upon application of the **CM** closure condition may be represented (after truth-functional reclamation) as $\mathrm{A\,A}\,\alpha\,\bar{\alpha}\,\mathrm{A(X)}$, where 'd' represents the dummy deleted by application

. .
d
. .

of the closure condition, and the indices of α and $\bar{\alpha}$ are otherwise closed by application of the **CMn** closure conditions. The cases involving the occurrence of 'd' or other modal indices in X may be suppressed as before. The modal reclamation of the closure elements before **CM** closure is a formula of the form $\mathrm{A\,O}\,\alpha\,\mathrm{O'\,M\,O''}\,\bar{\alpha}$, where the occurrence of 'M' among the modal operators governing $\bar{\alpha}$ results from the reclamation of 'd'. The modal reclamation after **CM** closure is applied is then $\mathrm{A\,O}\,\alpha\,\mathrm{O'\,O''}\,\bar{\alpha}$, which by hypothesis is a theorem of **Mn**. It is worth noting that the number of modal operators represented by O must then be equal to the number of modal operators represented by O′ and O″ taken together. $\mathrm{A\,O}\,\alpha\,\mathrm{O'\,M\,O''}\,\bar{\alpha}$ is proven from $\mathrm{A\,O}\,\alpha\,\mathrm{O'\,O''}\,\bar{\alpha}$, using Axiom B(1′) as necessary. The axiom is most conveniently used in the equivalent form $\mathrm{A}\,\alpha\,\mathrm{M}\,\bar{\alpha}$ in the proof.

4.4 *If a branch is closed by the application of the* **CS4** *or* **CS5** *closure conditions, then the wff resulting from the modal reclamation of that branch is a theorem of, respectively,* **M′**(**S4**) *or* **M″**(**S5**). The strategy for this demonstration is by now obvious. The reclamation of a branch *after* **CS4** or **CS5** closure has been applied is a theorem of **M** (more often of **Mn**)—i.e., if the branch closes after **CS4** or **CS5** closure conditions have been applied, it closes under the rules of **CM** or **CMn**, and it is already apparent that a formula which closes in **CM** or **CMn** is a theorem of **M** or **Mn**. It is sufficient to show, then, for the cases where **CS4** or **CS5** closure conditions will close a branch, that if the reclamation of the branch after these closure conditions have been applied is a theorem of **M**, then the reclamation of the branch before the closure conditions are applied is a theorem of **M′**(**S4**) or **M″**(**S5**), as appropriate. A somewhat stronger result can be shown by appeal to the **S4** and **S5** equivalences: that the formula reclaimed before **CS4** or **CS5** closure is equivalent in **M′**(**S4**) or **M″**(**S5**) to the formula reclaimed after **CS4** or **CS5** closure, and since the latter is by hypothesis a theorem of **M**, then the former is a theorem of **M′**(**S4**) or **M″**(**S5**). The demonstration, like the previous ones, involves a consideration of the cases in which the **CS4** and **CS5** closure conditions may be applied in such a way as to effect closure. The details are left to the reader.

5 SUMMARY

The results sketched in §3 and §4, taken together, provide the demonstration that the cancellation systems and the axiomatic-deductive systems are indeed equivalent in the sense that any iwff that is a theorem of a given axiomatic-deductive system is also a theorem of the corresponding cancellation system, and vice versa.

Appendix F

COMPLETENESS AND SOUNDNESS RESULTS

A general completeness and soundness result will be argued for in this appendix which not only covers the formal systems of modal logic developed in the book, but also covers other formal systems that may be developed using cancellation techniques and the Quantificational Analogue. Using the vocabulary of Appendix B, §4, the result may be stated as follows.

A cancellation system of modal logic, **CS**, *is complete and sound provided that:* (i) *the conjunction of antecedent assumptions permissible in* **QACS** *is consistent, and* (ii) *a formula,* α, *of* **CS**, *is a theorem of* **CS** *if and only if* **QACS:**$\boldsymbol{\alpha}$ *is a theorem of* **QACS.**

1 OBJECT LANGUAGE AND QUANTIFICATIONAL ANALOGUE

Completeness and soundness properties have been demonstrated by others for axiomatic **Mn**, **M**, **S4** and **S5**. Since the cancellation systems **CMn**, etc., are equivalent to their axiomatic counterparts, we may safely claim completeness and soundness for these systems in a derivative way. But the four "mainstream" systems are only the core of the many systems of modal logic discussed in this book. For these, and for the systems that can be constructed employing the same resources, we need the more general result stated above. This, in turn, demands some discussion of the way in which the cancellation systems of modal logic are, and are to be, developed, at least to the extent that we must confront the chicken-egg question concerning object language modal cancellation and the Quantificational Analogue.

The crucial point of information concerning the Quantificational Analogue is simply this: QA, as a "mechanization" of Hintikka's model system semantic technique, is prior to the object language cancellation systems of modal logic. Historically, the development ran something like this: Hintikka's model system technique and the Binkley-Clark cancellation techniques for elementary logic were developed independently. The two were brought together by Binkley and Clark in 1963, and the Quantificational Analogue for monadic modal logic was developed. From this, the cancellation techniques for monadic modal logic in the object language were developed in such a way that the cancellation of a modal formula reproduced, in abbreviated form, the cancellation of its QA translation. Compare, for example, the cancellation of 'CLpMp' with the cancellation of its QA translation. Moreover, the closure conditions for the object language systems of modal logic may be viewed as "carrying out instructions" given in the antecedent assumptions for those systems.

In designing further object language systems of modal logic, beyond those presented here, the relationships between object language cancellation and QA translation, and between object language closure conditions and QA closure conditions, should be preserved. Apart from the fact that object language manipulations without reference to the semantic devices often lead to confusing results, the preservation of the close "physical" relationship between object language cancellation and QA interpretation is vital to assuring completeness and soundness results of newly developed systems.

Properly, the definition of *truth of logic* for each object language formal system may be relativized to the Quantificational Analogue rather than to the model system described in Chapter IV. It seems obvious by now that to define a truth of logic in **Mn**, for example, as a formula which is true under every interpretation on the **Mn** model system described in Chapter IV is tantamount to defining a truth of logic in **Mn** as a formula whose Quantificational Analogue is a truth of logic in **QAMn**, as described in Chapter IV and Appendix B. A demonstration of this fact will not be attempted here. Instead, for our general result, we redefine *truth of logic* as follows, using '**S**', as before, to represent any given formal system and '**QAS**' to represent the Quantificational Analogue of that formal system as described in Appendix B, §4. '**QAS:α**' represents the Quantificational Analogue of a formula α of **S**, i.e. the QA conditional having the conjunction of the antecedent assumptions of **S** as antecedent and the QA translation of α as consequent.

A formula, α, is a truth of logic with respect to a given formal system, **S**, *if and only if* **QACS:α** *is a truth of logic with respect to* **QACS**.

This result can be shown, rather than accomplished by fiat, by attending to the definition of "truth of logic" given in Chapter IV for the modal systems and developing a similar definition in an obvious way for elementary logic, using the model *set* of Chapter IV plus the conditions on quantifiers given in Chapter V. It is perhaps easier to see in terms of provability of **QACS:α** in **QACS**. That is, we may show that α is a truth of logic in **S** if and only if **QACS:α** is provable in **QACS**, and then, relying on the completeness and soundness of **QACS**, develop the result above.

2 THE QUANTIFICATIONAL ANALOGUE AND ELEMENTARY LOGIC

The machinery of the Quantificational Analogue is merely elementary logic (with existence presupposition), with the addition of certain items of vocabulary (H, B) and abbreviation to allow for what seems to be quantification over model sets. The nature of these abbreviations is discussed in the note at IV.7.1. The Quantificational Analogue of any particular system of modal logic, as described in Appendix B, is a subset of the set of formulas of elementary logic. That is, for any system **S** of modal logic, **QAS** is the subsystem of elementary logic that has as its iwffs the conditionals of elementary logic that have as their antecedents the conjunction of antecedent assumptions associated with **S**, and as their consequents the QA translations of iwffs of **S**.

Let us assume that elementary logic, in its cancellation version, is complete and sound. This has been proven by Binkley and Clark, and the proof will be discussed in a later section. This means that all and only truths of logic eventually cancel to closed branches. Consider now the subsystem of elementary logic which comprises **QAS**, the Quantificational Analogue of some formal system **S** of modal logic. Clearly, if every theorem of elementary logic is a truth of logic, then every formula of the subsystem **QAS** that is a theorem of elementary logic is a truth of logic; and if every truth of elementary logic will eventually cancel to closed branches, then every truth of elementary logic which is a formula of the subsystem **QAS** must cancel to closed branches. That is, **QAS**, as a subsystem of a complete and sound system of elementary logic, is itself complete and sound.

All the formulas of a Quantificational Analogue system **QAS** are conditional in form; the formula designated **QAS:α** is the Quantificational Analogue conditional having the translation of the object language formula α as its consequent and the conjunction of antecedent assumptions associated with **S** as antecedent. Suppose now that **QAS:α** cancels to closed branches. There are three circumstances to consider: (1) all closure elements come from the consequent of **QAS:α**; (2) some closure elements come from the antecedent of **QAS:α** and some from the consequent; (3) all closure elements come from the antecedent of **QAS:α**. In case (1), the translation of α is itself a theorem of **QAS** and is a truth of logic. In case (2), the translation of α is a relevant consequence of the antecedent assumptions of **S**, meeting Binkley's Test for entailment in at least some of its branches. That is, given the assumed truth of the antecedent assumptions, α is a truth of **S**. Case (3), however, would allow the QA translation of any formula to cancel to closed branches in **QAS**. To prevent this—i.e., to keep the system sound—we must assure ourselves that the antecedent of every formula in **QAS** is of such a nature that case (3) cannot occur. The antecedent, β, of every formula $C\beta\gamma$ of **QAS** is of course the same formula, and what we must assure ourselves of is that $\bar{\beta}$ (as it would occur after the cancellation of 'C') does not cancel to closed branches. In short, we must be sure that the conjunction of antecedent assumptions of a given formal system is consistent. If this condition is met, then we can be sure that **QAS:α** will cancel to closed branches only when the translation of α is either itself a truth of logic or is, in the sense of Chapter VII, entailed by the antecedent assumptions of **S**.

It remains then to discuss the relationship between the cancellation of α in **S** and the cancellation of **QAS:α** in **QAS**.

3 MODAL CANCELLATION AND THE QUANTIFICATIONAL ANALOGUE

We have shown mapping rules for moving from iwffs of the object language to formulas of the Quantificational Analogue (Appendix B). To tighten the connection between object language formulas and the quantificational metalanguage, we can give two additional procedures relating the two: a set of reclamation rules for moving from an object language branch chart to the Quantificational Analogue translation of the formula from whose cancellation the branch chart resulted, and a set of rules for moving from an object language branch chart to the Quantificational Analogue branch chart of the translation of that formula.

The first procedure is a simple variant of the reclamation procedure for modal logic, and begins where modal reclamation begins: after truth-functional reclamation of an object language formula has been performed. At this stage, we have a formula containing occurrences of 'A' and 'K', with negation confined to atomic formulas and index matrices occurring below the atomic formulas. The following rules for Quantificational Analogue reclamation are used in conjunction with certain of the modal reclamation rules, as specified below.

Rule QAR-1. Each predicate of degree n in a basic formula is changed to a predicate of degree $n+1$. (A statement constant is considered to be a predicate of degree zero, and is changed to a capital letter at this step.)

The additional argument-place after each predicate is to the right of all other argument-places.

Individual dummies and constants from the index matrix are inserted in the argument-places directly above their occurrences in the index matrix. Modal dummies or constants which occur in the bottom position of the modal index columns of the bsfs are inserted, following the predicate constant, in the new argument-place to the right of the individual dummies and constants. (See Figure 8 for an example.)

A	F	G	→	A	F2a	G34
	0	0			0	0
	12	13			12	13
	a	4			a	4

FIGURE 8

Modal index columns are grouped as for modal reclamation, following reclamation rules Mod-1 and Mod-2 from Appendix D, §2. These rules are used alternatively with the following rules, as each becomes applicable.

Rule QAR-2. When a modal dummy or constant has precisely one occurrence in the modal index columns of the entire formula, and that occurrence is in the bottom position of a modal index column, it is reclaimed as follows, provided that there are no individual dummies or constants to the right of it.

A modal dummy, 'd', is reclaimed by inserting, immediately to the left of the sign in whose index the dummy appears, a subformula of the form '∃s K Hsm', where 's' is a model set variable that has no prior occurrence in the scope of '∃', and 'm' is an occurrence of the modal dummy or constant that occurs immediately above 'd' in the modal index column from which 'd' is being reclaimed. All occurrences of 'd' in bsfs are then replaced with 's'. The quantifier now acquires the remainder of the modal index column, with 'm' in the bottom position. A modal constant is reclaimed in the same manner, by inserting a subformula of the form '∀s C Hsm' (see Figure 9).

A	$\bar{P}$2	Qa	→	A	∀λ	C Hλ1 $\bar{P}$λ	∃λ	K Hλ1 Qλ
	0	0			0		0	
	1	1			1		1	
	2	a						

FIGURE 9

Indices are grouped, using Mod-1 and Mod-2, and reclaimed, using QAR-2, until the only remaining modal index is a single occurrence of '0', which is then deleted. The following two steps are continuations of the example in Figure 9.

A ∀λ C Hλ1 Pλ $\bar{\exists}$λ K Hλ1 Qλ
0
1

∀μ C Hμ0 A ∀λ C Hλμ $\bar{\text{P}}$λ ∃λ K Hλμ Qλ
Ø

Compare this result with the cancellation and the QA translation of 'L C M p M q' and its confined-negation equivalent 'L A L $\bar{\text{p}}$ M q'.

When individual dummies and constants are present in a partially reclaimed formula as well as modal dummies and constants, the following reclamation rule is used in conjunction with QAR-2.

Rule QAR-3. Individual dummies and constants are considered to be part of the index matrix and are grouped with modal dummies and constants in applications of the modal grouping rules Mod-1 and Mod-2. When two index matrices are alike with respect to the modal index column, but one contains occurrences of individual dummies or constants to the right of the modal index column where the other does not, the individual dummies or constants are retained in the new index formed under Rule Mod-2. An individual dummy or constant is reclaimed when it has precisely one occurrence in the index matrices of the entire formula, and when that occurrence is in the bottom position of an index matrix. Under Rule QAR-2, an individual dummy or constant must be reclaimed before the modal dummy or constant to the left of it is reclaimed. If two individual dummies or constants should occur next to each other on the same horizontal line of an index matrix, the Q-chart is used, as in quantifier reclamation (Appendix D), to choose which is to be reclaimed first.

An individual dummy, 'd', is reclaimed by inserting, immediately to the left of the sign in whose index the dummy appears, a subformula of the form '∃x K Bxm', where 'x' is an individual variable that has no prior occurrence within the scope of '∃', and 'm' is an occurrence of the modal dummy or constant which occurs on the same horizontal line of the index matrix as the occurrence of 'd' which is being reclaimed. All occurrences of 'd' in bsfs are then replaced with occurrences of 'x'. The new quantifier now acquires the remainder of the index matrix. An individual constant is reclaimed in the same manner, by inserting a subformula of the form '∀x C Bxm'.

Figure 10 is a step by step example of reclamation to QA, beginning with a simple object language branch, '$\bar{\text{F}}$, F '

0 0a
12 b

Step 1. A$\bar{F}$ F — Rule TF-1
0 0a
12 b

Step 2. A$\bar{F}$21 Fab — Rule QAR-1
0 0a
12 b

Step 3. A$\bar{F}$21 ∃λ K Hλ0 Faλ — Rule QAR-2 (reclaiming 'b')
0 0a
12

Step 4. A$\bar{F}$21 ∃x K Bx0 ∃λ K Hλ0 Fxλ — Rule QAR-3 (reclaiming 'a')
0 0
12

Step 5. A ∀x C Bx1 $\bar{F}$x1 ∃x K Bx0 ∃λ K Hλ0 Fxλ
0 0
1

Rule QAR-3 (reclaiming '2')

Step 6. A ∀λ C Hλ0 ∀x C Bxλ $\bar{F}$xλ ∃x K Bx0 ∃λ K Hλ0 Fxλ
0 0

Rule QAR-2 (reclaiming '1')

Step 7. A ∀λ C Hλ0 ∀x C Bxλ $\bar{F}$xλ ∃x K Bx0 ∃λ K Hλ0 Fxλ
0

Rule Mod-2

FIGURE 10

The formula is of course the Barcan Formula, and the result in Figure 10 should be compared with the cancellation, modal reclamation, QA translation and confined-negation equivalent of the Barcan Formula.

The Quantificational Analogue reclamation rules just given reproduce the First Interpretation of basic formulas. For the Standard Interpretation, Rule QAR-1 is expanded to include the insertion of 'C Bxm' before bsfs, where 'm' is the modal constant or dummy in the bottom position of the index matrix of the subformula, and 'x' is the individual dummy or constant brought up from the matrix.

A procedure for moving from an object language branch chart to the branch chart of the Quantificational Analogue of the object language formula can be set out. Such a procedure, however, involves grouping within the branch chart and is rather long and complicated. With the procedure just described, we have, in two steps rather than one, the means for moving from object language branch chart to Quantificational Analogue branch chart. The remainder of the procedure is, of course, to cancel the QA reclamation just derived.

The connection between object language branch chart and QA branch chart is such that we can claim the following to be obvious: Dummies and constants will be the same, and will appear in the same relative positions, if the object language matrix is interpreted in the manner of II.4.6. Beyond this, we will claim several results without laying out formal proofs—the results are at least apparent, if not demonstrated, once one has cancelled several object language formulas and compared these cancellations with the QA translations of these formulas. We recall the three cases mentioned in §2 above, where the formula **QAS:α**, associated with an object language formula α, cancelled to closed branches. The first case had to do with the formula where **QAS:α** cancelled to closed branches in such a way that all closure elements came from the consequent —i.e., from the translation of α. For this case, we claim the following result.

An object language formula, α, cancels to closed branches without application of any modal closure conditions other than **CMn** I *if and only if* **QAS:α** *cancels to closed branches in such a way that all closure elements come from the QA translation of* α (i.e., the consequent of the **QAS** formula).

The second case where **QAS:α** might close is where closure elements come from both the antecedent and consequent, i.e., where the antecedent assumptions associated with **S** must be used to obtain closure. Antecedent assumptions in QA provide the justification for modal closure conditions in the object language. If the object language branch chart is regarded as an "abbreviation" of the Quantificational Analogue branch chart, as it may be for this purpose, given the connection established above, we can now set down the condition that modal closure conditions in the object language must meet. For the systems in the text, as well as for additional systems that may be constructed along the same lines, a modal closure condition in the object language must produce the same effect on indices that an antecedent assumption in QA produces on basic formulas in the branch chart. That the modal closure conditions for **CMn**, **CM**, **CS4** and **CS5** accomplish this can be assured by arguments similar to those in IV.5. That is, the QA assumption *Rf*, as it would affect a QA branch chart, accomplishes what the **CM** closure condition accomplishes in an object language branch chart (the constant-repetition version of **CM** closure is the easier one to see), **CS4** closure matches *Tr*, and **CS5** closure matches the combination of *Rf*, *Tr* and *Sm*. To a certain extent, the QA assumptions may be read as "instructions" for manipulating object language indices: *Tr* allows jumping over a constant that appears between two other entries, for example, and *Sm* allows the reversing of two entries.

For other assumptions not considered in the text, care must be taken that the index manipulation produces just the results that are produced in QA by the antecedent assumption, and this is best assured by working back and forth between QA and the object language, as has been done throughout the text.

Once it is assured that object language closure conditions have the same effect as QA antecedent assumptions—and this is a mechanical matter—the second case where **QAS:α** might cancel to closed branches is covered by the following claim.

An object language formula, α, cancels to closed branches upon application of the closure conditions of the formal system **S** *if and only if* **QAS:α** *cancels to closed branches in* **QAS**.

The third case, where the cancellation of **QAS:α** closes in such a way that all closure elements come from its antecedent, is effectively ruled out by our earlier stipulation that the conjunction of antecedent assumptions of **QAS** must be consistent. The last claimed result, then, is the general result we are after, and it may be restated as follows.

An object language formula, α, is a theorem of a formal system **S** *if and only if* **QAS:α** *is a theorem of* **QAS**.

Putting this result together with the result stated in §1 above, we have completeness and soundness for any cancellation system **S** of modal logic that meets the conditions stated in the introduction to this appendix: (i) consistency of the antecedent assumptions of **QAS** and (ii) provability of α in **S** if and only if **QAS:α** is provable in **QAS**. If these conditions are met, we can claim the following.

An object language formula, α, is a theorem of a formal system of modal logic, **S**, *if and only if α is a truth of logic with respect to* **S**. This is the completeness and soundness result desired.

4 CANCELLATION SYSTEMS OF ELEMENTARY LOGIC

While we have discussed the use of the Quantificational Analogue for the interpretation of elementary logic itself (in VIII and Appendix B), a proof of the completeness and soundness of elementary logic cannot be carried out relative to the Quantificational Analogue, since the Quantificational Analogue simply *is* the cancellation system of elementary logic with all predicates of degree n replaced by predicates of degree $n+1$, and with the addition of 'B' and '0'. While the analogue is useful for spelling out existence presuppositions in the object language system of elementary logic, it cannot be used to show, as it were, its own properties relative to itself. Completeness and soundness results for the cancellation system of elementary logic, then, must be shown independently, and this can be accomplished in several ways. First, a model *set* of the sort described in Chapter IV, supplemented by conditions on quantifiers from Chapter V, can be laid out and used to define what is a truth of logic with respect to elementary logic. Then, by induction on the cancellation and closure rules, it must be shown that all and only such formulas cancel to closed branches.

A more interesting sort of result, however, is to show that the cancellation system is equivalent to a system of elementary logic that is known to be complete and sound. Such a result generally provides insight into both the cancellation system and the known system that is more instructive than the sort of proof described above.

For either sort of proof, the cancellation system should be re-expanded into a reduction system, so that the effect of each cancellation rule can be considered separately. That is, the procedure of rewriting an entire sequent after each rule is applied, as was done in the reduction system of Chapter II, must be reinstated in such a way that the restrictions on replacement of dummies by constants can be observed. The completeness and soundness results are then shown by working on the transition from a given sequent to a sequent that is derived from it by a cancellation rule, relative to the rules of the known system to which cancellation is being compared.

For soundness, it must be shown that if a formula cancels to closed branches, then it is a theorem of the known system to which cancellation is being compared. This is shown by showing first that a closed sequent is a theorem of the known system (or that the sequent alternation is, if the known system makes no provision for sequents). Then the soundness result proceeds by showing that if a sequent, X, is derived from another, Z, by a cancellation rule, and X is a theorem of the known system, then so is Z. For the truth-functional rules, this has effectively been shown in Chapter II. For the quantifier rules, it must be shown that the rules for inserting quantifiers in formulas within the known system produce the same results as the quantifier cancellation rules applied backwards (remember that we are working from cancelled/reduced branches back to the original formula). The restrictions on "Universal Generalization" and "Existential Generalization" from the known system are shown to have the same effect as the restrictions on choosing dummies and constants for quantifier cancellation and for replacement of dummies by constants (again, working backwards through the cancellation rules). While restrictions vary from one known system to another, the restriction on Universal Generalization generally has the effect of assuring that one may infer from a formula of the form 'F1' to a formula of the form '∀xFx' only when it can be established that it did not make any difference that '1' was the constant involved—i.e., that 'F1' would hold for any arbitrarily chosen constant. The rule for choosing an individual constant in the cancellation ∀-rule assures precisely this, of course, since the constants are chosen in an arbitrary (numerical) order. The Uniformity and Order Requirements and the rule governing the substitution of constants for dummies can be shown to correspond in a similar way to the rules of known systems. With each cancellation rule thus justified, the soundness result is at hand: A closed sequent is a theorem of the known system, and if that sequent is, then so is the one from which it was derived, and so on until the original formula is shown to be a theorem of the known system.

The completeness result for cancellation relative to a known system follows a similar pattern. It must be shown that if a formula α is a theorem of the known system then it eventually cancels to closed branches. This is best shown by demonstrating that if the cancellation of a α does not eventuate in closed branches, then α is not a theorem of the known system.

A branch (sequent) which is not closed can be shown to have a counterexample, i.e., an assignment of truth-values to the basic formulas of the branch which falsifies the branch. The proof of this is derived in a straightforward way from the Law of Infinite Conjunction,[1] in the form of a proof that unless the sequent is closed, then there is an assignment of truth-values to the members of the sequent which falsifies the sequent alternation. Then, by moving from the falsified sequent up the cancellation/reduction to the original formula in much the same way as in the demonstration of soundness, it can be shown that the falsification is transmitted to the original formula. In short, if the branches do not eventually close, they can be falsified and so can the original formula α, so α could not have been a theorem of a known complete and sound system.

[1] See W. V. Quine, *Methods of Logic*. (Holt, Rinehart and Winston, New York, 1950), pp. 254–256.

SELECTED BIBLIOGRAPHY

Binkley, R. W., and R. L. Clark. "A Cancellation Algorithm for Elementary Logic," *Theoria* **33** (1967), pp. 79–97.

Carnap, Rudolph. *Meaning and Necessity.* (University of Chicago Press, Chicago, 1947)

Church, Alonzo. Review of Quine's "Whitehead and the Rise of Logic," *Journal of Symbolic Logic* **7** (1942), pp. 100–101.

Feys, Robert. *Modal Logics* (Edited with some complements by Joseph Dopp) (E. Nauwelaerts, Louvain, Belgium, 1965)

Hintikka, Jaakko. "Form and Content in Quantification Theory," *Acta Philosophica Fennica* **8** (1955), pp. 7–55.

———. Review of E. J. Lemmon's "Modality as Referential Multiplicity," *Journal of Symbolic Logic* **24** (1959), p. 179.

———. "The Modes of Modality," *Acta Philosophica Fennica* **16** (1963), pp. 65–81.

———. "Studies in the Logic of Existence and Necessity," *The Monist* **50** (1966), pp. 55–76.

———. "Individuals, Possible Worlds and Epistemic Logic," *Nous* **1** (1967), pp. 33–62.

———. "A Program and A Set of Concepts for Philosophical Logic," *The Monist* **51** (1967), pp. 69–92.

Hughes, G. E., and M. J. Cresswell, *An Introduction to Modal Logic.* (Methuen, London, 1968)

Kanger, Stig. *Provability in Logic.* (Almqvist and Wiksell, Stockholm, 1957)

Kneale, William, and Martha Kneale. *The Development of Logic.* (Clarendon Press, Oxford, 1962)

Kripke, Saul. "Semantical Considerations in Modal Logic," *Acta Philosophica Fennica* **16** (1963), pp. 83–94.

———. "A Completeness Theorem in Modal Logic," *Journal of Symbolic Logic* **24** (1959), pp. 1–15.

———. "Semantical Analysis of Modal Logic," *Journal of Symbolic Logic* **24** (1959), p. 323.

Lemmon, E. J. "New Foundations for Lewis' Modal Systems," *Journal of Symbolic Logic* **22** (1957), pp. 176–186.

Lewis, Clarence, I. *A Survey of Symbolic Logic.* (University of California Press, Berkeley, 1918)

Lewis, Clarence, I, and Cooper H. Langford. *Symbolic Logic.* (Century, New York, 1938)

Pap, Arthur. *Semantics and Necessary Truth.* (Yale University Press, New Haven, Conn., 1958)

Prior, Arthur N. *Formal Logic.* (Clarendon Press, Oxford, 1955)

Russell, Bertrand, and Alfred N. Whitehead. *Principia Mathematica.* (Cambridge University Press, Cambridge, 1925)

Schilpp, Paul A. (ed.) *The Philosophy of Rudolf Carnap.* (Open Court, La Salle, Ill., 1963)

Snyder, Donald Paul. *Causation, Predication and Modal Logic.* (University Microfilms, Ann Arbor, 1964) (Dissertation: Duke University)

Strawson, Peter. *Introduction to Logical Theory.* (Methuen, London, 1952)

von Wright, G. H. *An Essay in Modal Logic.* (North Holland, Amsterdam, 1951)

———. *Logical Studies.* (Routledge and Kegan Paul, London, 1957)

INDEX

INDEX